ELECTRICAL AND ELECTRONIC DRAWING

FIFTH EDITION

CHARLES J. BAER
Professor of Engineering
The University of Kansas

JOHN R. OTTAWAY
Senior Engineer
AT&T Technologies, Inc.

McGraw-Hill Book Company

New York Atlanta Dallas St. Louis San Francisco
Auckland Bogotá Guatemala Hamburg Johannesburg Lisbon
London Madrid Mexico Montreal New Delhi Panama Paris
San Juan São Paulo Singapore Sydney Tokyo Toronto

Sponsoring Editor: **Paul Berk**
Editing Supervisor: **Alfred Bernardi**
Design and Art Supervisor: **Nancy Axelrod**
Production Supervisor: **Laurence Charnow**

Text Designer: **Edward Butler**

Library of Congress Cataloging in Publication Data
Baer, Charles J.
    Electrical and electronic drawing.

    Rev. ed. of: Electrical and electronics drawing.
4th ed. c1980.
    Bibliography: p.
    Includes index.
    1. Electric drafting.   2. Electronic drafting.
I. Ottaway, John R.   II. Baer, Charles J.   Electrical
and electronics drawing.   III. Title.
TK431.B3   1986   604′.2′6213   85-7808
ISBN   0-07-003028-6

ELECTRICAL AND ELECTRONIC DRAWING, Fifth Edition

2 3 4 5 6 7 8 9 0   HALHAL   8 9 2 1 0 9 8 7 6

ISBN 0-07-003028-6

Contents

	Preface	v
Chapter 1	General Drawing Techniques	1
Chapter 2	Computer-Aided Design	46
Chapter 3	Device Symbols	81
Chapter 4	Wiring, Cabling, and Chassis Drawings	113
Chapter 5	Printed-Circuit Boards	159
Chapter 6	Flow Diagrams and Logic Diagrams	203
Chapter 7	The Schematic Diagram	229
Chapter 8	Microelectronics	278
Chapter 9	Industrial Controls	321
Chapter 10	Drawings for the Electric Power Field	376
Chapter 11	Electrical Drawing for Architectural Plans	411
Chapter 12	Graphical Representation of Data	440
Appendix A	Glossary of Electronics and Electrical Terms	480
	Electrical Part (Device) Reference Designations	485
	Control-Device Designations	487
	Abbreviations for Drawings and Technical Publications	487
	The Frequency Spectrum	490
Appendix B	Symbols for Electrical and Electronics Devices	491
	Symbols for Electrical and Electronics Devices for JIC-Oriented Drawings	500
	The Relationship of Basic Logic Symbology between Various Standards	504
	Electrical Symbols for Architectural Drawings	508
	Signaling-System Outlets	512
Appendix C	Component Identification	517
	Resistor Identification	517
	Capacitor Identification	519
	Inductor Identification	523
	Width of Copper-Foil Conductors for Printed Circuits	526
	Color Code for Chassis Wiring	526
	Circuit-Identification Color Code for Industrial Control Wiring	527
	Transformer Color Codes	527

Appendix D   Approximate Radii for Aluminum Alloys for 90° Cold
                 Bend   529
                 Thickness of Wire and Metal-Sheet Gages (in Inches)   530
                 Metric Conversion Table   530
                 Minimum Radius of Conduit (in Inches)   531
                 Decimal Equivalents of Fractions   531
                 Small Drills — Metric   532
                 Twist-Drill Sizes   534
                 Standard Unified Thread Series   536
                 Metric Screw Threads   538
                 Wire Numbers and Sizes   540

Appendix E   Product and Manufacturers' Directories, Catalogs,
                 and Other Sources of Design Information   541

Bibliography   545

Index   547

Preface

The fifth edition of this book has undergone extensive revision, in response to major changes in the fields of electrical and electronic drawing. For example, many of the new drawings were prepared using computers or systems associated with them. Thus, a new Chapter 2, Computer-Aided Design, has been written to help the student more easily read these drawings and understand the differences between computer-aided and manually created drawings.

Most students do not have easy access to CAD systems that will produce complete electrical drawings. Therefore, Chapter 1, General Drawing Techniques, has been retained and strengthened so that a student who has not had previous drafting experience can quickly learn to make technical drawings that are not highly complex. In other words, he or she should be able to work most of the problems in the book after receiving proper instruction.

New subject material includes surface-mounted devices (Chapter 5), the learning curve (Chapter 8), solar power (Chapter 10), robotics (in several chapters), and printed-circuit board drawing (Chapter 6). However, much of the material from the previous edition that is still appropriate has been retained. It should also be noted that much of the material on instrumentation and controls in this book will not be found in any other book of its kind. In addition, no other text contains as complete a listing of standard symbols and tables as those appearing in Appendixes B and D.

These important standards have been established by agencies such as ANSI (American National Standards Institute), a private agency founded by industry many years ago, and the IEEE (Institute for Electrical and Electronics Engineers, Inc.), a professional engineering society. Other standards have been formed as needed by JIC (Joint Industrial Council), NMTBA (National Machine Tool Builders' Association), NEMA (National Electrical Manufacturers Association), and agencies of the U.S. government. The reader may obtain lists of the available standards by writing the organizations listed below.

ASME   American Society of Mechanical Engineers, 345 East 47th Street, New York, NY 10017

American National Standards Institute, 1430 Broadway, New York, NY 10018

EIA   Electronic Industries Association, 2001 Eye Street, N.W., Washington, DC 20006

IEEE   Institute of Electrical and Electronics Engineers, Inc., 345 East 47th Street, New York, NY 10017

JIC   Joint Industrial Council, 7901 Westpark Drive, McLean, VA 23101

NEMA   National Electrical Manufacturers Association, 2101 L Street, N.W., Suite 300, Washington, DC 20037

NMTBA   National Machine Tool Builders' Association, 7901 Westpark Drive, McLean, VA 23101

Superintendent of Documents, U.S. Government Printing Office, Washington, DC 20402

The authors wish to make known their gratitude to the officials and engineers of the many companies and firms who were interested and thoughtful enough to assemble drawings and photographs for use in this book. We have endeavored to reproduce these drawings as closely as possible to the way they were originally drawn.

The authors would also like to acknowledge the suggestions for improvements and additions that came from several of the users of the textbook's fourth edition. These instructors are:

Jack E. Bernard, Colorado Technical College, Colorado Springs, Colorado
Joseph Cutrone, Island Drafting & Technical Institute, Amityville, New York
Edgar Hund, Los Angeles Pierce College, Los Angeles, California
Kenneth Markowitz, New York City Technical College, Brooklyn, New York
James Merrigan, Brookdale Community College, Lincroft, New Jersey
William Moore, Tidewater Community College, Portsmouth, Virginia

CHARLES J. BAER
JOHN R. OTTAWAY

1

General Drawing Techniques

Because drawing for the electrical and electronics industry includes so many types of drawing, it utilizes all the various graphical techniques in one place or another, at one time or another. In the beginning stages of some projects, rough freehand sketches may be the most appropriate way to show ideas graphically. On the other hand, the final stages of some projects may require precision drafting that necessitates the use of sophisticated and expensive equipment.

This book can provide only an *overview* of the various ways to make technical drawings. A small amount of instruction is provided in the use of the time-honored method of making drawings with triangles or T squares. In addition, the subject of freehand sketching (often a prelude to the construction of a drawing by instruments or computer) is presented from time to time in the book. Most students will be able to develop a method to complete the drawing assignments made by their instructor, given the type of equipment available to them.

Computer-aided graphics is treated in Chap. 2. The bibliography contains a list of several excellent texts in engineering drawing, computer graphics, and related subjects. School libraries and college bookstores are likely to have books on technical or engineering drawing in case the reader wants to learn more about drawing techniques.

1-1 Line Work

Figures 1-1 and 1-2 are examples of drawings done in the electrical and electronics industry. Figure 1-1 is an older drawing that was made on a drafting machine and by use of a mechanical lettering device. Figure 1-2 is a computer-generated drawing that was laid out on a cathode-ray-tube (CRT) console and then plotted on a belt-bed plotter. The lettering is from a *font* (alphabet) provided by the manufacturer of the computer graphics equipment. Figure 1-3 is a drawing that was made by a student. The three orthographic views were obtained by use of drawing board, T square, and triangle. The lettering and arrows were drawn freehand. Freehand sketching was used to draw Fig. 1-4. It was made on a sheet having $\frac{1}{4}$-in. (6.35-mm) grid lines. Sketches such as this are used in several different ways; one is a preliminary layout for an instrument-produced drawing.

1

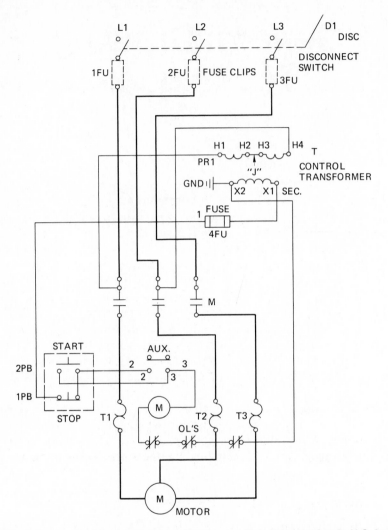

FIG. 1-1 An electrical drawing made by a professional drafter in U.S. industry.

    The drawings just discussed have one thing in common. The lines are almost all horizontal or vertical. There are quite a number of ways to make parallel lines, depending on the accuracy required and the equipment available. Some of these methods are:

1. With a plotting device such as the drum or belt-bed plotter
2. With a drafting machine
3. With a parallel straightedge, or bar

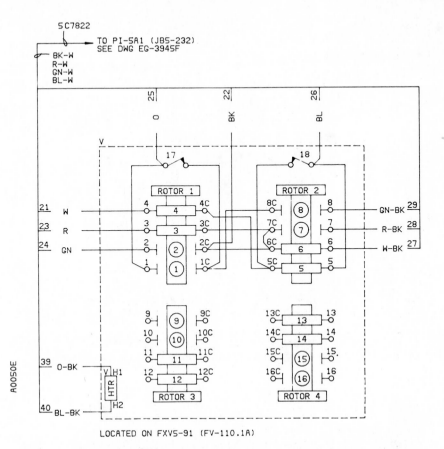

FIG. 1-2 An electrical drawing made with computer and related equipment.

4. With special section-lining devices

5. With a T square (and triangles)

6. With two triangles

7. With tape appliqués

8. Freehand

In general, the cost of equipment and the accuracy obtained decrease as we go from 1 to 8 in the above list. Figures 1-6 to 1-8 show some of the equipment listed. For small drawings, a second triangle can be substituted for the T square, although this is not as convenient as using the latter.

Different types of line are used for different purposes in electrical drawing. These are shown in Fig. 1-5. Some of the equipment that is used to produce electrical and electronic drawing is illustrated in Figs. 1-6 through 1-11.

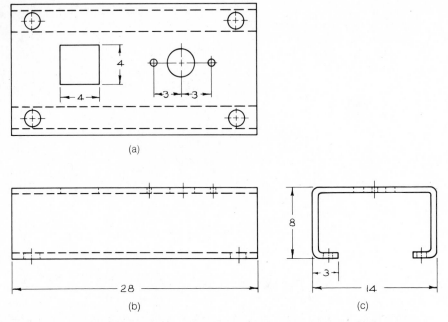

(a)

(b)   (c)

FIG. 1-3 A drawing made by a student. It contains three views of a chassis: *top view* at upper left, *front view* at lower left, and *right-side view* at lower right.

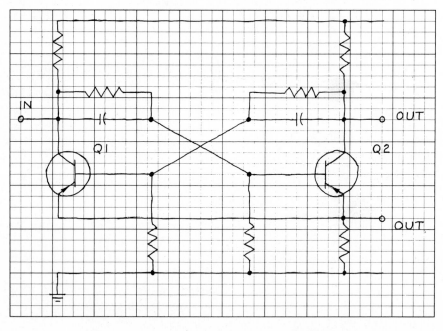

FIG. 1-4 A freehand sketch of a circuit schematic.

———————————— Medium	*General purpose. Outlines.* *Circuit path. Symbols*
— — — — Medium — — —	*Shielding. Mechanical connection.* *Hidden line. Future circuits.*
—— —— Medium —— ——	*Bracket – connecting line.*
———————— Thin ———————	*Dimension line. Leader. Bracket.*
——— Thin — — ——— — —	*Alternate position. Adjacent part.* *Mechanical grouping boundary.*
━━━━━━━━ Thick ━━━━━━━━	*For emphasis.*

FIG. 1-5   **Different lines and their uses in electrical and engineering drawings.**

1-2 Circles, Symbols, and Other Shapes

Circles are generally made with the compass or with templates. Because circles in most electrical drawings are smaller than 1 in., templates are used more frequently than the compass for circle construction. Templates are flexible, thin (0.020- to 0.060-in.) pieces of plastic with generally accurate shapes punched in them. Many models are designed for the purpose of making geometric shapes

FIG. 1-6   **Two styles of drafting machine.** (Keuffel & Esser Co.)

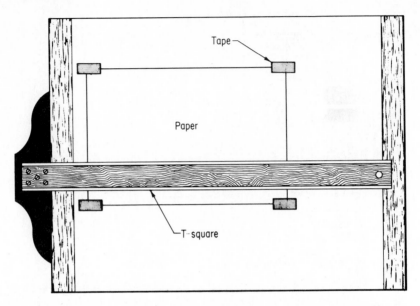

FIG. 1-7 Positioning of a small-size sheet of drawing paper on a drawing board. The bottom of the sheet should align with the top of the T square.

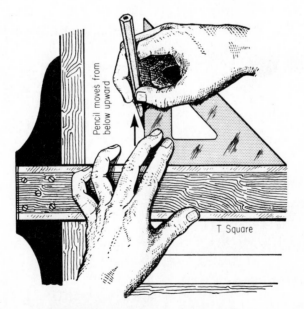

FIG. 1-8 Using a T square and triangle to draw vertical lines. The left hand holds the T square so that the head is tight against the edge of the board or desk.

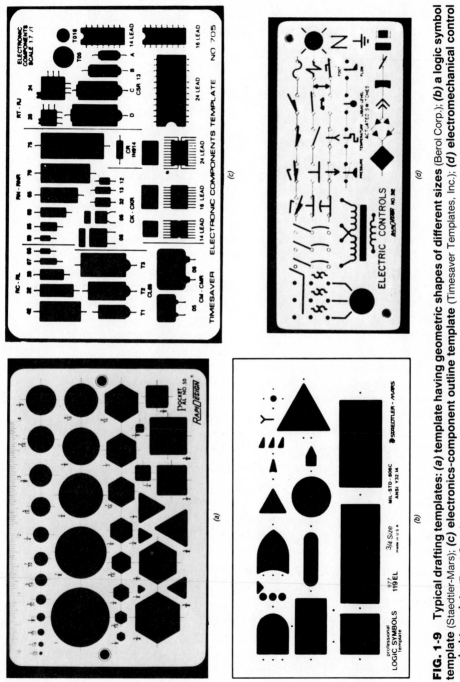

FIG. 1-9 Typical drafting templates: (*a*) template having geometric shapes of different sizes (Berol Corp.); (*b*) a logic symbol template (Staedtler-Mars); (*c*) electronics-component outline template (Timesaver Templates, Inc.); (*d*) electromechanical control symbol template (Berol Corp.).

such as circles, rectangles, triangles, etc.; while others, for use in electrical and electronic drawings, are for making symbols, component outlines, and wiring diagrams.

Over the years, designers and drafters accumulate a large collection of templates to do their drawings. Templates are a more convenient, faster, and more accurate method of drawing shapes and electrical symbols compared with conventional methods. For example, drawing a transformer symbol as in Fig. 3-11 would take several minutes by conventional methods, whereas drawing it using a template would take only 15 s. (The times mentioned are those expended by one of the authors.)

Care should be used in selecting templates in terms of the quality, accuracy, symmetry, arrangement of the shapes, and the shapes and symbols required. Many template manufacturers will make special templates upon request. Figure 1-9 shows several templates that could be used in doing drawings. Figure 1-9a is a template having a variety of geometric shapes. Figure 1-9b is a logic-symbol template that is used for making logic diagrams. Figure 1-9c is an electronic-components template used for making printed-circuit-board layouts and wiring diagrams. Figure 1-9d is an electromechanical-control-symbol template used in making control schematics. These are just a few examples of the many varieties of templates on the market today.

Several concepts should be followed in using a template.

1. In order to draw any geometric shape at the precise location desired, construct centerlines (along the axes, at least as long as the longest diameter of an equivalent circle) on the drawing before the template itself is positioned for the drawing of the shape. Many times, for small quick modifications to existing drawings, experienced designers use one line and an "eye-balled" centerpoint; however, this is not recommended for beginners.

2. For proper alignment, the template must be held firmly against a T square, drafting machine arm, or some suitable (horizontal) straightedge.

3. The pencil must be held *vertically* as the figure is being drawn. This is the only way to be certain of the correct alignment and shape of the figure. It may also be necessary to use a slightly duller pencil point than normal in order to achieve the desired line thickness of the figure.

4. During inking, the template should be slightly raised off the drawing medium by a template guide and a sheet of Mylar or tape should be fastened to the back of the template to prevent the ink from *feathering*. Figure 1-10 shows the general procedure on how to use a template.

Figure 1-11 shows a drafter using an ellipse template. The drawing paper is taped to a drafting table, and the person is holding the template snugly against the top edge of the parallel bar straightedge. Short construction lines (horizontal and vertical) have been made on the drawing to correspond with marks printed on the template at each ellipse. The marks on the template are placed directly

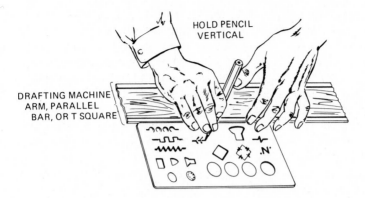

FIG. 1-10 Relationship of a template to a horizontal straightedge.

FIG. 1-11 Use of an ellipse template to make ellipses on a pictorial drawing.

over the marks on the drawing. When this has been done the drafter does not have to guess where to position the template. Just about all templates have marks that enable the user to position the template accurately on a drawing.

In the case of circle templates these marks correspond to centerlines. Since the centerlines are usually used to locate holes in the first place, the marks are very useful. In electrical and electronics templates there are marks which correspond to the signal path, wire, and so on. These marks are helpful to the user in positioning the template correctly when making symbols or outlines.

Templates are very convenient for making shapes and electrical symbols; however, a designer should also know how to construct symbols with the basic drafting tools because the necessary templates might not be available. These procedures are covered in Chap. 3.

1-3 The Importance of Lettering

Electrical drawings utilize all the conventions and shortcuts of engineering drawings in order to present a graphical concept of devices or systems. However, the line work, symbols, and other patterns of a drawing are not sufficient to give the complete picture. Considerable lettering is required on most electrical and electronics drawings. Figures 1-1 through 1-3 show the amount and kind of lettering required on an electrical drawing. Some drawings require even more in proportion to the amount of line work than these drawings. In many cases, as much time is used in doing the lettering on the drawing as is used in doing the line work. In drawings which are to be used for the manufacture, construction, or installation of equipment, lettering is often done freehand. In drawings that are to be used over and over again or in those which for certain reasons are to be printed in books, manuals, or journals, lettering is usually drawn with mechanical lettering devices.

In every case, lettering must be good. Poor lettering not only ruins the appearance of a drawing that is otherwise good but improves the chances of costly mistakes being made in the reading of the drawing. It so happens that the lettering in Fig. 1-1 was done mechanically, mainly because it was a part of a set of printed installation instructions. But before this final drawing was made, another one using freehand lettering had been drawn. Although mechanical and computer-aided lettering is becoming more prevalent, many electrical drawings still have freehand lettering.

1-4 Types of Engineering Lettering

Practically all the alphabets used in technical drawings are single-stroke Commercial Gothic. Four types of these single-stroke letters in use today are:

1. Vertical uppercase (capitals)
2. Inclined uppercase
3. Inclined lowercase ("small" letters)
4. Vertical lowercase

The alphabets above are arranged in what the authors believe is the order in which each type is used in U.S. industry, with type 1 used the most, type 2 the next, etc. One who wishes to become expert in the making of electrical or electronics drawings should be proficient in doing freehand lettering of type 1 and possibly types 2 and 3.

The person who has had no lettering instruction and wishes to become proficient at freehand lettering should practice one type of alphabet at a time. A logical sequence of alphabets would be (1) vertical uppercase, (2) inclined uppercase, and (3) inclined lowercase. With each alphabet the student should develop a set of numbers that appear to match or fit with the letters. As a rule, numbers are usually made narrower than the letters of the companion alphabet. Thus the number zero should be narrower than the letter O of the same alphabet.

1-5 Standard Alphabets

Lettering for drawings is standard throughout the United States. Three standardized alphabets are shown in Fig. 1-12. There may be slight differences between alphabets authorized by one agency and another, and also between

FIG. 1-12 Typical American National Standard (ANSI) alphabets.

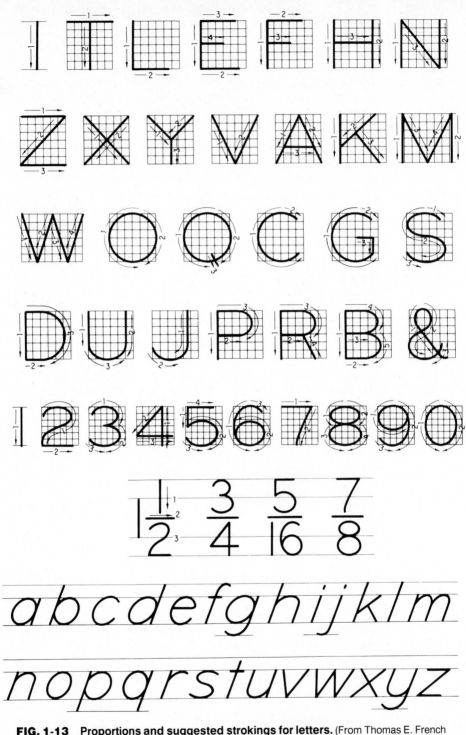

FIG. 1-13 **Proportions and suggested strokings for letters.** (From Thomas E. French and Charles J. Vierck, *Fundamentals of Engineering Drawing,* McGraw-Hill Book Company, New York, 1960. Used by permission.)

alphabets shown in drawing textbooks, but these differences are minor. The shapes of the letters are, in general, the same.

Figure 1-13 shows vertical uppercase letters placed in blocks that are six small squares high. Borrowed from a famous textbook, this illustration shows the comparative widths of all letters and the suggested order of strokes for right-handed persons. This order does not have to be followed, but it is used by many drafters. Left-handers will probably have to develop their own order of stroking, using whatever is most comfortable and effective.

1-6 Guidelines

Practically all freehand lettering must make use of guidelines. These are very light, thin lines that locate the tops and bottoms of capital letters, numbers, etc. They should be just dark enough to see, because erasing them is not practicable after letters are penciled in; therefore, they should be as nearly invisible as possible.

There are a number of ways to draw guidelines. The most convenient way is to use a device especially made for that purpose. Two such devices are shown in Fig. 1-14. In this illustration, guidelines for capital letters have been constructed with both the Braddock and the Ames lettering devices. In the lower figure, lines have also been partially drawn for the bodies of lowercase letters but were stopped near the letter C and number 3. Guidelines may also be made with T squares and triangles in conjunction with scales.

Different combinations of letters and figures require different arrangements of guidelines. If only uppercase letters are to be used, just the top and bottom guidelines are necessary (see Fig. 1-15a). For lowercase letters, the intermediate, or "waist," line must be used with the other two lines, as shown in Fig. 1-15b. A "drop" line may be placed below the baseline to facilitate making the descenders. This is not done very often, probably because the normal spacing provided by lettering devices does not allow enough space for drop lines. Occasionally it is desirable to have a line exactly centered between the upper and lower guidelines. This would be helpful, for example, when many fractions are part of a note (see Fig. 1-15c). Because fraction heights are usually made twice those of whole numbers and capital letters, additional lines may be added above and below, as shown at the right end of Fig. 1-15c. This would produce a series of lines all equally spaced. This type of spacing can be obtained with the Ames lettering instrument for nine different sizes of letters and with some models of the Braddock-Rowe lettering triangle for one size of letter.

Figure 1-16 shows another type of freehand lettering guide that is easy and simple to use. Lettering guides of this type have a number (usually four) of guide slots for various heights of letters—ranging from $\frac{3}{32}$ to $\frac{3}{8}$ in. Either inclined or vertical freehand lettering may be done using this type of guide with the assurance of uniform-height letters due to the upper and lower edges of the guide slots. A drafter can letter with great speed using this type of guide; however, care should be exercised to avoid flattening rounded letters against the edges of the guide.

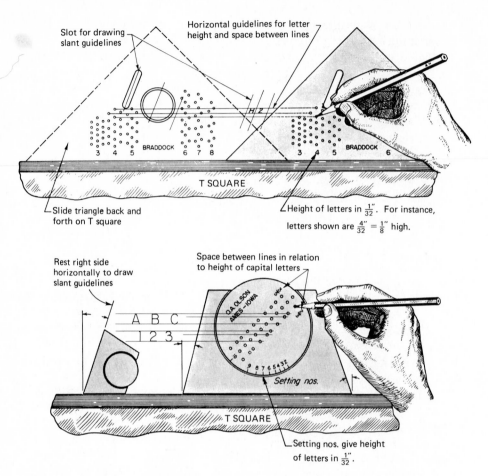

Slot for drawing slant guidelines

Horizontal guidelines for letter height and space between lines

BRADDOCK

T SQUARE

Slide triangle back and forth on T square

Height of letters in $\frac{1}{32}''$. For instance, letters shown are $\frac{4}{32}'' = \frac{1}{8}''$ high.

Rest right side horizontally to draw slant guidelines

Space between lines in relation to height of capital letters

A B C

1 2 3

O.A. OLSON AMES — IOWA

Setting nos.

T SQUARE

Setting nos. give height of letters in $\frac{1}{32}''$.

FIG. 1-14 **Use of lettering guideline devices.** (From Frank Zozzora, *Engineering Drawing,* McGraw-Hill Book Company, New York, 1958. Used by permission.)

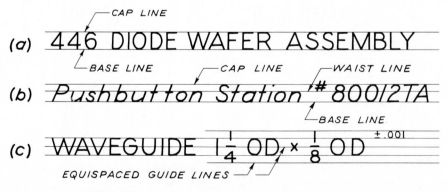

(a) 446 DIODE WAFER ASSEMBLY

CAP LINE

BASE LINE

(b) Pushbutton Station '# 80012TA

CAP LINE

WAIST LINE

BASE LINE

(c) WAVEGUIDE $1\frac{1}{4}$ OD $\times$ $\frac{1}{8}$ OD

±.001

EQUISPACED GUIDE LINES

FIG. 1-15 **Use of guidelines for different situations.**

FIG. 1-16 Another device for making guidelines for freehand lettering. (Timesaver Templates, Inc.)

Another way to align the tops and bottoms of letters is to use guidelines already drawn on a piece of paper. This paper is positioned under the sheet of tracing paper or transparent plastic material on which the drawing is to be made. The guidelines show through the drawing medium and are followed just as if they were drawn on the medium itself.

1-7 Parts Lists and Tables

Figure 1-17 shows a part of a typical parts list, or bill of materials. In practice, the heavy horizontal dividing lines are usually spaced $\frac{1}{4}$, $\frac{5}{16}$, or $\frac{3}{8}$ in. apart, and the letters are made not more than one-half as high as this vertical spacing. One convenient way to organize such a list is to use equally spaced guidelines, as shown in the lower part of Fig. 1-17. For instance, if the lines are spaced $\frac{1}{16}$ in. apart (using a lettering guide set for No. 4 letters, as in Fig. 1-14), letters and numbers will be $\frac{1}{8}$ in. high and centered between the horizontal dividing lines, which will be $\frac{1}{4}$ in. apart. Abbreviations are frequently used; examples are CRS for cold-rolled steel and AL for aluminum. Capital letters are almost always used. Parts lists and other tabular arrangements often accompany, or are part of, electrical drawings. Their exact arrangements (formats) are not standardized.

ITEM	REFERENCE	DESCRIPTION	MAT'L
1	310-19506	CHASSIS	
2	35A-19472	BRACKET, MOUNTING	
3	DD-6040A	CAPACITOR, 1 MF 25 WVDC	
4	50C-19503	CABLE ASSEMBLY	
5	30103744	LOCK SHAFT	CRS
6	60104321	SCREW #4-40 x $\frac{1}{4}$	STEEL

FIG. 1-17 Layout for a parts list, with suggested vertical spacing.

1-8 Mechanical Lettering Devices

A number of devices for making engineering letters mechanically have been successfully tried and used in the drawing room. They fall into two classes: (1) the stencil, or incised-letter, type, with which the drafter puts the letters on the drawing medium, and (2) the special typewriter, with which one puts the letters on the drawing after the drafter or engineer has written or otherwise indicated what material is to be typed. Examples of the first type are shown in Figs. 1-18 and 1-19. These can be used for penciled or inked letters. The fountain-pen type of lettering pen (the Rapidograph, for example) can be used with the stencil type such as that shown in Fig. 1-19.

Other ways to make lettering are the computer (or more specifically a computer and plotting system) and a special typewriterlike device called a *lettering system.* Figure 1-20 shows two computer-generated alphabets called *fonts* and a set of letters made by a lettering system instrument. Some of the computer equipment is shown in Chap. 2. *Helvetica* is one of several types of lettering that can be generated on the lettering typewriter, which is easy to use. The specified words, or combinations of words and numbers, are produced on strips of paper that have adhesive backing. Figure 1-21 shows a drafter putting some of this lettering on a block diagram. To make certain that the letters are horizontal, blue pencil lines (which don't show up well in the photograph) have been made with the horizontal straightedge. The best lettering on electrical, electronic, and engineering drawings today is done with this system.

The authors feel that most computer-generated lettering leaves something to be desired. The plotter that does the printing moves so very fast that small, unwanted marks sometimes appear when the pen puts down or lifts up, line weights are not always uniform, and smears are easy to make during handling.

FIG. 1-18 A mechanical lettering device consisting of a template and a scriber.
(Keuffel & Esser Co.)

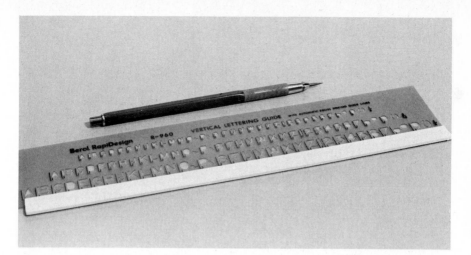

FIG. 1-19 A template used for making vertical uppercase letters $\frac{1}{8}$, $\frac{3}{16}$, and $\frac{1}{4}$ in. high.

(a) FONT 1   RESFT

ABCDEFGHIJKLMNOPQRSTUVWXYZ&
1234567890

(b) FONT 2   CONDFT

ABCDEFGHIJKLMNOPQRSTUVWXYZ&
1234567890

(c) TYPE FONT   HELVETICA

ABCDEFGHIJKLMNOPQRSTUVWXYZ
1234567890

(d) SPACING

RESFT PROVIDES THE SAME AMOUNT
OF SPACE FOR EACH CHARACTER.

RESFTP PROVIDES PROPORTIONAL
SPACING FOR EACH CHARACTER.

FIG. 1-20 Lettering produced by computer graphics equipment and the vari-typer. Alphabets produced by computer-driven equipment are called *fonts*. *(a)* RESFT is a standard engineering type letter. *(b)* CONDFT is a narrower letter. *(c)* Helvetica is a standard letter made on the Kroy lettering system. *(d)* Spacing produced by the computer-generated system. (Lines a and b by permission of Autotrol ® Technology Corp. Fonts and spacing samples made by Black & Veatch, Consulting Engineers.)

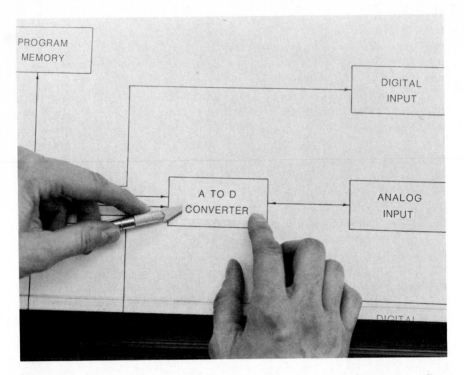

FIG. 1-21 Applying letters that were made with a lettering instrument to a flow diagram.

Some of the letter and number shapes depart somewhat from the standard alphabet, usually not for the better. We have had to discard quite a few computer-generated drawings for illustration in the book because of poor lettering, poor line width, or both. The fonts shown in Fig. 1-20 can be modified so that the crossbars can be removed from the *I* and the shapes of other letters such as the *W* can be changed. According to the user, a large consulting engineering firm, a simple modification in the program would produce these changes. However, after using this equipment for several years the company has not yet seen fit to make the changes.

Yet another way to place letters on drawings is by *transfer*. Sheets of transfer letters, including such styles as Helvetica, Roman, and Old English, can be purchased in most college bookstores. These preprinted letters can be positioned on a drawing and affixed to the drawing medium one by one by applying pressure on the desired letter with the blunt end of a pencil. They will stay on the drawing by virtue of their adhesive backing.

1-9 Orthographic Views (multiview drawings)

Among the many types of drawing that are produced and used within the electrical and electronics industry are those that show the front, top, and side views (or maybe just two, or as many as five views) of an object. In manufacturers' catalogs one often finds drawings of this type for certain devices or equipment. Or the designer of new products may sketch out or lay out with instruments several views of the object being designed, and the drafter may be required to make the finished scale drawing.

One such drawing is shown in Fig. 1-22. This is an outline drawing consisting of three views: a *left side* view at the left, the *front* view to the right of the left side, and the *top* view directly above the front view. This is the standard way these views are arranged (in the United States). Except for the front view of the notch, hidden lines are not shown. They are seldom shown in outline drawings. In this particular example dimensions are placed in the standard manner, using two-place decimals, with all digits horizontally oriented. It was drawn to a scale of 1 in. = 0.20 in. with a 20 scale. It appears in a reduced size in the book; therefore, it cannot be scaled, although the 30 scale comes close.

Another example is given in Fig. 1-23. This is a complete double-view drawing, consisting of the *front* and *right-side* views, or *left-side* and front views, whichever view the drafter and viewer consider to be the front view. (If the name of the view is important, each one could be labeled.) The writer has considered the view at the left to be the front view of the bracket; this means that the other one is the right side view. (Figure 1-24 shows the correct location of four orthographic principal views.) The bracket was drawn to full size using a metric scale, and the dimensions have been given in centimeters to two decimal places. In this book the bracket has been reduced and cannot be scaled. Hidden lines (dashes) show the thickness of the 16-gage (0.062-in.-thick) steel including holes (see the side view). All dimensions have been shown except for the bottom holes, which are designated in a note at the bottom. A third view (looking down

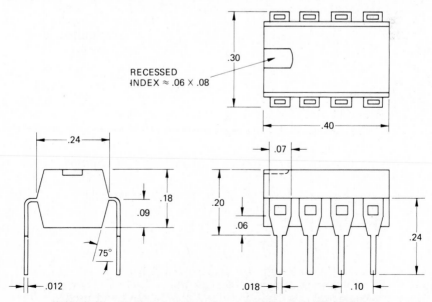

FIG. 1-22 Outline drawing of an integrated circuit package.

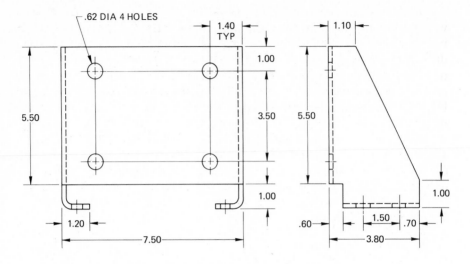

Note: Four Bottom Holes are .50 Dia.

FIG. 1-23 Orthographic views of a mounting bracket.

or up) might make the shape of the bracket a little more clear. But two views are all that are really necessary for the manufacture of the bracket.

Figure 1-24 shows four different orthographic views of an object. Each corner has been labeled in all four views. The reader should note that the front surface, *A B C D*, is placed nearest to the front view in the side and top views. If the object were a car or an aircraft, the right side, *B 2 3 C*, would be what we normally call the left side of the car or plane. The right side view in this type of drawing means the view is seen from the viewer's right, as if one moved to one's right 90° around the object and drew it as seen from that position.

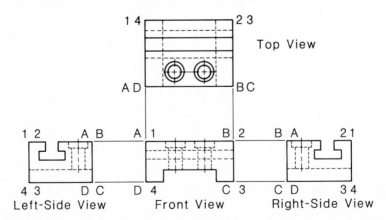

FIG. 1-24 Four principal views of a block with a T slot. Each corner has been lettered or numbered.

In drawing principal views one generally lines up the two or three views with a T square, parallel straightedge, or drafting machine arm or triangle and usually works back and forth between views. In other words, it is not efficient to draw one view completely, then another view completely, then another, and so on. The drafter should take advantage of the fact that certain features line up (or *align*) in two or three views. Note the T shot in the side views and front view of Fig. 1-24. The amount of space needed between the views must also be determined. This depends largely on how many dimensions are to be placed between views. In Fig. 1-22 the front and left side views were drawn $2\frac{3}{8}$ in. apart; in Fig. 1-23 they were $1\frac{3}{8}$ in. apart on the original drawing. If the objects are not to be dimensioned, a distance that appears pleasing to the eye should be used. For drawings on $8\frac{1}{2} \times 11$ in. paper an inch or a little more is usually about right.

1-10 Pictorial Drawing

As the electronics and electrical fields continue their phenomenal expansion, more and more devices that require pictorial representation, either in place of or as additions to standard orthographic projection, are encountered. Also, electrical drawings are being read by some technical personnel who have not had training in the reading and making of engineering drawings. These are two factors which have been responsible for the increase in the amount of pictorial drawing being done in the electrical industry.

Good examples of pictorial drawing applied in the electrical field are those used by the armed forces, maintenance personnel, assembly-line workers for assembling electronic equipment, and companies who manufacture do-it-yourself kits. Figure 1-25 is a typical example of the latter.

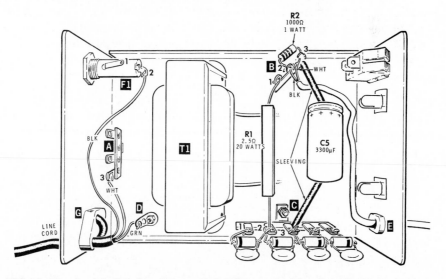

FIG. 1-25 **A pictorial drawing used for the assembly of a robot.** (Reprinted by permission of the Heath Company, Copyright © 1982.)

The types of pictorial drawing most often found in the electrical field are:

1. Isometric drawing
2. Oblique drawing
3. Dimetric drawing
4. Perspective drawing

Brief treatment of the first two categories of pictorial drawing will follow. References will be given for the more complicated subject of perspective drawings.

1-11 Isometric Drawing[1]

It is conveniently possible to draw an object at such an angle or position that all edges will be foreshortened equally. For example, all lines of the cube shown in Fig. 1-26 have been drawn to the same length; that is, they have been foreshortened equally. This drawing, called an *isometric* or *isometric projection,* was obtained by using three axes, as shown in Fig. 1-26*b*. Rectangular-shaped objects lend themselves readily to this type of drawing because all lines, or edges, are drawn along or parallel to the three isometric axes and can be scaled directly.

A simple and direct way to make an isometric drawing of an object composed of rectangular-shaped surfaces is to draw an outline of a cube or prism and then measure off distances from edges of this cube to the corners and edges of the object. Such step-by-step procedure is shown in Fig. 1-27.

The front and top views of a transistor are shown in Fig. 1-27*a*. The major dimensions (in millimeters) are included. Using the box method, we begin by constructing a rectangular box or prism that has the overall height, width, and depth dimensions (excluding the three pins) shown in Fig. 1-27*b*. We have chosen to orient the device as shown in the successive figures to show it to the best advantage, but it is possible to orient it differently. In Fig. 1-27*c* we have cut

[1] Liberal translation of the word *isometric* from the Greek gives "iso," meaning "the same," and "metric," meaning "measurement."

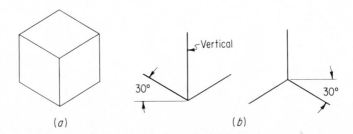

(a)   *(b)*

FIG. 1-26 An isometric drawing of a cube and the isometric axes.

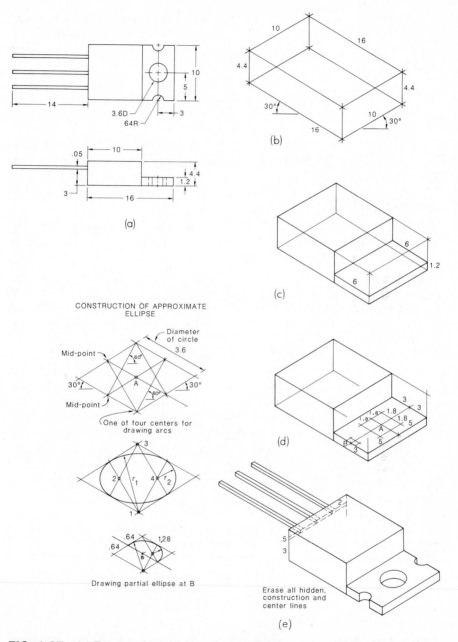

FIG. 1-27 *(a)* Front and top views of a transistor. *(b)* through *(e)* Step-by-step construction for making an isometric drawing. Lower left: drawing an isometric ellipse using the four-center method.

out part of the right side of the block leaving the lower 1.2-mm-thick (or high) part as shown. In Fig. 1-27d we have begun to locate and lay out the hole and the two semicircular cutouts. The isometric "diamond" within which the ellipse will fit has sides that are equal to the diameter of the 3.6-mm-diameter circle. But it is necessary to locate the center A by measuring "in" 5 mm from the front or back edges and 3 mm from the right edge, measuring along the isometric axes. (All measuring must be done along these axes.)

There are several ways of drawing ellipses, three of which are (1) freehand; (2) with a drafter's ellipse template; and (3) with the use of triangles, pencil, or pen and compass. Freehand curves appear unattractive on an instrument drawing; therefore, one should use an ellipse template that makes 35° ellipses, or triangles, compass, etc. Ellipse templates are very nice, but fairly expensive. For this reason, we have shown in the lower left part of Fig. 1-27 how to draw a horizontal ellipse and a partial ellipse by the four-center method. The main idea is to draw the diamond at the right location and to the right size, then to locate the midpoints of each side, and then to draw a line to each midpoint from the upper and lower corners to opposite midpoints. This construction completes the location of the four centers, 1, 2, 3, and 4 (see the second figure). The four arcs, with radius either r_1, or r_2, are then drawn tangent at the midpoints.

We have completed the isometric pictorial of the transistor in Fig. 1-27e, by drawing another set of ellipses 1.4 mm lower to show those parts of the curves on the bottom surface that can be seen. The leads have been added by measuring up 3 mm from the bottom on the left end and 2 mm "in" from the front and back sides for the outer two leads and then centering the middle lead. We have shown part of the leads with hidden lines. But these, as well as the construction lines, would be erased and should not appear in the final drawing.

Isometric drawings may be made with instruments (the 30-60 triangle is a natural choice) or drawn freehand. If an instrument drawing is to be made, any suitable scale may be used, and only the one scale selected will be used in making measurements along all three axes. Hidden lines are seldom shown. Dimensions may be added if necessary.

A freehand sketch of a chip carrier is shown in Fig. 1-28. It was made on specially prepared paper on which light grid lines had been printed in the direction of the three isometric axes. Most edges of a rectangular-shaped object can be drawn directly over the preprinted lines. Sometimes, however, a drafter might have to draw between — and parallel to — lines. These lines (usually in color) are generally $\frac{1}{4}$ in. apart and can be used as a sort of scale so that the person making the sketch can execute a sketch having the correct proportions.

To show the rounded edges on top, parallel broken lines have been drawn. Technically they aren't necessary, but they are often used. Other methods are available but are not discussed here.

Another example of isometric drawing is shown in Fig. 1-29. Shading has been added to make this object look a little more like a photograph. A good isometric drawing does not have to have shading, however. Note that the plus and minus signs have been drawn along isometric axes.

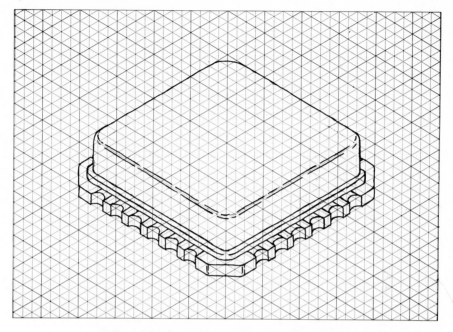

FIG. 1-28   Freehand sketch of a chip carrier.

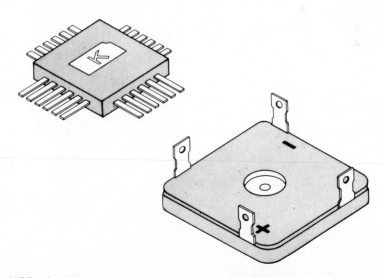

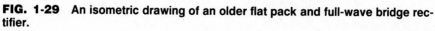

FIG. 1-29   An isometric drawing of an older flat pack and full-wave bridge rectifier.

1-12 Oblique Drawing

Another type of pictorial representation which is easily made is oblique drawing, or oblique projection. In this type of drawing, one face of the object is drawn in its true shape and the other visible faces are shown by parallel lines, or projectors, drawn at the same angle (usually 30 or 45°) with the horizontal.

In Fig. 1-30 a cube has been drawn in oblique pictorial projection with projectors going back at 30 and 45°. Any angle other than 0 or 90° can be used, but 30 and 45° are popular because they can be easily drawn with the use of triangles. There is a certain amount of distortion in an oblique drawing which is reduced when cabinet drawing is employed. This can be seen in Fig. 1-30c, in which the projectors have been made one-half as long as in the other two cube drawings. Figure 1-30c looks more like a cube than do Figs. 1-30a and 1-30b.

The rectifier (Fig. 1-31) is suitable for oblique drawing for two reasons: (1) the face drawn perpendicular to the line of sight (in the plane of the paper) contains circular parts which can be drawn with a compass or circle template; and (2) the other axis is comparatively short, thus making it unnecessary to shorten the projectors. The projectors of the rectifier were drawn 45° to the left and are not foreshortened. If the rectifier were drawn in the isometric manner, they would have been drawn at a 30° angle and all circles and arcs would have to be drawn as ellipses or parts of ellipses. The rectifier has also been drawn in cabinet projection in Fig. 1-31c. Like isometric drawings, oblique drawings may be made freehand or with instruments.

A freehand sketch of a seven-segment light-emitting diode (LED) is shown in Fig. 1-32. This sketch was made on a sheet of paper without any aids. An instrument drawing of a transistor with radiator is illustrated in Fig. 1-33. In Fig. 1-33 the circles and arcs were drawn with a compass because they are in the plane of the paper. Orientation of objects in oblique pictorial drawing can often circumvent the need for drawing circular parts as ellipses.

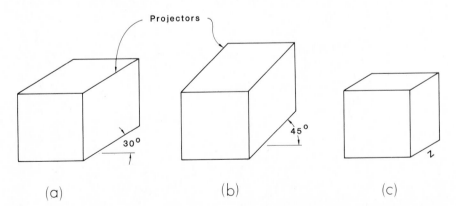

FIG. 1-30 Oblique view drawings of a cube: *(a)* with projectors to the right and 30°; *(b)* with projectors at 45°; *(c)* cabinet drawing with the projectors half as long as the other edges.

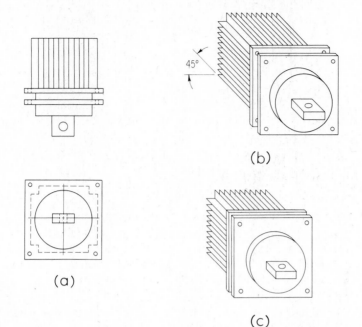

FIG. 1-31 Oblique drawing of a rectifier: *(a)* front and top views; *(b)* oblique drawing with projectors left and 45°; *(c)* modified oblique (cabinet) drawing.

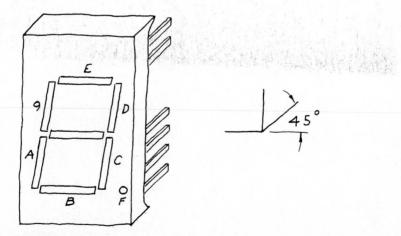

FIG. 1-32 A freehand oblique sketch of a seven-segment LED.

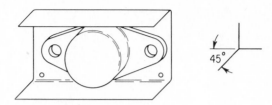

FIG. 1-33 An oblique instrument drawing of a transistor with a heat radiation attachment.

1-13 Dimetric Projection

Another form of pictorial representation similar to isometric drawing is dimetric drawing, or dimetric projection. In this type of drawing the reference cube is viewed from such an angle that edges on two axes are equally foreshortened. The third axis (the lines marked 3 in Fig. 1-34*a*) is shortened a different amount, however, and thus cannot be measured with the same scale that is used for the other two axes. Examples of dimetric drawing are shown in Fig. 1-34. Dimetric drawing is not as popular as isometric and oblique drawing for two reasons: (1) difficulty is encountered in drawing circular parts as represented dimetrically, and (2) it is necessary to use two different scales when laying out a dimetric drawing. However, a dimetric drawing will produce a less distorted, more pleasing effect than will either of the other two types of drawing.

1-14 Perspective Drawing

The three methods of pictorial representation described thus far in this chapter produce only approximate representations of objects as they appear to the eye. Each type produces varying degrees of distortion of any device or system so drawn. Because of the ease and quickness of their execution, however, they are the types of pictorial drawing most often used in the electrical industry.

On certain occasions, the exact pictorial representation of an object as it actually appears to the eye may be necessary or desirable. Such representation requires that the principles of perspective drawing be observed. An example of such a drawing is shown in Fig. 1-35.

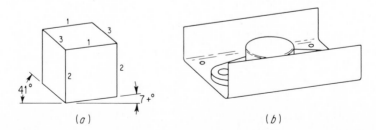

FIG. 1-34 Dimetric drawing: *(a)* cube; *(b)* TO-66 transistor can with radiator.

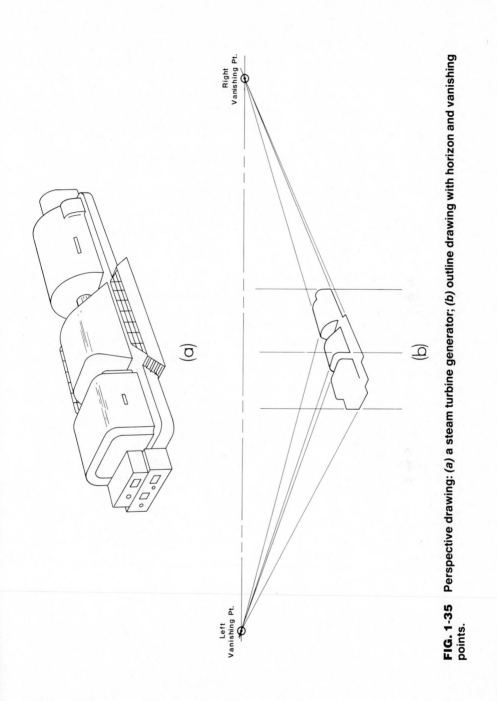

Left
Vanishing Pt.

Right
Vanishing Pt.

(a)

(b)

FIG. 1-35 Perspective drawing: (a) a steam turbine generator; (b) outline drawing with horizon and vanishing points.

29

In Fig. 1-35*b* an outline of the turbine generator has been drawn with the horizon and two vanishing points included. Several lines go to the vanishing points from the machinery. An experienced drafter can orient the object mentally and locate the horizon and vanishing points. Other ways to locate the horizon and vanishing points are to (1) use a commercially prepared perspective grid something like the isometric grid on which the chip carrier was sketched; (2) use an exact approach applying top and side views and a picture plane at 90° to the desired viewing direction; and (3) start with an isometric, oblique, or dimetric sketch and then modify it by making sets of lines converge at their respective vanishing points.

Perspective drawing is a time-consuming process which requires more explanation than can conveniently be given in this book. Most texts on engineering drawing (see Bibliography, p. 545) cover this subject.

1-15 Pictorial Sections

Occasionally, the interior detail of a device can be appropriately shown by means of a pictorial section. All types of pictorial views — isometric, oblique, and perspective — can be used to show unusual interior details with full sections, half sections, or broken-out sections.

Figure 1-36 is a section view showing the construction of a typical silicon point-contact diode. It is a full isometric section, in which the main axis is horizontally oriented. This is in contrast to the other isometric drawings in this chapter in which this axis has been drawn in a vertical position.

The same figure also illustrates the fact that standard section-lining (cross-hatching) symbols are not always adequate for electrical and electronics drawings. There is no standard section symbol for silicon. The silicon slice in this case was not section-lined. Each material has been labeled — a necessary, or at least highly desirable, feature when the American National Standard (ANSI) section symbols are not used throughout.

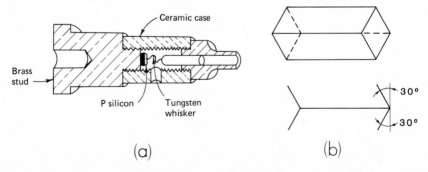

(a)

(b)

FIG. 1-36 *(a)* An isometric section of a diode; *(b)* views of a right prism from or with which the diode might have been drawn.

1-16 Hidden Lines and Centerlines

Hidden lines and centerlines are seldom drawn in pictorial drawings. This is because pictorial views are intended to show the object as it is actually seen, and hidden parts are not seen. However, hidden lines may be shown if special hidden details are desired to be shown, and centerlines may be required if an object is to be dimensioned.

1-17 References

More detailed explanation of the construction required for oblique, isometric, and dimetric views, as well as perspective, may be found in the reference books listed in the Bibliography at the end of the book.

SUMMARY

The techniques of making drawings for the electrical and electronics industry are the same as those used in making other engineering or technical drawings. With proper study, practice, and equipment nearly any person who has the desire can make most types of electrical drawing. The student having the least amount of equipment will have to work harder and longer than the person who has more sophisticated tools.

Good freehand sketching can be developed and used as an aid to lay out a drawing in the early stages. One can learn to letter well by practicing fundamental strokes and learning the shapes of standard technical lettering. There are several types of mechanical equipment that make standard lettering; however, these are rarely available to the student.

Examples of typical drawings presented in this chapter are one-line wiring diagrams; schematic diagrams; material parts lists; orthographic (multiview) drawings; and isometric, oblique, and other pictorial views. Many of these areas are covered in more detail in later chapters. Problems at the end of this chapter are of several types and have been developed with the beginning drafting student in mind. But there are some that are appropriate for the person who has had previous drafting instruction or experience.

QUESTIONS

1-1. List five different ways of drawing parallel horizontal lines.

1-2. Why is it necessary to have different line weights and different types of lines in technical drawings?

1-3. List several ways to draw circles.

1-4. Show by means of a sketch the proper position of the top, left side, front, and right side views of a cube.

1-5. If the horizontal dividing lines in a parts list are $\frac{1}{4}$ in. apart, what would be an appropriate height for the letters and numbers to be put in that space?

1-6. What lettering do you think is used the most in electrical drawing: vertical (roman) lowercase, inclined (italic) lowercase, vertical uppercase, or inclined uppercase?

1-7. In addition to making capital letters of a uniform height, how do we strive to achieve uniformity in lettering?

1-8. What is a *font?*

1-9. Why are dimensions placed on a scale drawing of an item of equipment?

1-10. What determines how far apart the views should be placed on a multi-view (orthographic) drawing?

1-11. What is the advantage of making an isometric drawing over a dimetric drawing?

1-12. What is the advantage of a cabinet-type drawing over a standard oblique drawing?

1-13. Identify a situation in which there would be an advantage in making an oblique drawing instead of an isometric drawing.

1-14. Can dimension lines be added to an isometric drawing?

1-15. List three ways to draw an ellipse.

1-16. At what angles (with the horizontal) are the "projectors" of an oblique drawing usually drawn?

1-17. What is the purpose of vanishing points in a perspective drawing?

1-18. In terms of using a scale, what is the advantage of making an instrument isometric drawing?

1-19. Why are hidden lines seldom shown in a pictorial drawing?

1-20. What is the purpose of hidden lines in a multiview (orthographic) drawing?

PROBLEMS

1-1. Referring to Fig. 1-37, make a scale drawing of the cover for a car AM/FM radio to full size. Use the English system of measurement (upper figures) or the metric system (lower figures), whichever your instructor specifies. The small holes are 0.16 in. (4 mm) in diameter, and the large holes are 0.50 in. (13 mm) in diameter. Put dimensions on the drawing if your instructor requires this. Use $8\frac{1}{2} \times 11$ paper.

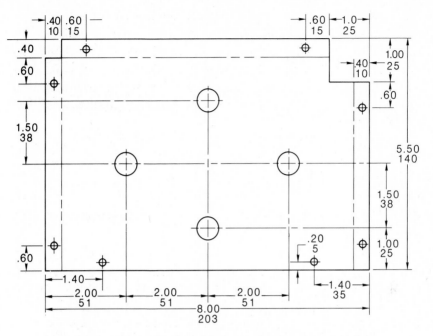

FIG. 1-37 (Prob. 1-1.) Scale drawing of car radio cover plate.

1-2. The major elements of a clock circuit are shown in Fig. 1-38, with scales along two edges having marks $\frac{1}{2}$ in. apart. These marks can be used to lay out the block diagram on a sheet of $8\frac{1}{2} \times 11$ paper. Make neat uppercase letters that will fit in the block and numbers and letters identifying the lines between output part and clock chip.

1-3. Two views of a robot are shown in Fig. 1-39. Marks around the edges are $\frac{1}{2}$ in. (1 cm) apart and may be used for drawing purposes. Draw either view or both, whatever your instructor assigns. Small crosses in the side view show the centers of the three wheels, one in front and two in the rear (one view on $8\frac{1}{2} \times 11$ paper).

1-4. An elevation of the receiving tower for a central solar collector is shown in the left part of Fig. 1-40. Dimensions are in meters. Two details are shown at the right. Draw an elevation of this tower and show the major dimensions that are indicated on the elevation. Use $8\frac{1}{2} \times 11$ paper.

1-5. Figure 1-41 is the layout of the cover for a smoke alarm. Make a scale drawing of this cover. Dimensions are in inches. Use the alternate position line (Fig. 1-5) for the fold lines. Use $8\frac{1}{2} \times 11$ paper.

1-6. Make a drawing of the photodiode shown in Fig. 1-42. Include the lettering (on $8\frac{1}{2} \times 11$ paper).

1-7. Make a drawing of the third rail section shown in Fig. 1-43. Where

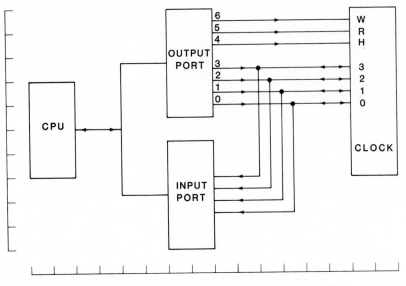

FIG. 1-38 (Prob. 1-2.) Diagram of a clock circuit.

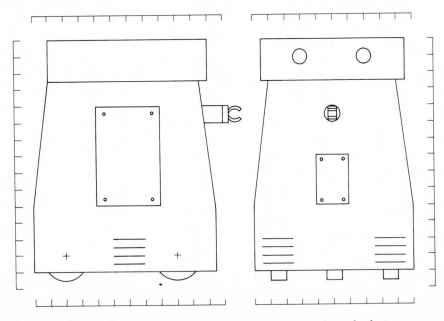

FIG. 1-39 (Prob. 1-3.) Left side view and front view of robot.

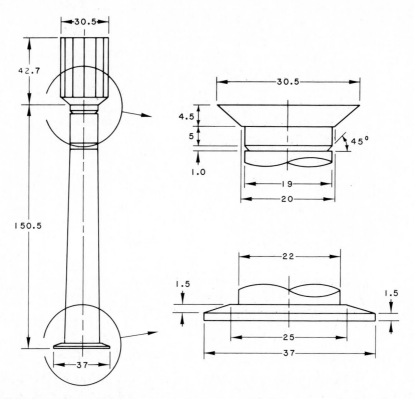

FIG. 1-40 (Prob. 1-4.) Elevation view of central receiving tower for solar electric power generating station.

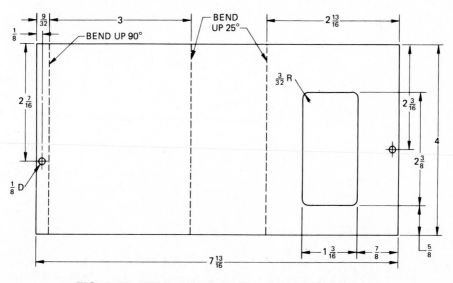

FIG. 1-41 (Prob. 1-5.) Cover for smoke alarm chassis.

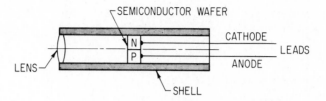

FIG. 1-42 (Prob. 1-6.) Diagram of a section of a photodiode. The length is 4 times the diameter.

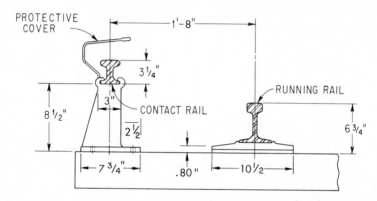

FIG. 1-43 (Prob. 1-7.) Section showing the third rail detail of a mass transit system.

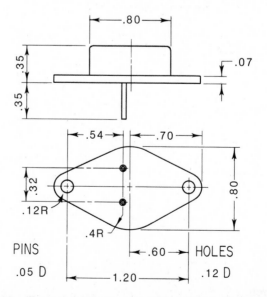

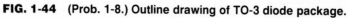

FIG. 1-44 (Prob. 1-8.) Outline drawing of TO-3 diode package.

dimensions are not given, make your shapes as close to those shown as possible. Add lettering and dimensions if required by your instructor. Use $8\frac{1}{2} \times 11$ paper.

1-8. Make an outline drawing of the diode shown in Fig. 1-44. Use an enlarged scale so that this drawing will fit on an $8\frac{1}{2} \times 11$ sheet or so that it and Prob. 1-9 can both be put on an $8\frac{1}{2} \times 11$ sheet. Dimensions are in inches.

1-9. Make an outline drawing of the transistor shown in Fig. 1-45. Use an enlarged scale so that this will fit on an $8\frac{1}{2} \times 11$ sheet by itself, or with the diode from Prob. 1-8.

1-10. Make a two-view outline drawing of the microprocessor socket shown in Fig. 1-46. Dimensions are in inches. The pins are 0.20 in. apart (across) and pins and holes are 0.10 in. apart (center to center) in the long dimension. Use $8\frac{1}{2} \times 11$ paper.

1-11. Make a two- or three-view drawing of the integrated circuit package shown in Fig. 1-47. Use an enlarged scale so that the drawing will fit on $8\frac{1}{2} \times 11$ paper. Dimensions and title may be added if the instructor so specifies. Tops of pins are 0.015 in. wide.

1-12. Make two or three views of the subchassis shown in Fig. 1-48. It may be shown as a development or formed as shown. Hole diameters are: No. 1, 40; No. 2, 7.5; No. 3, 18; No. 4, 7.5; and No. 5, 7.5 mm. Material: SAE 1020 steel. Add notes: *Break all sharp edges, remove all burrs, degrease per Spec. 51606, bright dip per Spec. 51606.* All dimensions are in millimeters. (Two views, $8\frac{1}{2} \times 11$.)

1-13. The dimensions of the mounting bracket shown in Fig. 1-49 are in inches. It is made of 16-gage (0.062-in., 1.6-mm) SAE 1020 steel. The thickness may be exaggerated as you draw two orthographic views on $8\frac{1}{2} \times 11$ paper. Dimensions may be added if your instructor so directs. The bottom four holes are 0.50 in. in diameter.

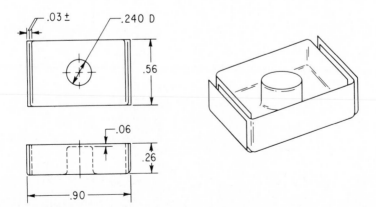

FIG. 1-45 (Prob. 1-9.) Front and top views of a TO-5 transistor "can" with heat radiation attachment.

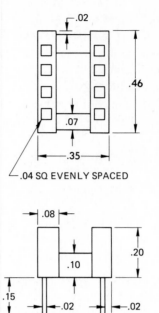

.02

.46

.07

.35

.04 SQ EVENLY SPACED

.08

.20

.10

.15

.02

.02

FIG. 1-46 (Prob. 1-10.) Outline drawing of a socket for a microprocessor.

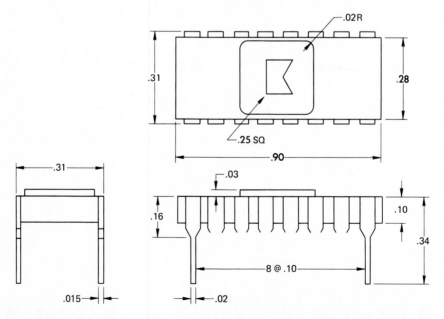

.02R

.31

.28

.25 SQ

.90

.31

.03

.16

.10

.34

8 @ .10

.015

.02

FIG. 1-47 (Prob. 1-11.) Outline drawing of 16-pin DIP ceramic case.

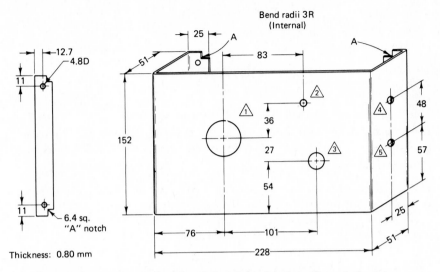

FIG. 1-48 (Prob. 1-12.) Pictorial and partial view of a subchassis.

1-14. In $\frac{1}{8}$-in. uppercase letters, letter the following terms or notes:

a.   0–099   Parts on display board
b. 101–199   Parts on arm drive board
c. 201–299   Parts on motion board
d. 301–399   Parts on power supply board
e. 401–499   Parts on CPU board

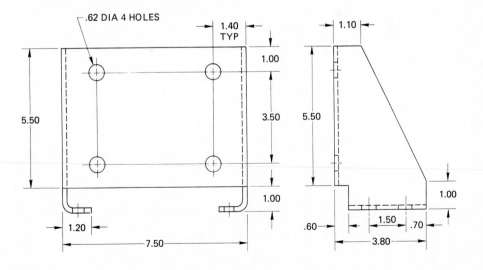

Note: Four Bottom Holes are .50 Dia.

FIG. 1-49 (Prob. 1-13.) Two views of a mounting bracket.

1-15. With $\frac{1}{8}$-in. uppercase letters, letter the following:
a. Clock enable
b. 90 mW across 50-Ω load
c. MC 1488 pin 14
d. All resistors are $\frac{1}{4}$ W $\pm$ 5 percent
e. $V_{CC} = 4.75$ V

1-16. In a space of about 6 $\times$ 12 cm make a set of general notes as follows: Resistance values in ohms; Capacitance values less than one in microfarads, one and above in picofarads; Direction of arrows at controls indicates clockwise rotation; Voltages should hold within $\pm$20 percent with 117 V ac.

1-17. With uppercase letters $\frac{3}{32}$ or $\frac{1}{8}$ in. high, letter the following specific notes:
a. Knockout for $1\frac{1}{2}$-in. conduit may enter top or bottom of cabinet
b. $\frac{9}{16}$ diameter (4 holes) for wall mtg.
c. $5\frac{1}{2}$ $\times$ 6 cutouts for secondary cable conn.
d. Tie-bolt holes $\frac{1}{2}$ maximum
e. All pins 0.093 $\pm$ 0.003 DIA
f. 187 $\pm$ 0.003 DIA, 4 pins

1-18. Make a parts list with the following headings: Item No., Company No., Description, No. Required. Place the following items in the six spaces:

A_3	5260	Drive motor	1
F_3	241-1	3-A fuse	2
C_1	12-60	0.05-μF ceramic capacitor	1
SW_1	62-1	Switch	1
U_{301}	556	Integrated circuit	3
Q_{704}	768	Transistor	2

Use uppercase letters, using Fig. 1-17 as a guide.

1-19. Using inclined upper- and lowercase letters, letter the following equipment descriptions, one below the other, as they would appear beside a vertical one-line diagram:
a. 1/0 ACSR
b. 3.30KV Line-type Lightning Arresters
c. 23KV 200 AMP S&C Hook-operated Disconnect
d. 3000KVA SC-3917KVA FAC West. Transformer 22000Δ $-$ 4160Y / 2400 volts
e. 3-CT's 1000/5
f. Watthour Meter W/Demand
g. 5KV 600A Air Circuit Breaker
h. 400-AMP 5KV Enclosed Disconnect @ 3$-$4/0 cu.

1-20. Using the instructor's selection from among the following, letter the notes or sentences:
a. 40-pin DIP package
b. $\frac{9}{16}$ diameter (4 holes)

c. Unless otherwise noted, all resistances are in ohms

d. 144KV surge arresters

e. 556.5 MCM ACSR

f. $16.1 - 13.2$KV transformer

g. $0.01\ \mu$F ceramic capacitor

h. A_{12} 473-34 sonar transducer

i. U_{502} 442-734 LM388 IC

In many of these problems, one or more types of pictorial drawing can be used to depict the object or system satisfactorily. In only a few cases, therefore, has a specific type of pictorial view been required. Many of the objects can be drawn pictorially as freehand sketches or as mechanical drawings. Your instructor may, in addition, require you to put dimensions on the pictorial drawing.

1-21. Make a pictorial drawing of the bracket shown in Fig. 1-50. Position the holes as closely as you can to the way they appear in the figure. Use $8\frac{1}{2} \times 11$ paper.

1-22. Make an isometric drawing of the automatic guided vehicle (AGV) shown in Fig. 1-51. Orient it so that the imaginary point 0 is at the bottom and corner 1 is to the left and above 0.

1-23. Make a pictorial drawing of the microcircuit waveguide accessory shown in Fig. 1-52. Use the rough scales along the left edge for correct proportions. Use $8\frac{1}{2} \times 11$ paper.

1-24. Make a pictorial drawing of the IC package shown in Fig. 1-53.

1-25. Make a pictorial drawing of the bracket shown in Fig. 1-49. It is made of 16-gage (0.062-in.) steel. Thickness may be exaggerated.

1-26. Draw an oblique view of the subchassis shown in Fig. 1-48. Orient so that holes 1, 2, and 3 may be drawn as circles. *Optional problem:* Draw the

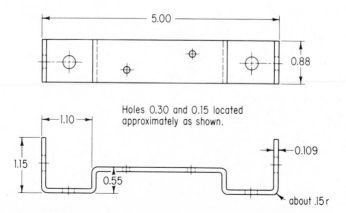

FIG. 1-50 (Prob. 1-21.) Front and top views of a bracket.

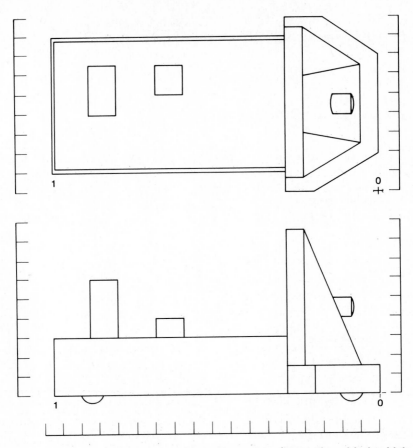

FIG. 1-51 (Prob. 1-22.) Front and top views of an automatic guided vehicle.

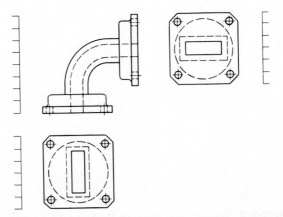

FIG. 1-52 (Prob. 1-23.) Three orthographic views of a waveguide bend or elbow.

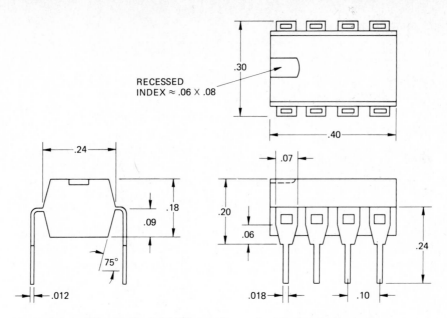

FIG. 1-53 (Prob. 1-24.) Three views of an 8-pin plastic package.

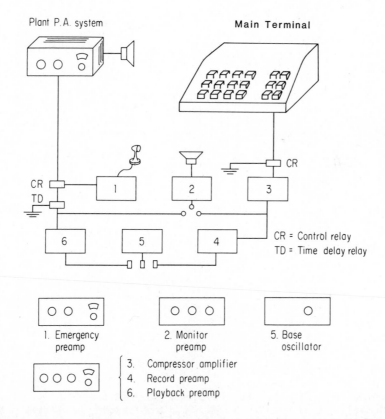

FIG. 1-54 (Prob. 1-27.) A partially completed pictorial diagram of an automatic paging system.

bracket in the isometric pictorial. *Optional problem:* Draw an oblique pictorial and an isometric view of the bracket.

1-27. Construct a complete pictorial drawing of the automatic paging system shown in Fig. 1-54. The front panel of each of the six blocks is shown in the figure. The rest of each unit can be shown as the plant public address (PA) system, upper left, appears. If oblique projection is used, the emergency preamplifier (block 1) will appear almost the same as the plant PA, upper left. Label each unit and the relays. Assume all items (1 through 6) are the same size. Use 11×17 or 12×18 paper.

1-28. Make a pictorial drawing of the personal or business computer system shown in Fig. 1-55. Add the plotter (shown at lower right) if your instructor so indicates. Maintain the relative size of the terminal, monitor, and peripherals,

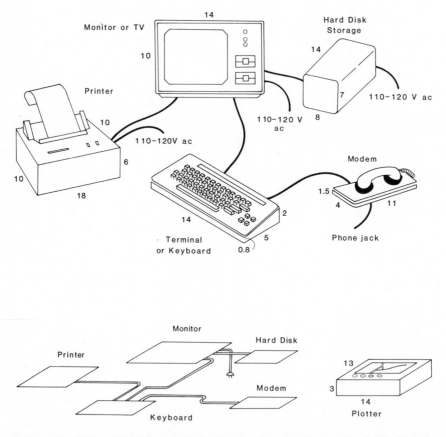

FIG. 1-55 (Prob. 1-28.) Elements of a personal computer system. Overall height, width, and depth dimensions are shown except for depth of monitor. A partial oblique pictorial is shown below.

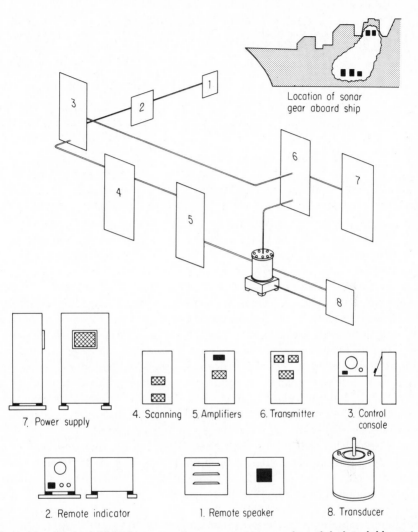

Location of sonar
gear aboard ship

7. Power supply

4. Scanning 5. Amplifiers 6. Transmitter 3. Control console

2. Remote indicator   1. Remote speaker   8. Transducer

FIG. 1-56 (Prob. 1-29.) Elements of a sonar system and partial pictorial layout.

using the overall dimensions shown in the figure. Use 11×17 or 12×18 paper.

1-29. Make a pictorial drawing of the sonar system shown in Fig. 1-56. Show each unit at the location shown in the partial pictorial diagram. Maintain the proportional sizes of the units. Use 11×17 or 12×18 paper.

2

Computer-Aided Design

2-1 Introduction

Today, computer-aided design (CAD) is a reality. In the last few years major changes have taken place in the field of CAD. Systems have become more rapid in terms of both input and output, larger and — yes — even smaller, easier to use, and less expensive. With the invention of the microcomputer and the continuous compression of more memory per integrated circuit (IC), a great variety of systems have been developed, thus making CAD available to not only large businesses, but also small businesses and individuals. Presently CAD systems range from the large mainframe computer systems that support many workstations to the small, single personal computer-based workstation.

It is the intent of this chapter to acquaint the beginning drafter with CAD. We will examine basic systems, equipment, input devices and how to use them, output devices, and finally the output. Also, photographs of each type of CAD equipment that the beginner might encounter are shown. Throughout the chapter, CAD terminology and "buzz" words are defined as they are used. But before this is done, some general background is presented.

Computer-aided design, computer-aided drafting and design (CADD), automatic drafting systems (ADSs), computer graphics, and interactive design and drafting are essentially different names for the same thing — a design and drafting tool which, through the assistance of a computer, provides significant, more accurate, and better-quality productivity increases for the designer and drafter. Originally, ADSs were capable only of drafting, but with the advent of the interactive graphic CRT, they also incorporate design features. Some CAD systems do primarily design but also have some engineering software packages. Computer-aided design systems are being used in all types of engineering, including structural, mechanical, electrical, automotive, aerospace, and civil; however, because this text concerns electrical and electronic drawings, this chapter is orientated in that direction.

In general, the term *computer-aided design* implies that the computer assists the designer in modifying existing designs and creating new designs. Almost any designer or drafter can use a CAD system to create, modify, and distribute graphical data with only a limited or no knowledge of programming and no need to know that graphical data are being manipulated. This is because the designer and the computer are in a two-way communication or interacting, using the CAD system's predescribed symbols and text. While these symbols or text are just that to the designer, they are actually graphical software functions, which are processed by a computer to appear as symbols and text.

Almost any type of drawing associated with the electrical or electronic industry can be created and modified on some CAD systems. It should be stated here that most CAD systems cannot do *all* types of electrical and electronic drawing. They are generally specialized to perform one or more areas (e.g., IC artwork, printed-circuit (PC)-board layout of design); however, some extremely large mainframe computer systems come very close to creating all types of electrical and electronic design drawing. Computer-aided design systems associated with fields of IC and PC-board layout are generally the leading edge of CAD electrical and electronic design. Some typical design capabilities of a CAD system are:

1. Integrated or hybrid IC layout (artwork)
2. Printed-circuit layout and component arrangement
3. Logic diagrams, schematics, or block diagrams
4. Plan views, general arrangements, and wiring diagrams
5. Perspective and isometric drawings
6. Two-dimensional (with multiple views) and three-dimensional views
7. Perspective and isometrics
8. Exploded views
9. Material and parts lists with associated legends
10. PERT charts, flowcharts, and maps
11. Plots of numerical data
12. Text with multiple fonts
13. Direct manufacturing information, such as paper tape for driving numerical control (N/C) equipment

A CAD system has many advantages. It is faster, more accurate, and less expensive in comparison to the conventional way of doing design and drafting. For example, a CAD system can do dimensional accuracy of an IC layout that is beyond the drafter's ability. So because of this ability, as others, a CAD system is an absolute necessity. A CAD system can reduce overall costs up to 50 percent. For certain types of drawing, especially repetitious ones, labor costs can be cut 90 percent. A CAD system can increase a drafter or designer's productivity up to 3 or 4 times. With the use of a CAD system, drawings are usually better and revisions are faster and easier. Symbols, lettering, line weight, notes, and other style and format factors are more consistent. CAD systems can give the designer the option of looking at multiple views of the same design, rotating the image, making design changes, instantly seeing these changes, and being able to store or output them. In addition, a CAD system will take over the drudgery of simple, repetitive drafting chores, leaving the designer free to do more important aspects of the design. A CAD system also allows the owner to store design drawings on magnetic disks or tape, rather than keeping and storing the actual drawings.

2-2 The Basic System

Years ago, CAD systems were fairly typical in that they were all driven by a central computer. Today CAD systems range from large central host computer-driven systems to stand-alone microprocessor-based systems or a combination of both. Figure 2-1 shows many of the components of a CAD system. Some of this equipment is obsolete, but this figure shows a large selection of CAD equipment. In the center there is an alphanumeric terminal (CRT and keyboard), a graphics monitor with a function tablet, and a magnetic stylus. From left to right, there is a photoplotter, an ink plotter, a graphics processor with a keyboard and tape drive, and a digitizing table with a hockey puck-type digitizer.

Figure 2-2 is a photograph of some of the state-of-the-art CAD equipment. In the foreground there are two graphics terminals, each equipped with a digitizing (graphics) tablet, a hockey puck-type digitizer, a magnetic stylus, and a hard-copy unit. In the left rear of the photograph is a mainframe computer and on the right, a graphics processor. In front of the graphics processor is an alphanumeric terminal.

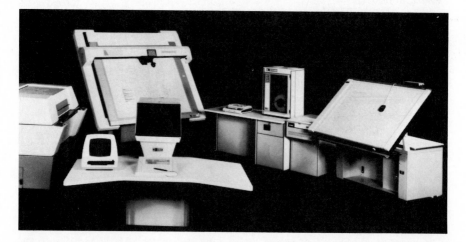

FIG. 2-1 Photograph of many of the CAD system components. In the center is an alphanumeric terminal (CRT and keyboard), a graphics monitor with a function tablet, and a magnetic stylus. From left to right, there is a photoplotter, a belt-bed plotter, a graphics processor with keyboard and tape drive, and a digitizing table with a hockey puck-type digitizer. (Computervision.)

FIG. 2-2 Some state-of-the-art CAD equipment. In the foreground are two graphics terminals, each equipped with a digitizing tablet, a magnetic stylus, and a hard-copy unit. In the background on the left is a mainframe computer. On the right is a graphics processor, and in front of it is an alphanumeric terminal. (Computervision.)

From a designer's point of view, however, there are some typical aspects of all CAD systems. First, all CAD systems revolve around a workstation where the designer or drafter operates. The CAD workstation consists of an alphanumeric keyboard, a graphics CRT or graphics monitor or two, possibly an alphanumeric terminal, some type of graphic CRT cursor moving or locating device (keyboard keys, thumbwheels, mouse, joystick, trackball, light pen, or magnetic pen-digitizer tablet), and some type of graphic input devices (keyboard keys, mouse function buttons, light pen or magnetic pen-digitizer tablet, or hockey puck digitizing tablet). As will be discussed in the next section, some of these input devices can do more than one function.

If the CAD system revolves around the workstation, the workstation revolves around the graphics CRT or monitor. Please note that when the term *monitor* or *CRT* is used in this chapter, it refers to only the CRT and not CRT and keyboard. The term *terminal* refers to both keyboard and CRT. There are three basic types of CRT: the vector, the raster, and the storage tube. The vector CRT utilizes the *XY* coordinate system. The computer first locates the points on the CRT and then connects them. The process by which vector images are written is called *refresh.* An image is vector-refreshed with each change in the drawing. The raster-type CRT is similar to a television screen. Displays are made in a vertical and horizontal grid. The area within each area between the grid lines appears as a dot and is called a *pixel.* The closer the grid lines are, or in other words, the more pixels per unit area there are, the sharper or clearer the graphics picture is. This is called *resolution.* The storage tube CRT allows a drawing to be plotted without the refresh of a vector CRT. The inside surface of the tube is coated with phosphor and when hit with electrons, continues to give off picture. Some storage tubes have the ability to hold a picture without flickering. Vector and storage CRTs are available in monochrome, while raster are available in both monochrome or color. Graphics CRT are manufactured in all sizes ranging from 9 to 25 in. The raster and storage type CRTs are the most popular today.

If the workstation is part of a stand-alone system, it will also have a micro- or minicomputer and one or more disk drives. The disk drives will be a combination of hard and floppy. Hard disks may be removable or nonremovable; the floppy disks are always removable. If a workstation is part of a large system with multiple workstations, there will be a large host computer, multiple disk drives, and graphics conditioning and manipulating equipment associated with the workstation. The host computer and associated equipment may be located adjacent to or remotely from the workstation.

A plotter or a hard-copy unit printer may be associated with the workstation. In some installations, however, the plotters are remote from the workstation while a hard-copy unit may be located near the workstation. A plotter produces a full-size drawing while the hard copy produces a reduced scale (in range of 12 X 14 in.) drawing to be used for quick reference. Figure 2-3 shows a stand-alone CAD workstation which has a portable microcomputer with two associated floppy disk drives and one hard disk drive, a 19-in. color monitor for viewing the graphic design on an adjustable arm, a mouse for moving the monitor's cursor, a full-function alphanumeric for inputing graphical data, and a D-size drawing ink pen plotter.

Most workstations are ergonomically designed, which means that they are designed to accommodate the physical features and motions of the operator. Most ergonomically designed workstations have removable keyboards. The CRTs have adjustable viewing angles to prevent glare and eyestrain. All the controls are within easy operator reach and designed for optimum comfort levels. The work surfaces are large to accommodate the necessary design or reference material. Workstations have comfortable chairs and spacious legroom.

The CAD system is usually housed in an area that has a controlled environment of both temperature and humidity. Because of the intricate nature of the

FIG. 2-3   A standalone CAD system having (from left to right) a portable computer with two floppy and one hard disk drives, an ink pen plotter (behind table), a mouse for monitor cursor positioning (on table), a 19-in. color monitor (attached to the table), and an alphanumeric keyboard (on table). (Bruning CAD.)

equipment, this area is generally secured or access-controlled to prevent unauthorized personnel from tampering with the components or accidently damaging them or from having access to a company's proprietary or secret material.

A CAD system uses normal electric power — typically 115 V ac, 50 A, single-phase. However, current requirements may exceed 80 A, depending on how much peripheral equipment is used. Large current requirements will sometimes dictate the use of larger multiphase voltage of 240 and 208 V ac. Some CAD systems, because of the criticality of the work being done, require uninterruptible power systems.

A CAD system consists basically of four functional elements: the input devices, the processing equipment, the output devices, and the software (or programs). Figure 2-4 shows a pictorialized block diagram of a CAD system emphasizing the four major parts. These basic parts and their operation and interaction are discussed in the following sections.

2-3 Software

Because a CAD system uses a computer, it is controlled and run by a set of instructions called a *program* or *software* (to distinguish it from the physical

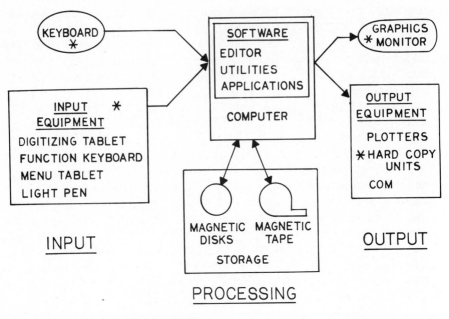

✱ THIS EQUIPMENT IS LOCATED
AT THE WORK STATION.

FIG. 2-4 Block diagram of the major components of a CAD.

equipment, such as a computer, which is categorized as hardware). The designer or drafter seldom writes or sees the software. By acting on the input and output devices, however, the designer executes it and knows its capabilities. This process is known as *interactive* communications. Therefore, a cursory knowledge of the workings of software and software packages is necessary.

Essentially, the computer is a giant calculator, but it must be told precisely what, how, when, and where to do these calculations. It is told what to do by a *program*. The program is written in either a high-level language, an assembly language, or a machine code. However, no matter which language the program is written (or coded) in, it must be converted into machine code or the operating code of the computer. Original programs are entered into the computer by a programmer using an alphanumeric terminal; if the program has been previously converted to machine code, it will probably have been stored on magnetic disks (hard or floppy) or magnetic tape (in some cases). The programs can be retrieved from the devices as necessary when required by the system. In some CAD systems the applications software allows users to develop their own process-oriented functions or use their own written programs. In essence, the application software is the user-machine communications tool whereby the designs are created.

Custom-made or standard ("canned") programs are available from CAD system manufacturers, software-developing houses, major manufacturers, and the government. Software packages are available for electrical design for all the items mentioned in the introduction of this chapter, including electrical schematics, logic diagrams, layouts, IC and PC design and artwork, detailing, general arrangements, wiring diagrams, and bill or materials, plus a host of programs in other disciplines, including structural, architectural, mechanical, and civil engineering.

High-level languages such as FORTRAN (FORmula TRANslation) or Pascal are easier to write programs in. They require fewer statements than do programs written in assembly languages or machine codes for the same tasks. High-level language programs are converted by the computer into machine code by a program called a *compiler.* Assembly languages are resident to only one manufacturer's type or series of computer. Compared with higher-level languages, assembly languages are more difficult and time consuming to write programs in; however, when they are executed, they are generally faster and more efficient. Assembly-language programs are converted into machine code by a program called an *assembler.* It is the program that allows the designer or drafter to communicate the drawing information (data) to the computer by permitting the data to be entered, manipulated, verified, corrected, and outputted. The input and output devices are discussed in the following sections.

The applications software which the designer uses should be of English-language format. It will contain all the necessary commands for graphics design, editing, plotting, displaying, and view manipulation. In some CAD systems, the applications or graphics software permits designers to design and implement their own function menus for a digitizing tablet, hockey puck-type digitizer function keys, or function keyboard. The applications software will probably offer the user help or tutorial instructions that assist in the understanding of the applications software. The applications software permits the designer to set up, label, and protect files for the designs to be created and store these files of designs on a disk.

2-4 Input Equipment

The workstation operator uses the input equipment to communicate with the CAD system. With this equipment the operator creates the graphical drawing with and issues instructions to the CAD system to have the drawings outputted in some form (storage, hard copy, finished drawings, etc.). In most CAD systems, input equipment is associated with the CRT or a digitizer table or tablet.

It is difficult to understand the input equipment without understanding the input processes. To do any type of CAD drawing, an operator must perform two basic input operations: (1) communicate to the system "what" to put in the graphics drawing and (2) indicate "where" to put "what" on the graphics drawing.

Let's discuss the "where" operation first. In the use of the CRT this is determined by the location of the cursor, which is the movable bright mark on the screen. The cursor may be a dot, a cross hair, a check, etc. The position of the

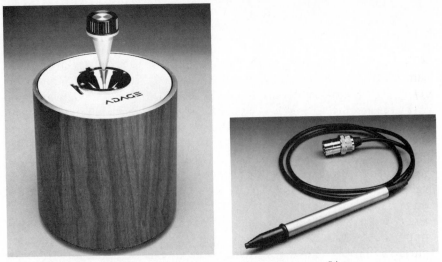

(a)

(b)

(c)

(d)

FIG. 2-5 CRT cursor positioning devices: *(a)* a three-axis input joystick (ADAGE, Inc.); *(b)* a light pen (ADAGE, Inc.); *(c)* a mouse and associated digitizing (graphics) tablet (Summagraphics Corp.); *(d)* keyboard thumbwheels (Tektronix, Inc.).

CRT cursor may be controlled by any number of devices, which include keyboard keys, a joystick, a trackball, a mouse, a light pen, a magnetic stylus, or a hockey puck with digitizer tablet.

Figure 2-5 shows several of these cursor positioning devices. Figure 2-5*a* shows a joystick, which is not unsimilar to those joysticks found in video arcades, but of much greater precision. This joystick is a three-axis device that can also be used for analog data input. The joystick is moved in the same direction as the user desires the cursor to move on the screen. Figure 2-5*b* shows a light pen. The light pen is a hand-held, light-sensitive, high-speed photopen used to detect visible positions on a display monitor. The light pen assists in operator interaction with image data. It must be noted that the CRT or video monitor must be designed for use with a light pen, or a light pen cannot be used. Figure 2-5*c* shows a mouse. Like the joystick, the movement of the mouse corresponds to a cursor movement on the monitor screen. The pushbuttons and call function buttons on the mouse are discussed later. Figure 2-5*d* shows keyboard thumbwheels. Two thumbwheels, one for a horizontal line movement (top to bottom) and one for a vertical line movement (left to right), move to form a cross hair, where it is desired to enter graphical data. Trackballs (not shown) operate in a similar manner to the joystick.

Figure 2-6 shows a graphics tablet and a hockey puck and magnetic stylus (pen) to the right of the tablet. The tablet is a unique input device. A rough sketch may be placed on the table, the stylus or puck is used to trace the sketch, and the traced sketch appears on the monitor. This is done in the following manner. The tablet has a grid pattern beneath the surface. The puck or stylus is pressed on the tablet surface at a particular location, creating electrical pulses. This location (horizontal and vertical coordinates) or digital signal is transmitted to the computer. This process, on creating the digital signals, is called *digitizing,* and the tablet is called the *digitizer.* Thus the assembling of many close locations or coordinates actually creates a sketch.

Figure 2-7 shows a small graphics tablet and its associated magnetic stylus and controller. This tablet performs in the same manner as the one in Fig. 2-6, except that it can be used at a desk or a workstation.

It should be stated that some of the "where" input devices can also be used as "what" input devices, as will be discussed soon.

What a workstation operator puts on the graphics drawing is determined by software functions. Whether this is a line, a circle, or an electrical symbol, its graphic representation is determined by a software function. The variety of ways to input graphical data to the computer are many, as we shall soon see; but this graphical data or function must reside in the CAD system storage so that it can be retrieved and then inputted on a particular drawing. Each CAD system has many libraries or groups of associated functions, such as libraries of electronic symbols, geometric shapes, etc. Some devices used for graphical data input are a function keyboard, a menu tablet, a light pen, and a keyboard. These devices are explained in the following paragraphs.

Figure 2-8 shows a programable function keyboard. Each key is programmed to represent a different software function. These functions may be

FIG. 2-6 A large digitizing (graphics) tablet and hockey puck-type digitizer used for inputting graphical data. (Summagraphics Corp.)

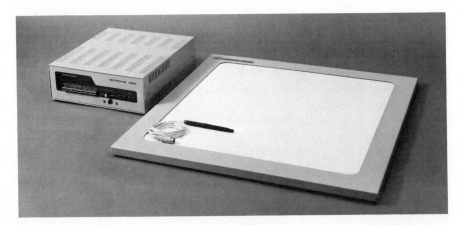

FIG. 2-7 A digitizing tablet and its associated controller and magnetic stylus (pen). (Tektronix, Inc.)

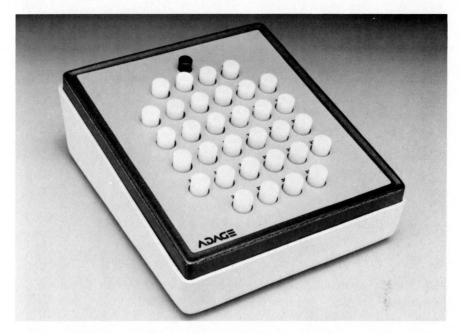

FIG. 2-8 **A programmable function keyboard.** (ADAGE, Inc.)

system commands, symbols, text, etc. Figure 2-9 shows a workstation operator doing a graphical drawing of a PC board. The operator is using a light pen for cursor position location and a functional keyboard for graphical data input.

Remember earlier that we said that when the stylus touches the tablet, an electrical pulse is sent back to the computer signifying coordinates; these discrete coordinates can also correspond to software functions if the computer is so programmed. By putting a diagram or *menu* of symbols and functions on top of the graphics tablet, a menu tablet has been created. A menu tablet is an integral part of most CAD systems. In many CAD systems, the tablet may be programmed for many different function menus on the number of software functions libraries available in the system. On some menu tablets the tablet covers that physically show the menu symbols and commands are changeable. Figure 2-10 shows a closeup view of a menu tablet. Figure 2-11 shows a workstation with an alphanumeric terminal, a graphic monitor, and a menu tablet with a stylus for graphic data input.

Another device that we have discussed for cursor positioning—the light pen—may also be used for graphical data input. In order to use the light pen for graphical data input, the menu of software functions (commands or symbols) must appear on the screen of the graphic monitor.

The alphanumeric keyboard is equipped with a special cursor positioning and graphical data inputting device. Cursor keys are used for cursor positioning, while the keyboard graphical data input is usually text of alphanumeric commands that represent graphical data. In most instances, the alphanumeric keyboard is too cumbersome for use with graphical data input.

FIG. 2-9 A workstation operator using a light pen for position location and a function keyboard for graphical data input. (ADAGE, Inc.)

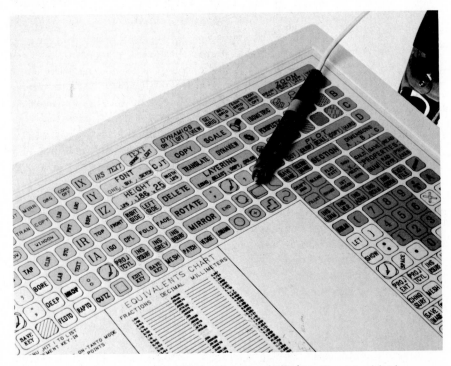

FIG. 2-10 A menu tablet and magnetic stylus. (Computervision.)

FIG. 2-11 **A workstation with an alphanumeric terminal, a graphics monitor, and a menu (graphics) tablet and associated stylus.** (Cascade Graphics Development.)

Now that we have examined the CAD system input devices, we will discuss the input process.

2-5 Input Process

It should be emphasized that the input process differs from CAD system to CAD system. These systems range from those that have two input devices (keyboard and some other) to systems with a multitude of input devices and ways to input data and from systems that have a limited amount of functions to systems that have many libraries of functions of symbology augmented by many functions of display manipulation and control. The following is a generalized example on how to use a CAD system.

One must first log in on the system by typing one's name and then a password at the keyboard. In large systems, the system manager is responsible for assigning user names, user identification codes (UIC), and passwords. An operator who doesn't know what to do next may type in the name of a help or tutorial program. Some systems are also self-prompting—or asking the operator what to do.

At this point, in some systems the operator may have the option of selecting the menu for the type of design that will be performed (e.g., architectural, civil, electronic, electrical). This menu will be associated with an input device such as a function keyboard, a graphics (menu) tablet, screen menu, etc. For example, a graphics tablet menu may contain all or a portion of command functions such as symbology, graphic elements, three dimensions (3-D), dimensioning, view controls, display controls, digitizing, and text controls.

Command functions such as those of symbology (resistor, capacitor, logic gate, etc.) and graphical elements (e.g., line, arc, rectangle) determine what is entered in the graphical drawing. Dimensioning functions might include dimension placement, justification (right or left hand or center), and control of the witness line. The view control commands are commands that manipulate the screen view that the operator sees, such as multiple views of the same drawing, zooming the drawing in and out, provide and rotate the view as in 3-D, move selected areas of the screen, etc. The display controls will allow the operator to control the basic drafting functions such as drawing line weight, patterns, line construction, dimensional data, and construction. Digitizing, as we described earlier, will allow the operator to digitize parts of the drawing. The text controls will include letter font, display, change, and justification. Note that text is entered at the keyboard itself.

Now that menu is set, the operator may have to modify an existing drawing by typing in the drawing identification, which retrieves it from storage. To create a new drawing, the operator types in an identification of that drawing. In typing a drawing identification, the operator has told the system the drawing name and to reserve storage for this drawing on the disk.

The operator may next have an option to set up drawing grid symbology. This may consist of light lines, dots, or plus signs. It may appear or disappear as the operator requests. The grid may change with change of views.

During the design session itself, the operator can place, delete, modify, move, rotate (if 3-D), and copy various drawing symbols to a new or existing drawing. On some CAD systems the operator may have dual monitors, one alphanumeric for tutorial commands and instructions and one graphic screen for display of the drawing being modified or created. Remember that the operator, while creating a drawing, is telling the system "what," or which graphical data to use, and "where," or in which position to place this data. The operator creates the drawing by the use of the graphical data and cursor positioning input devices discussed in the previous section.

Some helpful hints, based on experience, in regard to creating CAD drawings are:

1. Save or store the drawing about every 10 to 15 min. This will assure minimal time loss in the event that a workstation or system ceases to operate.

2. If possible, begin a new drawing by copying and editing another drawing that has been completed.

3. Use a standard drawing as a starting point unless you are doing item 2 above.

4. Use tablet menus whenever possible. (Graphics tablets seem to be the most predominant method of data input on most CAD systems.)

5. Use a symbol library where possible, rather than creating your own symbols.

6. If you create a symbol for a library, make sure that it has been properly stored. Check to see if it exists.

7. Avoid time-consuming special line styles unless they are a part of the menu.

8. Use coordinates when accurate positioning is required.

9. Minimize changing tasks; for example, place all symbols, and then enter text.

10. Know your CAD system and all its features so that you may make best use of them.

2-6 Processing Equipment

The processing equipment is the "brains" of the CAD systems, consisting of a computer [central processing unit (CPU)] and the necessary data conditioning equipment and storage. The computer is not really a brain, as most of us know it, but a very large and fast calculating machine that must be programmed. In CAD work voluminous amounts of data must be handled, manipulated, and processed to create the graphic displays and outputs. The computers used in CAD systems, whether they are the large mainframe host computers, the mini-computers, or the microcomputers used in the standalone system, usually are the state-of-the-art design in terms of architecture and speed.

The mainframe central computers usually have a 32-bit processor (or processors) and a large, expansive operating system, which is a software program (or programs) governing the operation of the machine and the remote workstations. Figure 2-2 shows a small mainframe computer. The operating system is capable of directly accommodating approximately 20 to 30 workstations. Mainframes can perform a multitude of other types of processing. They have the largest amount and greatest variety of software associated with them. Mainframes were the original source of processing for the first CAD systems. The mainframe has an on-board memory (RAM) of up to 8 megabytes and extremely large, mass (disk) storage of up to 160 gigabytes. (A byte is 8 bits of information.) Mainframes, unlike mini and micro systems, are usually located in a location different from the workstation.

Minicomputer systems are the stand-alone or dedicated systems. They utilize 16- to 32-bit processor architecture. The term *stand-alone* means that the system doesn't have to feed its data into a mainframe computer. More than one but generally not more than three terminals may be connected to a mini-computer system. These systems have up to 4 megabytes of on-board (RAM) storage and the capability of using 300 megabytes of mass storage. Figure 2-12 is a "super" minicomputer-based workstation. The computer is shown in the left-hand portion of the photograph. It has 32-bit architecture. The system supports both computational and color graphics process with a palette of 16 million colors and the ability to conduct 24 simultaneous processes. The system can also support several languages (e.g., FORTRAN, Pascal, C).

The microcomputer system is used for small, dedicated applications. Many of these systems are desk-top or personal-computer-based systems. These

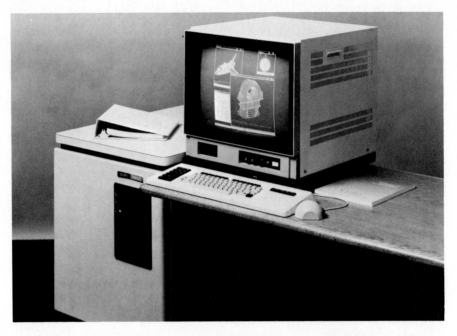

FIG. 2-12 A minicomputer-based CAD workstation. (Apollo Computer, Inc.)

units generally have computers with 16-bit architecture, but some have 32-bit architecture. The systems have a floppy disk drive for the operating system and hard or floppy disks for mass storage, which can range up to 80 megabytes. Most of these systems are equipped with the bare essentials for input and output. Some personal computer systems now offer PC-board and IC design. Figure 2-13 shows a desk-top-computer-based CAD workstation. This system has two associated floppy disk drives. The operator is using a hockey puck-type digitizer.

As mentioned, all these systems have some type of disk storage capacity for storing temporary or permanent drawing data. They may also use magnetic tape for storing and filing drawings. Disks are easier and faster for the computer to search and retrieve data from, while tapes are more convenient and economical for bulk storage and filing of data, although the current trend is toward the use of disks. Figure 2-1 shows a typical magnetic tape drive. Figure 2-14 shows a typical large magnetic disk drive used in CAD systems. The disk drive is top-loading, and a disk case for the removable disk is sitting on top of the disk drive.

2-7 Output Equipment

Obviously the output is the end product of the CAD system, whether in the form of final drawings, microfilm, photographs, or storage of information on magnetic disk, magnetic tape, or paper tape. Without output, a CAD system would be worthless.

FIG. 2-13 A desk-top-computer-based CAD workstation. (Bruning CAD.)

FIG. 2-14 A top-loading magnetic disk drive and associated disk setting on top. (Digital Equipment Corporation.)

The prime output device is the line plotter, which converts data directly from the computer or storage into a finished drawing. Essentially there are four types of plotters: flat-bed, drum, belt-bed, and electrostatic. Some of these plotters have stand-alone control units. Drawing media used by plotters, except the electrostatic, include vellum, paper, Mylar; the electrostatic plotter uses a special electrographic paper. A flat-bed plotter is also capable of using, depending on the plotter, film, metal, plastic, or material with a scribe coat.

With the exception of the electrostatic plotters, which have stationary print heads, ink pens, ballpoint pens, Rapidograph pens, and pencils may be used for making the drawing; however, pressurized liquid in pens gives better results at higher plotting speeds. All plotters can plot in black and white, and many can plot in color.

Figure 2-15 shows a flat-bed plotter in an automated drafting system. The bed of a flat-bed plotter is horizontal. In this particular plotter the bed remains stationary and print heads move in both axes. On some flat-bed plotters, the bed moves and the print heads remain stationary or both the bed and the print heads move.

The plotter shown in Fig. 2-15 is capable of making plots of 5×6 ft expandable to 6×24 ft. Flat-bed plotters may be used as photoplotters or scribe

FIG. 2-15 **An automated drafting system with a flat-bed plotter.** (Gerber Scientific Instrument Co.)

plotters. A flat-bed plotter can be converted to a photoplotter or scribe plotter by simply removing the printing head and replacing it with a photoplotting head or a cutter head.

Photoplotting is done using a computer controlled light source for producing graphic representations of patterns directly on photographic media, such as film or glass. There are two types of photohead on a flat-bed plotter: (1) one that utilizes continuous visible light and draws a vector of the photographic media and (2) a raster laser beam type in which the beam is turned on and off. The vector type is more accurate and smoother, while the raster type is faster. The plotter shown in Fig. 2-15 can be converted to a vector-type photoplotter by changing the print heads to photoheads. Its light is not turned off, but is accurately blocked by shutters in the photohead. The light is also increased or decreased by a varying voltage. In general, photoplotters produce some very accurate artwork patterns for printed circuits, ICs, and chemical etching.

Figure 2-16 shows a drum plotter. On the drum plotter, the drawing medium rests on a drum and moves in a vertical axis while the pens move in a horizontal axis. Notice that the pens are plotting at the top of the drum. This unit has four pens and thus the ability to plot in four different colors. Drum plotters usually are loaded with a complete roll of drawing media, such as paper.

Figure 2-17 shows a belt-bed plotter. The bed on belt-bed plotters is either

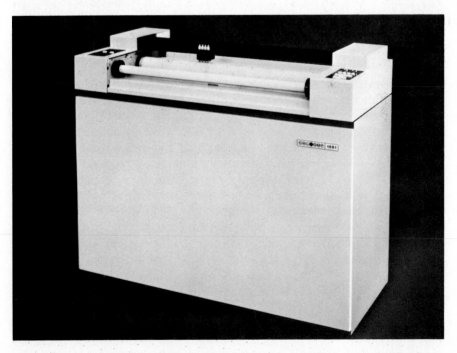

FIG. 2-16 A drum plotter. Notice that the pens are plotting on the top of the drum. (Calcomp.)

FIG. 2-17 A belt-bed plotter. This plotter has the ability to use either continuous-roll or cut-sheet drawing media. (Calcomp.)

vertical or almost vertical. On a belt-bed plotter, either the pen and belt bed move in opposite directions or the pens move in both axes. This particular belt bed is microprocessor-controlled and utilizes a four-pen block. It has the ability to use either continuous-roll or cut-sheet drawing media.

Figure 2-18 shows the top view of an electrostatic plotter. Electrostatic plotters have been developed in the last few years and constitute one of the most ingenious innovations in the plotter field. Unlike the other types of plotter discussed previously, the electrostatic printing head does not move. The stationary head consists of an array of densely spaced writing nibs. A voltage is applied to the nibs, and on digital commands, the nibs selectively create tiny electrostatic dots on the electrograph paper that is passing over the writing head.

Figure 2-19 shows a hard-copy unit. This hard-copy unit happens to be an ink jet color copier; many are black and white and use other printing processes. The hard-copy unit reproduces the CRT pictures quickly and conveniently. It is initiated by pushing a button on the keyboard or by a software command. The primary advantage of a hard-copy unit is that a copy of the CAD drawing can be made quickly, more conveniently, and less expensively than with a plotter. The hard copier is usually located at the workstation for convenience of the operator. Hard copies, especially black and white, are often used as check prints for the operator. Hard-copy units use the electrostatic, the ink jet, photoplotting, the dot matrix, and thermal printing processes.

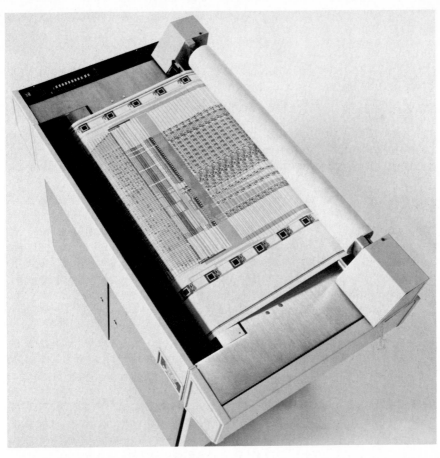

FIG. 2-18 Photograph (top view) of an electrostatic plotter. (Versatec.)

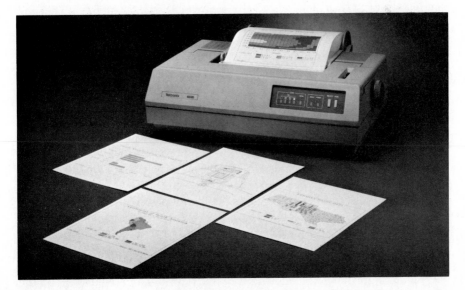

FIG. 2-19 Photograph of a color hard-copy unit. (Tektronix, Inc.)

Other output devices might include a teletype writer, which could be used for outputting textual material information such as XY coordinates, messages, and parts lists. A computer output microfilm (COM) is used to take the output from the computer and directly generate microfilm drawings in the form of aperture cards, microfilm, or roll film. Magnetic disks and tape are used to store the drawing data for future use. A paper or Mylar tape punch may be used to produce punched tape for driving plotters or numerically controlled machines.

2-8 The Output

This section of the chapter shows and explains a number of examples of drawings and artwork that have been produced by CAD systems. All outputs are produced in basically the same steps. The initial conceptions were inputted into a computer by one or more of the input devices that have been discussed, processed by the computer, stored in bulk (disk or tape) storage, instructed to be outputted by a command entered at the workstation keyboard, and outputted on a plotter or hard-copy unit.

Figure 2-20 is part of a drawing showing the one-line diagram of a large ac generating station. The regulator and control circuits, which are enclosed in a boundary line, are shown as block diagrams. This drawing was done on a belt-bed plotter, using pressurized-ink pens and drafting vellum as the drawing medium. First, notice the overall flowing symmetry and easy readability of the drawing, which is typical of most computer-generated drawings. The line weights are of such an even consistency—imagine yourself doing the same drawing in ink. The heavy lines are done by multiple passings or strokes of the pens, usually after everything else is completed. This drawing has been done on a CAD system that uses the dot-type wire connections preferred by the authors, but other types of connection are available. Notice how the lettering is neat and even; either vertical or 22.5°-slant letters are available. The authors watched the complete drawing (the figure shown is approximately one-fourth of the drawing), and it took approximately 10 min to complete.

Figure 2-21 shows part of a one-line diagram of a 4160 V-480Y/277 V substation. This drawing was done on the same machine as the previous figure. Again notice the symmetry and easy readability. Notice how well the mechanical grouping boundaries isolate but do not confuse the electrical components. The lettering is again vertical. The authors believe CAD lettering should be vertical with a balance of open space within and between the letters so that if a drawing is reduced, which happens often, the lettering will not look like an ink spot.

Figure 2-22 shows the control schematic of a recycle pump. The original drawing had the diagrams for two pump and valve connections. The diagram is easily readable; however, the schematic is turned 90° from the conventional arrangement. This arrangement is typical of circuit control schematics which may be encountered in the industrial power or electric power fields.

Figure 2-23 shows the valve limit-switch development, which was on the same drawing as Fig. 2-22. This development explains the operation of the limit

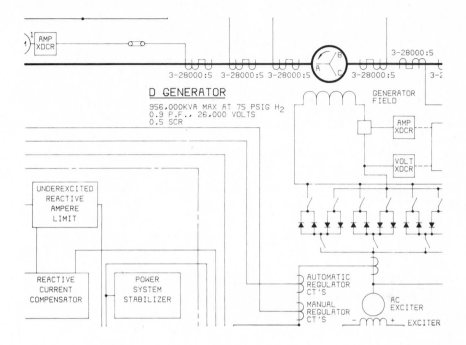

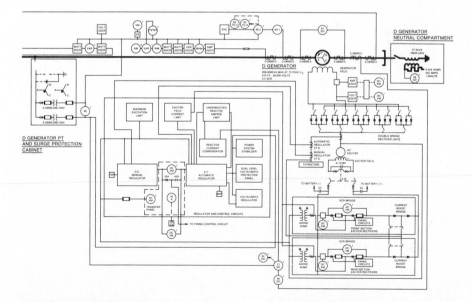

FIG. 2-20 **Part of a one-line diagram of a large ac generating station.** (Black and Veatch, Consulting Engineers.)

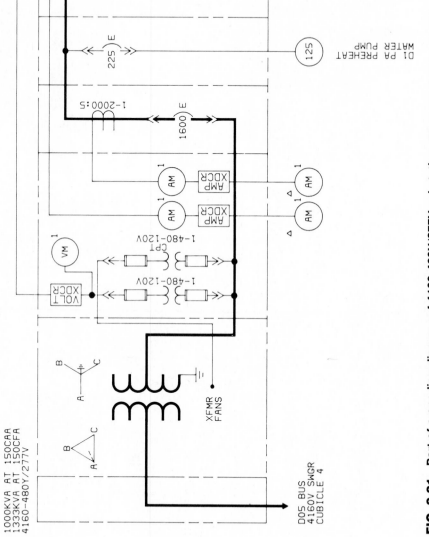

FIG. 2-21 Part of a one-line diagram of 4160-480Y/277V substation. (Black and Veatch, Consulting Engineers.)

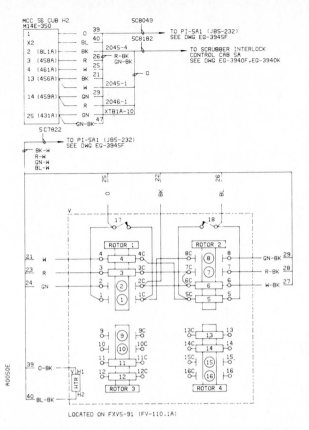

FIG. 2-22 The control circuit schematic of recycle pump and discharge valve on a flue gas treatment system. (Black and Veatch, Consulting Engineers.)

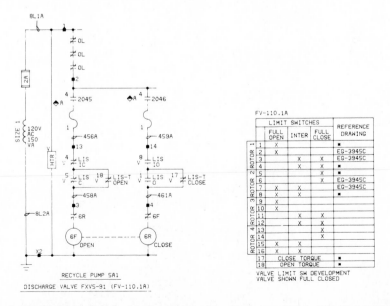

FIG. 2-23 A limit switch development of a discharge valve. (Black and Veatch, Consulting Engineers.)

switches on the valve. The printing of the development is easy to read with little possibility of error. The development is an example in which the printing must be done in a small area ($3\frac{1}{2}$ in. high $\times$ $2\frac{7}{8}$ in. wide, full size). Again, without belaboring a point, the authors would have preferred that this figure have more white space between and within the letters.

Figure 2-24 shows part of a schematic of a silicon-controlled rectifier (SCR) circuit. The SCR circuit is discussed in more detail in Chap. 9. This

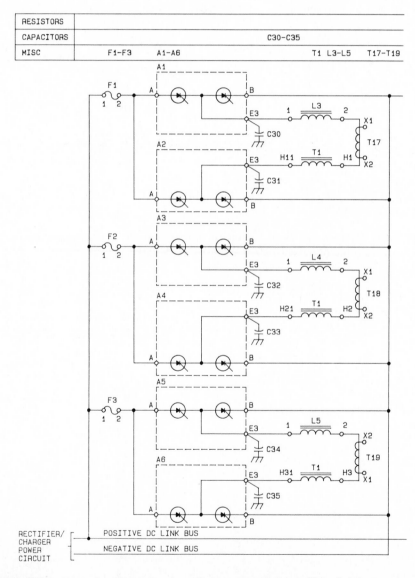

FIG. 2-24 **Part of schematic of a silicon-controlled rectifier circuit.** (Information Displays, Inc.)

schematic is a more conventional schematic (as discussed in Chap. 7), with the reading from left to right. Note the spacing between components and between the lettering and components — the spacing is sufficient for reading the schematic without any undue hardship.

Figure 2-25 shows part of the same schematic; Fig. 2-25a was plotted on a drum plotter, and Fig. 2-25b was plotted on a hard-copy unit. When the drawing was being developed at the workstation terminal drawing as a basis of 100 percent scale, the hard-copy drawing was done at 55 percent of scale, which was the largest scale to permit copying, while the drum plotter was done at 84

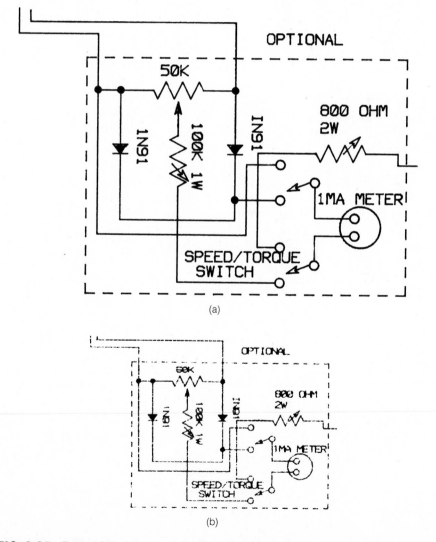

(a)

(b)

FIG. 2-25 Parts of the same schematic: *(a)* plotted on a drum plotter; *(b)* plotted on a hard-copy unit.

percent of scale to permit copying of the complete schematic (of which Fig. 2-25*a* is a part). In comparing Figs. 2-25*a* and *b*, note that as reduction occurred, lettering approached an ink spot, especially the legends "50K" and "100K." This was not the fault of the hard copier but was due to lack of sufficient space between and within the original lettering. Now notice the line quality between the two figures (symbols and text). The drum plotter drawing (Fig. 2-25*a*) is of much better quality than the hard-copy unit drawings, which are used as check prints and not final drawings as the drum plotter drawing is.

Figure 2-26 shows part of a PC-board component layout, showing only the components. Notice how the components and lettering are so closely spaced and imagine doing this by traditional drafting methods. Notice the quality of the arcs and circles—this is one way to determine both the quality of the software (good functions with clear data) and the plotter (no zigzags on arcs).

Figure 2-27 shows the artwork of a two-sided PC board with feedthrough and component layouts. After the design of the PC board was done, it was sent to an artwork generator, in this case a photoplotter, where the artwork was done on 15- × 20-in. photoetchable Mylar. The artwork for the printed circuits has been optimally designed, and, therefore, it is laid out with good organization for circuit spacing and component placement. Thus it assures a functional PC board. The circuits, feedthrough, and component terminals are very accurately

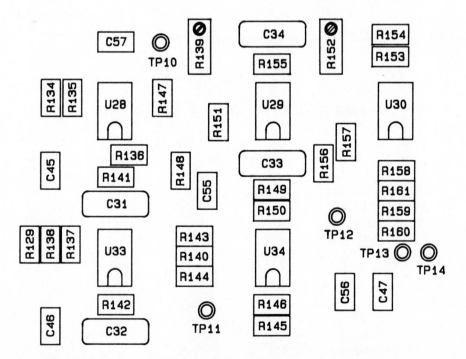

FIG. 2-26 **Part of a PC board component layout.** (Information Displays, Inc.)

done; again this assures good operation of the PC board. The accuracy of artwork is a major reason for using a CAD system. To support the design of the PC board, the designer had a number of design options available, namely:

1. Board outline
2. Pads and feedthroughs
3. Component side runs
4. Solder side runs
5. Component outlines
6. Drill-hole data
7. Annotation and signature block
8. Parts list
9. Schematic drawing
10. Functional flowchart
11. Ground plane (if required)
12. Signal plane (if required)
13. Assembly hardware

These options of drawings and artwork could also have been designed, and many would act like movable overlays on the PC-board artwork. If the designer had changed a component on one drawing, the others would have been automatically changed.

2-9 Computer-Aided Manufacturing

Any discussion of CAD would not be complete without discussing computer-aided manufacturing (CAM). Usually the CAD and CAM terms are linked together as CAD/CAM. Today, the CAD functions and CAM functions are being brought increasingly closer together to form one integrated unified entity. Computer-aided manufacturing builds on or utilizes the data base developed in CAD operation to do manufacturing functions. To date, these functions are concentrated in four main areas: numerical control (N/C), robotics, process planning, and factory management.

Numerical control is one of the oldest CAM technologies. It is, in essence, the process of controlling a production machine with prerecorded, coded information, such as a paper tape, to fabricate a part. Numerical control instructions are written in one of several languages and stored on paper tapes or magnetic tape. These instructions can be developed from data developed on a CAD system. Larger and more advanced systems use a mini- or microcomputer directly connected to a machine tool(s). This process is called *computer N/C*. The largest and most sophisticated systems use *direct N/C,* which connects several minicomputers to a large mainframe computer.

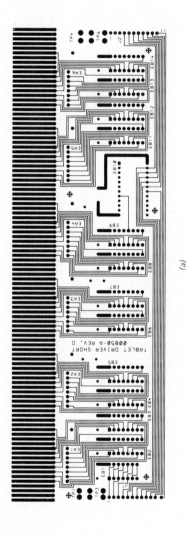

(a)

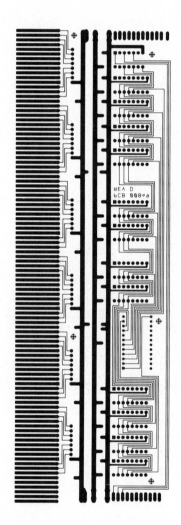

(b)

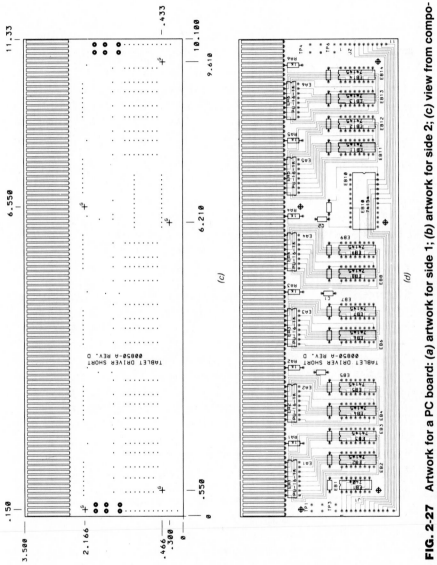

(c)

(d)

FIG. 2-27 Artwork for a PC board: (a) artwork for side 1; (b) artwork for side 2; (c) view from component side; (d) component layout. (Courtesy of Applicon, Inc.)

Traditionally, N/C instructions were written from engineering drawings by engineers or programmers. The instructions were then reiterated and refined several times before the manufactured part was made correctly. The reiteration and refining process took large amounts of operating time and substantially increased the cost of the machined part. Advances in computer technology such as computer simulation have made the creation and checking of N/C programs faster and more efficient. These N/C programs require less reiteration and refinement, thereby requiring less actual machine time for proving.

The N/C programs describe the movement of a machine. The machine can move, with the direction of the CAM instructions, in three axes (X, Y, Z) or six degrees of freedom, i.e., forward-backward, left-right, or up-down. Machines that are commonly controlled in industry by N/C instructions include the milling machine, punch presses, lathes, and drilling machines.

Another area of CAM is robotics. In robotics, automated manipulator arms or robots perform a variety of functions. Essentially robots, like N/C-controlled machines, have six degrees of freedom, which are called "waist," "shoulder," "elbow," "wrist rotation," "wrist bending," and "flange." In many factories, robots work in conjunction with N/C machines by doing material handling functions and tool selection for the N/C machines. In the automobile manufacturing industry, for example, robots are used to automatically weld the sheet metal to the unitized frames.

Most robots are programmed in a walk-through or teach basics by actually leading the robot through the required steps of the operation. This teach process is slow and error-prone, sometimes requiring complete reprogramming of tasks. Presently, there are programming languages being developed through a computer that will permit the robot to be programmed without this teach mode. There are other programs being developed to eliminate the detailed instructions and permit the robot program, by access to the data base, to determine the most efficient route to perform the required operation. The U.S. Air Force is developing a program to organize every step of a manufacturing operation around computer automation.

The process planning area of CAM considers the entire manufacturing process, rather than the control of a single machine or process as do numerical control or robotics. Process planning is not new; it has been in existence since manufacturing was first performed; however, process planning utilizing a computer is relatively new. Process planning can be as simple as planning and delivering the raw materials to complete monitoring and controlling of the process. Process planning can organize similar parts into families to allow standardization of fabrication steps. This process is called *group technology.*

For example, in semiconductor production, where chemistry and physics are pushed to the known limits, a CAM process planning system would follow the silicon wafer through the 160 steps of IC creation on a silicon wafer. At each step, the CAM system would collect quality data. The CAM system would feed both forward and backward data. In feeding forward this data, quality could be correlated to final yields; in the feeding backward of this data, trends could be established in any steps or series of steps along the process. These trends could be used to correct or modify the processes for better yields, more accurate resource (material, worker, and machine) scheduling. The data would also permit "what if" or "blue sky" modeling of a new or different process. The CAM system could communicate the process data through terminals to other areas within the factory or to other remote factories or offices by telephone communications. The process planning CAM system might also monitor and control the facilities where the semiconductor manufacturing is done.

Factory management is the umbrella for all the other CAM functions in that it ties them together to coordinate the operations of an entire factory. As can be guessed, factory management of a large factory producing a variety of products is, to say the least, an ambitious project. Factory management systems must rely heavily on group technology, with families of similar parts fabricated in individual manufacturing cells with each cell having one computer extracting information from the process planning computers. In factory management, there would be one large mainframe computer monitoring and controlling the cell computers, thus making factory automation a reality. Experts predict that factory management CAM will be in use before the end of this century.

SUMMARY

This chapter has described the operations of a CAD system from the input devices to the output. Computer-aided manufacture and its relationship to CAD have also been discussed. It should be emphasized that this chapter was oriented toward computer-generated electrical drawings and artwork and that CAD systems are being used very effectively in other design disciplines, such as automotive, civil, mechanical, and aeronautical. The future for CAD systems looks bright, especially now that there are many low-cost, single-discipline CAD systems available for the small specialized manufacturer. The large, multidiscipline CAD system supplies are continuing to offer more and better options to large industrial government and educational users at a lower cost per productivity ratio. It is extremely difficult for anyone to deny the CAD system advantages of speed, accuracy, productivity, and quality.

The integration of CAD and CAM under one unitized system utilizing a common data base is beginning but has a long way to go to meet the hopes of its initial fanfare.

QUESTIONS

2-1. What is CAD? Name three other terms that mean approximately the same thing.

2-2. Name at least five electrical and electronic drawings that a CAD system is used to create.

2-3. What are the four functional basic parts of a CAD system?

2-4. Describe a workstation, naming some of its components.

2-5. What is software? Name the two types of language that CAD software is written in.

2-6. Name two languages that CAD systems are programmed in. Does the designer or drafter usually do the programming?

2-7. Describe the two (2) basic input operations. Name several input devices that perform these operations.

2-8. What is digitizing? What is a digitizer?

2-9. What is interactive graphics?

2-10. Describe the basic input process.

2-11. What are some helpful hints when doing CAD work?

2-12. What functions does the computer perform during CAD?

2-13. Describe the functions of peripheral storage devices.

2-14. Name three output devices.

2-15. What are the four types of plotter? Describe them.

2-16. What is a photoplotter? What does it do?

2-17. What type of drawing medium is used by plotters? Name several drawing devices used by plotters.

2-18. What is COM?

2-19. Discuss the advantages and disadvantages of CAD.

3

Device Symbols

A very large portion of drawing in the electrical and electronics fields is of a diagrammatic nature. This diagrammatic drawing makes great use of symbols. Originally, these symbols were drawn to look something like the parts they were to represent. However, the parts and the symbols have changed considerably through the years, until now there is not much physical resemblance between symbols and the respective parts. In some cases, though, a symbol may give a suggestion of the shape of the part it represents. Examples are the inductor, resistor, headset, antenna, and knife switch.

In order to facilitate making and reading electrical drawings, representatives of industry and government have come together to make up lists of standard symbols to be used in drawings. Two important standards are the following:

> 76-ANSI/IEEE[1] Y32E, "Electrical and Electronics Graphic Symbols and Reference Designations." (This standard includes the most recent editions of ANSI Y32·2, IEEE 315, and CSA Z99.)
>
> IEC (International Electrotechnical Commission) 117, "Graphical Symbols."

Some of the commonly used symbols will be shown and explained in this chapter. Many more symbols will be shown in Appendix B. To improve coordination with IEC Pub. No. 117, IEC-approved versions of certain symbols have been added to the ANSI/IEEE standard as alternatives. We have included some of these, such as the capacitor symbol. Another set of symbols, approved by the Joint Industry Committee, will be explained later in the book and is also included in Appendix B.

Symbols of Common Nonelectronic Devices

3-1 The Battery

One of the simplest symbols is that which represents a single-cell battery, shown in Fig. 3-1a. (The horizontal line represents the path of the signal, or current,

[1] The Institute of Electrical and Electronics Engineers, Inc.

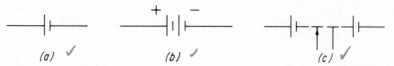

FIG. 3-1 Symbols for batteries: *(a)* single cell; *(b)* multicell with polarity marks added; *(c)* multicell with two taps. *(d) multicell without polarity*

and is not a part of the battery symbol itself.) The longer of the two lines always represents the positive terminal. The short line is about half the length of the long one. Multicell batteries can be shown with four or more lines (four is enough), as shown in Fig. 3-1*b*. Polarity symbols have been shown but theoretically are not necessary. However, these are sometimes used for emphasis or when it is believed the reader may be unsure. The third battery symbol shows two taps—one fixed and one (with arrow) adjustable.

3-2 The Capacitor

Another often used symbol that is easy to draw is the capacitor—sometimes called a *condenser;* its main function is to store electrical charge. The straight line can be made about the length of, or larger than, the long line of the battery symbol. The entire symbol is sometimes drawn twice the size of the single-cell battery symbol. (More about sizes will be found in Sec. 3-11.) The curved line represents the outside electrode in fixed-paper and ceramic-dielectric capacitors, the moving element in variable and adjustable types, and the low-potential element in feedthrough capacitors. Showing the polarity sign usually indicates an electrolytic capacitor. Good drawing practice should be observed in the drawing of the shielded symbol. Corners should be full and complete, and the dashed lines should not touch or intersect the signal path. In the case of Fig. 3-2*e*. the capacitance of one part increases as the capacitance of the other part decreases. But in the split-stator capacitor, the capacitances of both parts increase simultaneously.

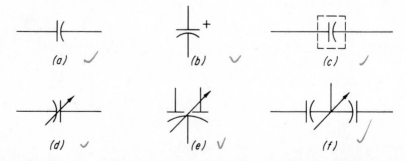

FIG. 3-2 Symbols for capacitors: *(a)* general; *(b)* polarized; *(c)* shielded; *(d)* adjustable or variable; *(e)* adjustable or variable differential; *(f)* split-stator.

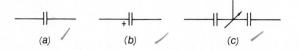

FIG. 3-3 IEC-approved symbols for capacitors: *(a)* general; *(b)* polarized; *(c)* split-stator. The distance between the plates should be between 20 and 30 percent of the length of the plates. This symbol has been used increasingly in the last decade.

A symbol that was not included in the standards now appears in the latest ANSI/IEEE Y32E. It is the parallel straight-line symbol listed as preferred by the IEC. Shown in Fig. 3-3, it is almost identical to the contact symbol, which has a wider gap between the lines. (See Fig. 3-7*f*.) This form of the capacitor symbol is experiencing greater usage in technical literature.

3-3 Chassis, Ground, and Circuit Return

It is usually necessary to connect parts of a circuit to a chassis, ground, frame, etc. If the conducting connection is to a chassis or frame that may have substantially higher potential than ground or the surrounding structure, the chassis symbol (Fig. 3-4*a*) should be used. If the conducting connection is to earth, a body of water, or a structure which serves the same function (such as a land, sea, or air frame), the symbol shown in Fig. 3-4*b* should be used. The common connection symbol should be used for common-return connections at the same potential level. This triangle symbol is used when a common conductor such as a ground bus or battery bus is used. Then, in the lower part of the diagram, a key is used in which the symbol is shown and the words GRD BUS or BAT BUS— 24 V or other appropriate terms are indicated. Sometimes a letter system is used if more than one type of common circuit return is shown on the same diagram. The triangle may be omitted and proper identification made where the asterisk appears on the right-hand symbol in Fig. 3-4*c*, although the triangle is more meaningful to the authors than the end of a line without a symbol.

3-4 Connections and Crossovers

Two systems are approved for showing connections (junctions) or crossovers. One is the dot system, as shown in Figs. 3-5*a* to *c*. The other is the no-dot system,

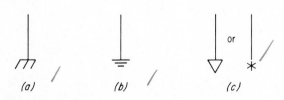

FIG. 3-4 Symbols for circuit return: *(a)* chassis connection; *(b)* ground (frame) connection; *(c)* common connections.

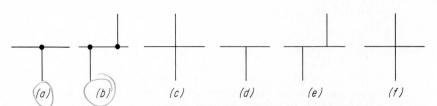

FIG. 3-5 Connections and crossovers: *(a)* and *(b)* connections for dot system; *(c)* crossover for dot system; *(d)* and *(e)* connections for use with no-dot system; *(f)* crossover for use with no-dot system.

used in Figs. 3-5*d* to *f.* The no-dot system has some weaknesses, but it has equality in the standards. However, many users (a majority in the authors' estimation) prefer the dot system for clarity.

3-5 The Inductor

The inductor, or induction coil, is used in a great many ways. In different situations it may be a transformer winding, a reactor, a radio-frequency coil, or a retardation coil. Some recent standards have shown the two symbols in Fig. 3-6*a* as approved. ANSI/IEEE Y32E stresses the more rudimentary symbol shown at the left. Many United States companies, however, still use the more complicated helical symbol. Because induction increases as the frequency increases, no cores (or air cores) are usually necessary at high frequencies. But at low frequencies, magnetic cores are often used to increase the induction. Ceramic cores are often composed of magnetic materials called *ferrites.* Inductors are also sometimes called *chokes, coils,* or *reactors,* depending on their function in a circuit.

3-6 The Relay Coil

Three symbols are approved for the relay coil, also known as the solenoid. These are shown in Fig. 3-7*a* to *c.* The asterisk in Fig. 3-7*c* indicates that a letter or value, such as the relay number, should be in the circle. The semicircular dot in Fig. 3-7*d* indicates the inner end of the winding. Figures 3-7*e* and *f* show combinations of relay coils with switches; the whole combinations are called *contactors.* The switches shown here are often called *contacts* and are turned off or on by action of the relay coils. In Fig. 3-7*f,* the left-hand contact is normally

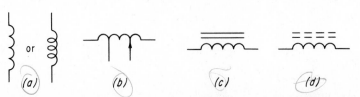

FIG. 3-6 Inductor symbols: *(a)* general symbols; *(b)* with fixed and variable taps; *(c)* with magnetic core; *(d)* with ceramic-type core.

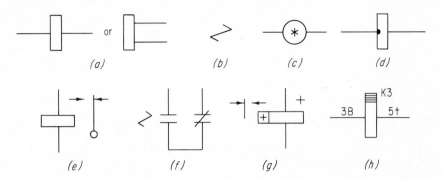

FIG. 3-7 Relay symbols: *(a)* through *(d)* approved relay coil symbols; *(e)* and *(f)* relays with transfer contacts; *(g)* polarized relay with transfer contact; *(h)* slow-release relay with letter-number designations.

open and the right-hand contact (with inclined line drawn at 60° to the horizontal) is normally closed. Action of the relay will close the left contact and open the right one. The term *slow-release* (Fig. 3-7h) is only relative. This relay closes one contact before it activates another, but the total action happens very quickly.

3-7 The Resistor

Two approved symbols for the resistor are shown in Fig. 3-8a. The older, zigzag symbol is probably used the most. The rectangle symbol, originally used in the electrical-controls field, has been adopted by a number of companies in other areas. Made with a 60° angle between adjacent lines, the zigzag symbol needs only three points on each side, unless extra taps or other special features require more. The asterisk within the rectangle means that identification should be placed within or near the rectangle. Typical values would be 210 (Ω) or 20 kΩ (20,000 ohms, often 20k, 20 kΩ, or 20,000Ω in drawings).

Resistors may be fixed, variable (rheostat), or tapped, with either fixed or variable taps (see Fig. 3-8d and e). They are usually linear. If not, a special nonlinear symbol is available. Resistors are used for such purposes as dividing voltage, dropping voltage, developing heat, and minimizing current and voltage surges. The shaft of the arrow in Fig. 3-8b is drawn at about 45° in this and other symbols that require variability.

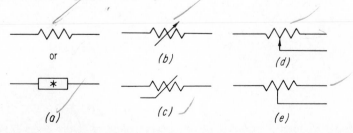

FIG. 3-8 Symbols for resistors: *(a)* general; *(b)* variable or adjustable; *(c)* nonlinear; *(d)* with adjustable contact; *(e)* tapped.

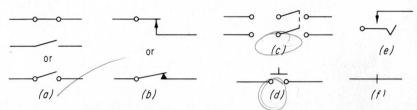

FIG. 3-9 Symbols for switches: *(a)* single-throw, general—closed and open; *(b)* nonlocking, momentary—opening (break); *(c)* double-throw, two-pole; *(d)* pushbutton—closing (make); *(e)* locking—closing (make); *(f)* switching-function symbol—closed contact (break).

3-8 The Switch

The purpose of the switch is to open or close circuits. The words *break* or *make* are often used instead of *open* and *close.* Mechanical-switch symbols are usually combinations of contact symbols, and they may be fixed, moving, sliding, nonlocking, etc. They are shown in the position in which no operating force is required. This is sometimes called the *normal*, or *initial*, position, in which the circuit is not energized. Figure 3-9*f* is not really a switch symbol; it simply defines the switching function. If, instead of a bar, an *X* were shown, we would have an open (break) contact. Many other devices—diodes, transistors, tubes, and cryotrons—also perform switching functions.

Rotary-switch symbols are viewed from the end opposite the control knob and with the operational sequence in the clockwise direction. The projection on each segment of the "wafer" represents the moving contact. When several functions are performed, the tabular form of presenting information, as in Fig. 3-10, is preferred. Here, dashes link the terminals that are connected. In position 2, for example, terminals 1 and 3 are connected (not terminal 2) and terminals 5 and 7 and 9 and 11 are connected.

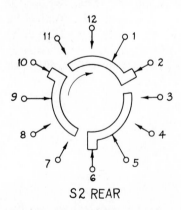

S2 REAR

POS.	FUNCTION	TERM.
I	OFF (SHOWN)	1-2,5-6,9-10
2	STAND BY	1-3,5-7,9-11
3	OPERATE	1-4,5-8,9-12

S2

FIG. 3-10 **Position-function relationships for rotary switches.** (From American Standard, "Drafting Manual," ANSI Y14.15, "Electrical and Electronics Diagrams." Used by permission of the publisher, The American Society of Mechanical Engineers.)

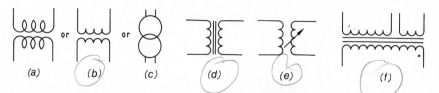

FIG. 3-11 Symbols for transformers: *(a)* through *(c)* general; *(d)* for showing a magnetic core; *(e)* with one winding having adjustable inductance; *(f)* power transformer.

3-9 The Transformer

The symbols used for inductors are also utilized in the drawing of transformers. The American National Standard Y32.2 approves the use of either the helix symbol or the more rudimentary symbol. Except for one instance, ANSI/IEEE Y32E shows only the simpler symbol. We shall use the simpler symbol in most of our drawings. Transformers are made with air cores (usually found in high-frequency circuits) or with iron or laminated cores (found primarily in low-frequency ac circuits). The standard, however, does not require that the two parallel lines be drawn for magnetic or metallic cores. A recent addition is the IEC-approved symbol shown in Fig. 3-11*c*. To date this symbol has not been used in drawings in the United States. However, it does appear as a constant current source in some illustrations (see Fig. 3-12*b*).

The power transformer supplies power (usually in different amounts) to two or more circuits. The secondary windings are often drawn with different numbers of loops, or cusps, to suggest the relative voltages going to these circuits. Sometimes the secondary circuits have additional taps.

3-10 New Device Symbols

In an expanding area such as electronics, new devices for which no symbols exist will be forthcoming. Two methods of solving the problem of portraying these devices are in common use: (1) a new symbol for each device may be invented or designed, or (2) a rectangular block with appropriate identification may be inserted in the diagram at the place where the new device would be placed. Figure 3-12*a* shows a symbol which was designed by someone to show a cryogenic switch.

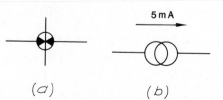

FIG. 3-12 Symbols invented to show *(a)* a cryotron and *(b)* a constant-current source.

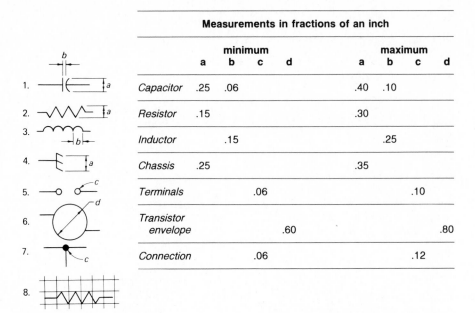

FIG. 3-13 Suggested sizes for electrical-device symbols. (The lowest figure shows a resistor symbol that has been drawn on cross-ruled paper.)

Sometimes a new symbol becomes the "industry standard" before it appears in the IEEE-ANSI standard. Such a symbol is that shown for a constant-current source in Fig. 3-12*b*.

3-11 Size of Symbols

Theoretically, size of a symbol is not important. But the relative sizes of symbols are an important matter. These relative sizes are shown in the standards, in Fig. 3-13, and in Appendix B of this book. Figure 3-13 suggests minimum and maximum sizes for several commonly used symbols. The minimum sizes are for small diagrams or for larger diagrams that are filled with many devices. The maximum sizes are for large drawings, say those on paper larger than 12 × 18. Other factors may dictate what size symbols should be drawn. A student or drafter using a template to make symbols is pretty much limited to the sizes that can be made with the template. Crowded conditions may require that the drafter draw symbols smaller than the suggested sizes. Lines are usually of medium weight — the same weight (width) used to draw the connecting paths and other lines in a drawing. However, symbols are sometimes made with heavy, thick lines for purposes of emphasis.

3-12 Drafting Aids

There are at least four ways in which symbols can be made more quickly than by drawing them with conventional drawing instruments. These methods use:

1. Templates
2. Preprinted symbol cutouts
3. Typesetting equipment with symbols
4. Computer graphics

There are many designs of electrical templates on the market, and no one template will make all the symbols that are in use today. Great care should be used in the selection and purchase of such a device. A template should be used with a T square or drafting machine, as described in Chap. 1. Some templates are so thin that they tend to slip under the edge of the T square and hence are not entirely satisfactory. Some users prefer to raise the template off the paper. One method is to place the template over a triangle and use the grooves that are over the open part of the triangle.

Preprinted symbols are manufactured as contact (pressure-sensitive) adhesives, or appliqués. These are cut out or lifted off a sheet of preprinted symbols and then positioned on the drawing as shown in Fig. 7-14.

Symbols that have been made with a personal computer are shown in Fig. 3-14c. These were constructed using BASIC language and printed on a dot-matrix printer.

3-13 The Diode

One class of semiconductors is the diode. Semiconductors do not allow current to flow as easily as conductors, and under certain conditions they may act as

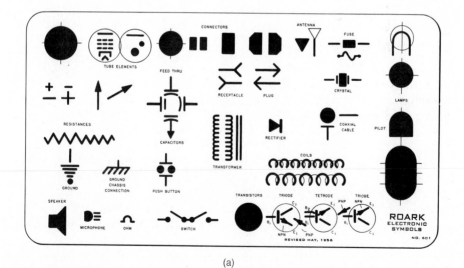

(a)

FIG. 3-14 Templates and appliqués (pressure-sensitive adhesives): *(a)* and *(b)* templates (the black areas represent open spaces); *(c)* symbols made on a personal computer.

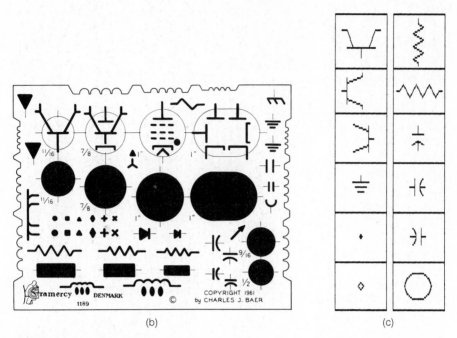

(b)

(c)

FIG. 3-14 *(Continued)*

insulators. Diodes are usually made of P- and N-type semiconductor material.[1]
Figure 3-15*a* shows a PN-junction diode. The *junction* is that region of the
crystal where the N-type material (doped with arsenic to provide free electrons)
ends and P-type material (doped with boron to produce *holes,* or missing
electrons) begins. After a diode is formed, some of the electrons cross the
junction to fill the holes. A region is formed as shown in Fig. 3-15*b* when the
electrons have crossed over and the holes are filled. This region is then said to be

[1] Semiconductor construction is discussed in greater length in Chap. 8.

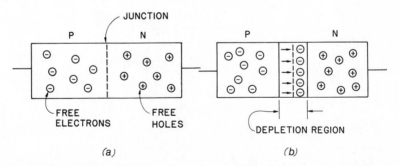

FIG. 3-15 **The structural representation of a PN-junction diode.** (With permission
from C. A. Schuler, *Electronics Principles and Applications,* McGraw-Hill Book Company, New
York, 1979.)

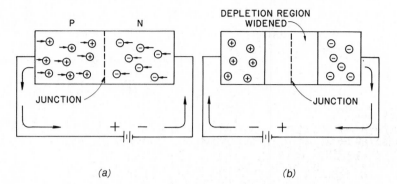

FIG. 3-16 A voltage, or current, applied to a PN-junction diode: *(a)* **forward bias;** *(b)* **reverse bias.** (With permission from C. A. Schuler, *Electronics Principles and Applications,* McGraw-Hill Book Company, New York, 1979.)

depleted. The action stops because a negative charge forms on the P-type side which repels other electrons that might try to cross. With the depletion region acting as an insulator, the diode is a very poor conductor.

Because it was initially formed by electrons moving and filling holes, the effect of the depletion region can be removed by applying a voltage. If a voltage is applied as shown in Fig. 3-16*a*, the depletion region is collapsed. The positive terminal of the battery repels the holes on the P-type side and the negative terminal repels the electrons, pushing them toward the junction. The electron current leaves the negative side of the battery, flows through the diode, and returns to the positive terminal of the battery. This *forward-biasing*, as it is called, *turns on* the diode, which then *semiconducts.* In Fig. 3-16*b* the direction of current flow is changed and the current flow is blocked except for a very slight leakage.

In Fig. 3-17 a junction diode is shown by means of its symbol. In electrical terms, the anode is the terminal or electrode that attracts electrons. The cathode is that electrode that emits (gives off) electrons. Depending on the function to be performed, the diode symbol may be oriented as shown in Fig. 3-17 or reversed.

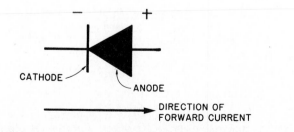

FIG. 3-17 Schematic of diode with forward direction of current.

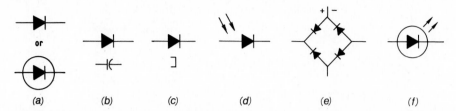

FIG. 3-18 Diode symbols: *(a)* general; *(b)* capacitive (varicap); *(c)* tunneling (tunnel); *(d)* photosensitive type; *(e)* bridge; *(f)* light-emitting (photoemissive) type (LED).

Figure 3-18 shows symbols for several types of diodes. The basic symbol is not usually circled, but it *may* be circled as shown in Figs. 3-18a and *f*, which was the older practice. Diodes are used for a number of purposes such as switching, rectifying, detection and amplification. The tunnel diode, for example, has a negative resistance characteristic that permits it to be used as an amplifier or switch. The bridge rectifier circuit provides a smoothed dc output.

It is important that diodes be properly placed (or replaced) in a circuit. Most packages identify the cathode or anode as shown in Fig. 3-19. In a few cases where identification is not clear, the technician may have to use a volt-ohm ammeter (VOM) or a vacuum-tube voltmeter (VTVM) to check a diode in a circuit and identify its leads. Manufacturers' catalogs also contain information on polarity. Package types — usually called *cases*, are generally labeled as DO1, DO5, etc. In diagrams diodes are usually identified as 1N followed by digits and letters as 1N914 or 1N758A.

The location and installation of semiconductors requires the following procedures:

1. Don't exceed their maximum voltage, current, or power ratings.
2. Use adequate heat sinks for power devices.

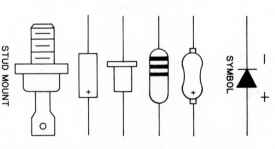

FIG. 3-19 Diode package styles. (With permission from C. A. Schuler, *Electronics Principles and Applications,* McGraw-Hill Book Company, New York, 1979.)

3. Avoid prolonged exposure to heat during soldering.

4. Don't locate sensitive circuit devices adjacent to heat-producing power devices.

5. Don't solder semiconductors into an electrically live circuit.

The silicon-controlled rectifier (SCR) is discussed in Chap. 8.

3-14 The Transistor[1]

Like the diode, the transistor is a semiconductor device and is usually made of the same material, silicon or germanium. Silicon is now extensively used.

Bipolar junction transistors are similar to junction diodes except that they have an additional junction, sometimes more. The word *bipolar* comes from the fact that both holes and electrons take part during the current flow through the device. Bipolar transistors are prevalent throughout the industry because of their low cost and good performance in many tasks.

A structural schematic drawing of an NPN transistor is shown in Fig. 3-20. The emitter region is very rich in current carriers. These carriers go (are emitted) into the base region and then on to the collector, whose job is to collect the carriers. The base region acts as a control; it can allow many or just a few electrons to flow from the emitter to the collector. It is very narrow and doesn't have many holes. If the base-emitter junction is forward-biased as shown in Fig.

[1] We have added material on the construction and operation of transistors in this edition. Students who are not interested do not have to read this material. However, we feel that many users should read this because it will help them to determine how a transistor symbol is to be placed in a schematic circuit diagram.

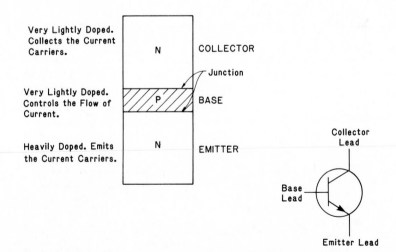

FIG. 3-20 Structural schematic of a NPN transistor. (With permission from C. A. Schuler, *Electronics Principles and Applications*, McGraw-Hill Book Company, New York, 1979.)

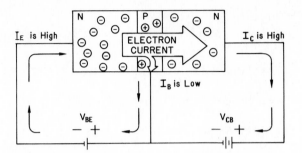

FIG. 3-21 NPN transistor currents. (With permission from C. A. Schuler, *Electronics Principles and Applications*, McGraw-Hill Book Company, New York, 1979.)

3-21, the resistance of the base-emitter junction is very low when compared to the resistance of the base-collector junction, which is reverse-biased. This large difference in junction resistances means that the resistor is capable of *power gain.* Although the collector is an N-type region, it is charged positively by V_{CB}. Since this positive field is quite strong, most of the electrons will not find holes in the base and will be attracted to and collected by the collector. A small current in the base controls much higher levels in the other two regions.

The PNP transistor differs from the NPN device in that the two outer regions are P-type material and the base is N-type material. Voltages V_{BE} and V_{CB} are reversed in polarity from that in a NPN transistor circuit. The emitter-to-collector current is *hole* current in the PNP transistor. Inasmuch as electron mobility is higher than hole mobility, NPN transistors operate faster than the PNP type. For this and other reasons more NPN transistors are in use than the PNP devices.

The bipolar transistor has the important feature of amplifying signals and therefore can be found in many circuits. Typical construction is shown in Fig. 3-22, and symbols for several types of bipolar transistor are shown in Fig. 3-23.

Another type of transistor is the *field-effect* or *unipolar transistor*. A unipolar transistor uses only one type of carrier. For example, in the N-channel JFET shown in Fig. 3-24 the only carriers are electrons. The channel contains enough electrons to support the flow of current from the source to the drain. Also, if a

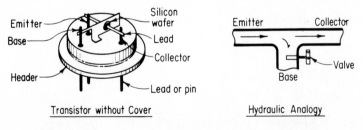

FIG. 3-22 Assembly and action of bipolar transistor.

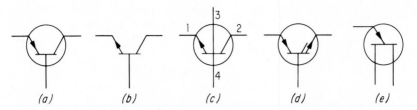

FIG. 3-23 Symbols for bipolar transistors: *(a)* PNP type; *(b)* NPN type; *(c)* tetrode; *(d)* PNPN type; *(e)* unijunction with P-type base.

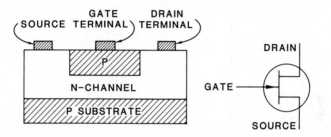

FIG. 3-24 An N-channel JFET. (With permission from C. A. Schuler, *Electronics Principles and Applications,* McGraw-Hill Book Company, New York, 1979.)

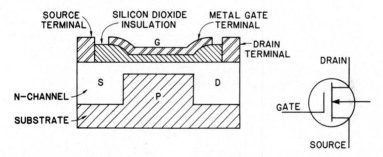

FIG. 3-25 An N-channel MOSFET of the depletion type. (With permission from C. A. Schuler, *Electronics Principles and Applications,* McGraw-Hill Book Company, New York, 1979.)

negative voltage V_{GS} is applied from gate to source, the drain current decreases. Whereas the bipolar junction transistor is *off* until base current is applied, the JFET is *on* until the gate voltage is large enough to remove the carriers from the channel.

Another FET is shown in Fig. 3-25. The metal gate terminal is insulated from the N-channel by a layer of silicon dioxide. (Hence the name *metal oxide semiconductor field effect transistor,* or MOSFET.) As in any FET, the arrow points *in for N-channel*, whereas for a P-channel FET the arrow points outward. Figure 3-26 includes symbols for five different FETS. Field-effect transistors have one advantage over bipolars: their gate terminal requires no current. This is fine when an amplifier with high input resistance is needed. With the enhancement-mode MOSFET (Fig. 3-26*e*), however, which is normally in the *off* mode, a proper gate voltage will attract carriers to the gate and form a conductive channel. The broken lines in the symbol between source and drain imply that the channel is not always present.

Several points should be made about transistor-symbol construction:

1. The circle (envelope) does not have to be drawn if no confusion arises or if no leads are attached to the envelope.

2. Orientation, including a mirror-image presentation, does not change the symbol meaning.

3. For the NPN and PNP transistors, the base symbol is drawn about one-third of the way "in" if the envelope circle is drawn.

4. Collector and emitter lines are drawn at about 60° to the base symbol, and the arrowheads do not touch the baseline.

5. The vertical line in the FET symbol, called the *channel*, is drawn through the center of the circle.

Although the envelope circle is not required, many engineers and drafters prefer or recommend that it be drawn. For this reason, we have shown most of the transistor symbols in Figs. 3-23 and 3-26 with envelopes. In Fig. 3-23*c*, the leads have been numbered, starting with the emitter, which is the standard order when numbering is desired. The student should become familiar with the symbols and current-flow characteristics of bipolar PNP and NPN transistors. Figure 3-27 supplies this information. The circular symbol with ~ inside indicates a signal source. We have only briefly touched upon the electron action of the transistor, a fascinating subject. There are good books that go into this subject fully. Several are listed in the Bibliography.

Transistors are available in several sizes and shapes of packages. Some examples are shown in Fig. 3-28. Many are identified on diagrams as *2* or *3* followed by *N* and several digits, such as: *2N918, 2N4393*, and *3N225*. But they may also include the manufacturer's number, which is usually an entirely different number. The body or case style is given as *TO3, TO5, TO10*, etc. One problem that confronts the technician and assembler is how to identify the

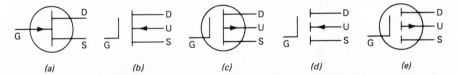

FIG. 3-26 Field-effect transistors: *(a)* N-channel junction-type FET, or JFET; *(b)* N-channel depletion-type metal-oxide FET, or MOSFET; *(c)* P-channel depletion-type MOSFET; *(d)* N-channel enhancement-type MOSFET; *(e)* P-channel enhancement-type MOSFET.

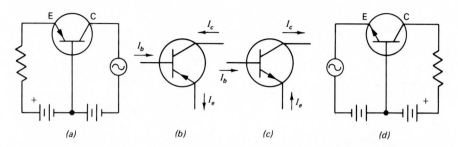

FIG. 3-27 Transistor biasing: *(a)* biasing of PNP transistor in an amplifier circuit; *(b)* and *(c)* electron flow in PNP and NPN transistors; *(d)* proper biasing of NPN transistor.

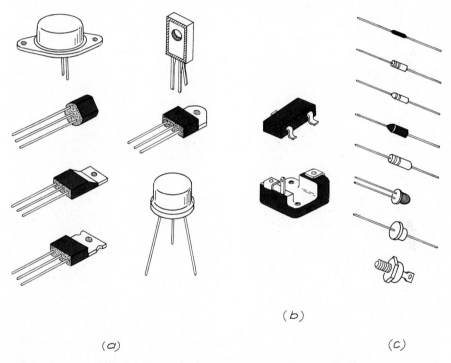

FIG. 3-28 Transistors and diodes: *(a)* transistor shapes; *(b)* gull-wing surface-mounted transistor, above, and power MOSFET, below; *(c)* diodes.

leads. These are not always shown on the case. Occasionally they appear as 1, 2, and 3 or E, C, and B. Sometimes one must refer to the manufacturer's catalog or a substitution guide for this information.

3-15 The Electron Tube

Before the transistor was invented, all or most of the functions now performed by semiconductors were performed by *thermionic devices* called *electron tubes*

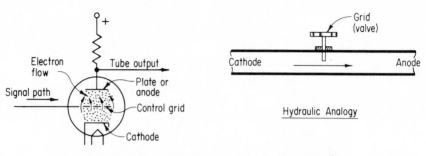

FIG. 3-29 Action of electron flow in vacuum tube.

or *vacuum tubes.* Today tubes are used only in special applications such as x-ray tubes, cathode-ray tubes (CRTs) (including television picture tubes), high-power RF amplifiers, and microwave circuits.

Thermionic emission involves the use of heat to liberate electrons from one element of a tube, called the *cathode.* When the cathode or a heater circuit within the tube is heated, the released electrons will flow toward the *anode* (or the *plate*) if the latter is made positive with regard to the cathode. Another electrode called the *grid*, which is a fine mesh or wire pattern, may be placed between the cathode and grid as shown in Fig. 3-29. If the grid is attached to a circuit that receives certain signals, the grid potential may rise and fall, causing changes in the flow of electrons toward the positively charged plate. Because of this throttling, or valvular, action the electron tube is sometimes referred to as a *valve.*

In most cases the tube envelope is drawn as a circle, although it is sometimes elongated, or even split. There are sometimes as many as five grids in a tube. The heater symbol (the inverted V, bottom of Fig. 3-29) is not a grid.

There are two types of video or cathode-ray tube, electrostatic and electromagnetic. Oscilliscopes generally have *electrostatic deflection*, while television picture tubes generally use *magnetic deflection* by means of coils around the neck of the tube. The cathode is heated to generate thermionic emission. Positive potential is applied to the anodes (dashed lines in Fig. 3-31*a, c,* and *d*) and to the posphor coating on the tube, thus accelerating the electrons toward the screen. As shown in Fig. 3-30*a*, the electrons are focused into a narrow beam, making a dot of light on the tube face. Positive voltage applied to vertical and horizontal deflecting plates can move the dot around on the screen. Each dot retains its brightness long enough for the eye to capture its image. If the process is repeated very rapidly, the effect of a moving picture is produced.

3-16 Qualifying Symbols

There are many qualifying symbols that, when added to a device symbol or conductor, indicate that that special characteristic is important to the function of the device. We have shown a few of these in Figs. 3-2, 3-8, 3-9, and 3-18. Because their number has grown so much in the last two decades, we have

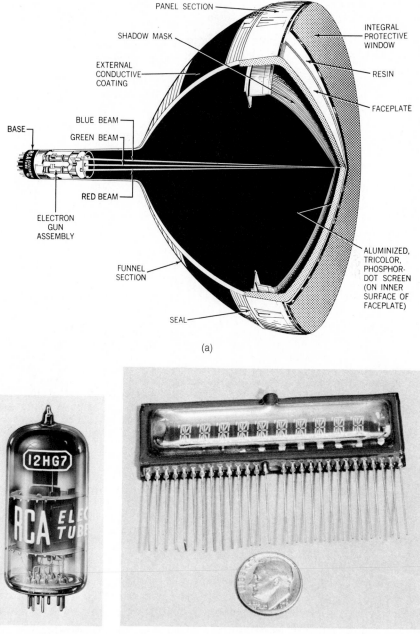

PANEL SECTION

SHADOW MASK

EXTERNAL
CONDUCTIVE
COATING

INTEGRAL
PROTECTIVE
WINDOW

RESIN

FACEPLATE

BLUE BEAM

BASE

GREEN BEAM

RED BEAM

ELECTRON
GUN
ASSEMBLY

FUNNEL
SECTION

SEAL

ALUMINIZED,
TRICOLOR,
PHOSPHOR-
DOT SCREEN
(ON INNER
SURFACE OF
FACEPLATE)

(a)

12HG7

RCA ELEC
TUBE

(b)

(c)

FIG. 3-30 Tubes: *(a)* cutaway section of a 70° color picture tube; *(b)* old-type electron tube; *(c)* vacuum fluorescent display popular in automobile digital displays.

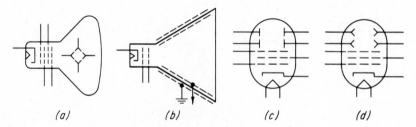

FIG. 3-31 Picture-tube symbols: *(a)* and *(b)* electromagnetic; *(c)* electrostatic; *(d)* electromagnetic.

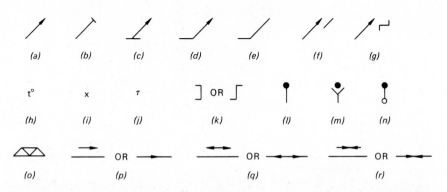

FIG. 3-32 Qualifying symbols: *(a)* adjustability, general; *(b)* preset adjustability; *(c)* linear, extrinsic; *(d)* nonlinear, extrinsic; *(e)* nonlinear, intrinsic; *(f)* adjustable, continuous; *(g)* adjustable, in steps; *(h)* temperature dependence; *(i)* magnetic-field dependence; *(j)* storage; *(k)* breakdown, overvoltage absorber; *(l)* test-point recognition symbol, general; *(m)* test point for a test jack; *(n)* test point for a circuit terminal; *(o)* electret (without electrodes); *(p)* direction of power or signal flow, one way; *(q)* either way; *(r)* both ways, simultaneously.

shown many of the qualifying symbols (listed in ANSI/IEEE Y32E) in Fig. 3-32.

3-17 Integrated-Circuit Modules

The symbols shown thus far have been for individual devices. The wide use of IC packages has made it necessary to somehow portray the entire circuit, which may have the equivalent of hundreds of transistors or other devices, as a part of a larger circuit. Present use includes two symbols that are employed to show the integrated circuit. They are the rectangle and the equilateral triangle, as shown in Fig. 3-33a and b. The triangle is usually reserved for ICs that are amplifiers because it is so listed in the standards. However, since many ICs are not amplifiers, the rectangle is widely used.

Unfortunately, the use of the rectangle is not spelled out in the standard. Usually, the manufacturer's number and the function are listed, as has been

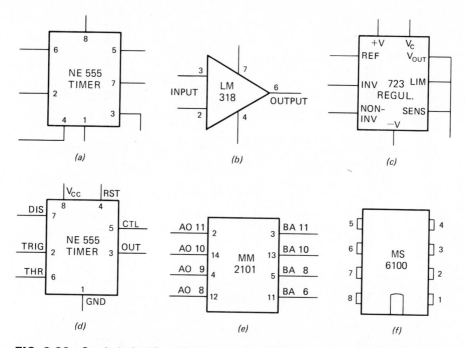

FIG. 3-33 Symbols for IC modules: *(a)* rectangle with pin numbers; *(b)* triangle, generally used for amplifier circuits only; *(c)* rectangle with functions assigned to each pin; *(d)* rectangle with pin numbers and assignments; *(e)* rectangle with pin numbers and circuit identification; *(f)* standard numbering of pins on 8-pin dual-in-line package.

done in Fig. 3-33*b* and *d*. Pin numbers are now generally shown. More information, such as the ground or input voltage, is helpful. Circuit destination or identification, as shown in Fig. 3-33*e*, is also helpful. Going on the assumption that a schematic or wiring drawing should be easy and quick to read, the authors favor the methods shown in Fig. 3-33*d* and *e*. Figure 3-33*f* shows the typical pin arrangement for an 8-pin dual-in-line package. The reader will notice that the other examples in Fig. 3-33 do not show this sequence. This is because the pins have been rearranged (on the drawing) to facilitate making the physical circuit. The manufacturer of the integrated circuit provides the information about the pin assignments.

Figure 3-34 illustrates the standard pin numbering system for a large-scale integrated-circuit (LSI) package and a dual in line integrated circuit package. Other chips have various numbers of leads up to 64, but the package size is growing. The functional diagram (Fig. 3-34*c*) shows what is in the 555 timer package. This is a very popular circuit. One of its several functions is to produce an accurately timed clock pulse for microcomputers. Figure 3-35 shows typical packages for integrated circuits. Figure 3-36 includes popular styles of other devices that are in great use.

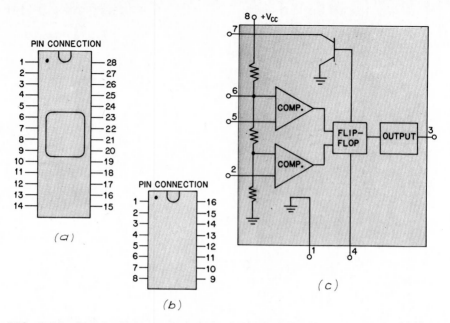

FIG. 3-34 Standard pin-numbering systems for LSI chip at *a* and dual-in-line package at *b*. A functional diagram for the NE 555 timer is shown at *c*.

FIG. 3-35 Integrated-circuit packages. At left and right: 40- and 28-pin packages. Below: an erasable read-only memory. Beside coin: surface-mounted packages. Below coin: Microprocessor chip, unpackaged. At top right is a relay.

CAPACITORS

SWITCH, SPST
OR DPDT

INCANDESCANT

ELECTROLYTIC

NEON   LED

LAMPS

SWITCH, ROTARY

VARIABLE   ADJUSTABLE

RESISTORS, FIXED

TRANSFORMER
LAMINATED IRON CORE

CRYSTAL, QUARTZ
(PIEZOELECTRIC)

RESISTORS
ADJUSTABLE,
POTENTIOMETER

TRANSFORMER,
POWER

TRANSFORMER
AIR OR MAGN. CORE

INDUCTORS
MOLDED   IRON CORE

SWITCH, ROCKER

TRANSFORMER
ADJ. IRON CORE

FIG. 3-36 **Drawings of some commonly used components.** (Artwork courtesy of the Heath Company.)

3-18 Reference or Device Designations

Schematic and other diagrams that include symbols for devices almost always include other information about the devices. This may include such items as a number arbitrarily assigned for each device in a circuit (e.g., R_1, R_2, D_1, and D_2)* and the capacity of such items as capacitors, fuses, and resistors. For diodes and transistors, an Electronic Industries Association (EIA) number may be shown. This is a standard number that is registered with the EIA and begins as 1N-, 2N-, 3N-, or 4N-. In some diagrams a "house" number is shown instead of the standard EIA number. House numbers are assigned by the manufacturer and often are not clearly related to the item's function or capacity. These numbers may resemble the following: LM 400, MR 1816, 10-390.

The most frequently used devices and their designations are shown in Table 3-1. This information is placed close to the symbol(s) in a drawing. It is

* In most schematic diagrams all letters and numbers are made with vertical uppercase characters of the same height.

TABLE 3-1 Device (Component) Designations

DEVICE	NUMBER IN CIRCUIT	VALUE OR CAPACITY	EIA OR HOUSE NUMBER
Capacitor	C_1	22 μF	
Crystal	Y_1	32.768 kHz	
Diode	D_2		1N419
Fuse	F_3	2 A, slow-blow	
Inductor	L_2	650 μH	
Integrated circuit	U_{201}		LM 255
Lamp, LED, or LCD	V_7	0.6 V	
Pin connector	P_{302}		
Resistor	R_5	680 Ω	
Resistor pack	RP_4	10 kΩ (4)	
Switch	SW_{401}		
Transformer	T_1		
Transistor	Q_5		2N4121

TABLE 3-2 Examples of Component Designations on Drawings

COMPONENT	LETTER SYMBOL	SAMPLE VALUE	CODE DESIGNATION
Capacitor	C_1	22 μF	
Diode	D_2		1N419
Inductor	L_2	650 μH	
Integrated circuit	U_{501}		LM 388
Resistor	R_{205}	680	
Transistor	Q_5		2N4121
Transistor with Function	Q_5		2N482 DETECTOR

either above, below, or at the side of some symbols and inside other symbols. We will show how this is done in Chap. 7, on schematic diagrams.

Some typical identifications of component devices are illustrated in Table 3-2. When a number such as 501 is shown for a device, it does not mean that the device is the 501st in a circuit. It indicates that the device is the first in a circuit board whose parts are numbered 501 and above. In a robot having 12 circuit boards, the motion circuit board might contain parts numbered 101 and above; the arm CB, 201 and above; the wrist, 301 and above; speech, 401 and above; and so on. At times the function and EIA or house number are shown for transistors and ICs. The detector Q_5 in Table 3-2 is such an example.

SUMMARY

Nearly every electrical and electronics device has a standard symbol which can be used to represent the device in a diagrammatic drawing of a circuit. These symbols are drawn with medium-weight lines but can be made with heavier lines, if it is necessary to highlight the symbol. Transistor symbols may or may not include the circle envelope. Theoretically, there are no particular sizes to which symbols should be drawn. Practically, though, there are minimal and maximal sizes for any symbol or set of symbols, and, within a drawing, symbols should be drawn in correct relative sizes. Usually a device is represented by a symbol of one size throughout a drawing. Not more than two sizes of a symbol are recommended for a drawing. There are certain basic electrical and electronics devices with which a reader or drawer of electrical drawings should become familiar. Orientation of a symbol depends on the direction of the conductor path along which it is placed, the polarity (if any) of the device, and other factors such as the collector, emitter, and drain leads of transistors. Symbols for the most commonly used devices or functions have been standardized on an American and international basis.

QUESTIONS

3-1. What authority (publication) or authorities would be good source material for the selection of symbols to be used in an electrical drawing?

3-2. How long would you make the resistor symbol in an average-size drawing?

3-3. What determines the position (orientation) of a symbol in a drawing?

3-4. In the diode symbol the bar that is perpendicular to the conductor path has a polarity. What is it?

3-5. What would be a typical designation that would be placed in the rectangular resistor symbol? In a circular relay-coil symbol?

3-6. What diameter would you make the termination symbols in a drawing? The connection dots?

3-7. Where more than one symbol is approved for an element (or component), how would you go about determining which symbol to use on an electrical drawing?

3-8. Assume that you are required to draw the line representing a signal path of a circuit approximately 0.5 mm wide. How wide a line would you use to draw the symbols in that circuit drawing? Why?

3-9. Do holes drift toward the negative or positive end of a semiconductor?

3-10. A statement in the standard indicates that not more than two sizes can be used for any one symbol in the same drawing. Describe a situation that, in your opinion, would require two different sizes of the same symbol in a drawing.

3-11. The IEC-approved capacitor symbol of two parallel lines has a disadvantage. What is it?

3-12. Name five different types of switch or device that can be used for switches in electric circuits.

3-13. How would you show a device in a circuit drawing if no known symbol for that device exists?

3-14. What happens to the depletion region of a junction diode when a forward bias is applied?

3-15. How do manufacturers mark the cathode lead of a diode?

3-16. The emitter of which type of transistor, NPN or PNP, emits holes?

3-17. When would you draw the circle around a diode or transistor symbol?

3-18. At what angle approximately is the collector lead drawn with respect to the base part of a bipolar transistor symbol?

3-19. How is a battery symbol oriented in a drawing?

3-20. What do the letters V_{GS} and V_{CB} stand for?

3-21. The symbol for an electron tube may contain three or more types of electrode. Name or describe three such electrodes.

PROBLEMS

The following problems will require that the student make use of the ANSI/IEEE Standard or some standard which indicates the correct symbols. The symbols shown in Appendix B are taken from several American standards. It would also be advantageous for the student to look at some of the schematic drawings shown at various places in the text, especially in Chap. 7, before beginning to work the following problems.

Cross-section paper with four or five divisions to the inch may be helpful, especially if the symbols are to be sketched freehand. It is not an absolute requirement, however. Any type of detail or tracing paper will be adequate. Exercises can be drawn on $8\frac{1}{2} \times 11$ paper, except where otherwise indicated. If they are to be done on a computer, your instructor will add additional instructions.

3-1. Draw three horizontal lines about 20 cm or 9 in. long and 8 cm or 3 in. apart. At 5-cm (2-in.) intervals starting 1 cm or 0.5 in. from the left end, draw the symbols for the following elements along the line(s) which represent the signal path.
- a. Diode (cathode on right)
- b. Battery (positive on right)
- c. Resistor (general)
- d. Voltmeter
- e. Speaker (at end of line)
- f. Inductor (general)
- g. Relay coil, ac
- h. Shielded capacitor
- i. Motor
- j. Fuse
- k. LED
- l. Zener diode
- m. Single-throw switch
- n. Current transformer
- o. Pushbutton switch, "break"

3-2. Draw three horizontal lines spaced about 8 cm or 3 in. apart. With about five symbols to each line, place the following symbols along the line(s), each representing a signal path. Additional short leads may have to be added for some devices (transistors, etc.).
- a. Shielded capacitor
- b. Chassis connection
- c. Transformer with core
- d. NPN transistor
- e. Contact (normally closed)
- f. Inductor with two taps
- g. Polarized relay
- h. LED
- i. Adjustable capacitor
- j. Pushbutton switch
- k. Variable resistor
- l. Switch, double-throw
- m. Tetrode transistor
- n. Split-stator capacitor
- o. Inductor with ceramic core

3-3. Sketch or draw the following device symbols, four to a row, in three rows. In general, the signal path will be left to right.
   a. PNP transistor, base at left
   b. Thermistor, general
   c. Zener diode, cathode right
   d. LED
   e. Thermocouple, current-measuring
   f. P-channel enhancement-type MOSFET
   g. Delay function, 2 s
   h. Incandescent lamp
   i. N-type FET
   j. Transformer, general
   k. Lightning arrester, general
   l. N-channel MOSFET, depletion type

3-4. Follow the instructions given in Prob. 3-1, but show the following symbols:
   a. Dipole antenna
   b. Schmitt trigger
   c. Separable connectors
   d. Slow-operate relay
   e. Breakdown diode, unidirectional
   f. Capacitor with split stator
   g. Pickup head, recording
   h. Switching function, transfer
   i. Thermistor, general
   j. Adjustable resistor
   k. Field-effect transistor, N-type base
   l. Twin-triode tube
   m. Four-conductor shielded cable
   n. Tunnel diode

3-5. Draw the following semiconductor or tube symbols; three lines of three each are suggested. In general, signal path is left to right.
   a. Zener diode, anode right
   b. PNP transistor, base right
   c. X-ray tube with grid
   d. Triode vacuum tube
   e. Varicap diode, anode left
   f. Tunnel diode, cathode left
   g. P-type unijunction transistor
   h. Tunable magnetron
   i. N-channel enhancement-type MOSFET, gate left

3-6. Redraw the symbols shown in Fig. 3-37 at about twice the size they appear in the book. (Use one 11 × 17 or 12 × 18 sheet, or two 8½ × 11 sheets.)

SYMBOL   IDENTIFICATION

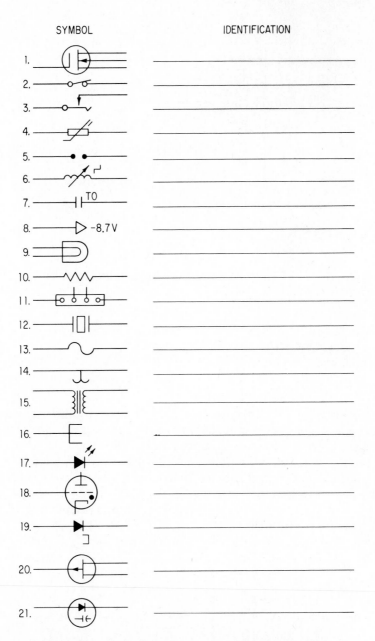

FIG. 3-37 (Prob. 3-6.) Symbol identification exercise.

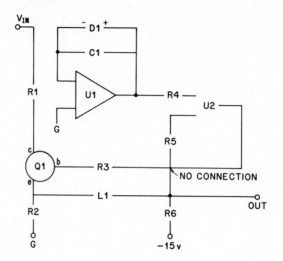

FIG. 3-38 (Prob. 3-7.) **Symbol exercise. Voltage-to-frequency converter circuit diagram.**

Then identify each device symbolized at the right. (*Suggestion:* Draw a horizontal line to the right of each symbol, as shown in Fig. 3-31, and letter the name along each line.) Identification, in many cases, should include more than just the name. If it is a transformer, for example, what type is it; if a switch, what type is it and what is its operating condition?

3-7. Redraw the circuit shown in Fig. 3-38 to about twice the size it appears in the book. Then place the standard symbols for the devices where they are indicated by letter-number designations such as R_1. The letter R stands for resistor, C for capacitor, Q for transistor (NPN in this circuit), D for diode, L for inductor, G for ground, and U for integrated circuit. The triangle symbol is used for integrators (U_1) and comparators (U_2), as well as for amplifiers; L_1 is continuously adjustable. Use $8\frac{1}{2} \times 11$ paper. Add dot connector symbol(s) where you think there should be one.

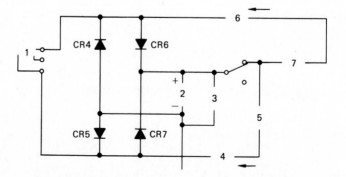

FIG. 3-39 (Prob. 3-8.) **Calculator charging circuit.**

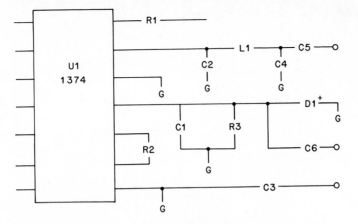

FIG. 3-40 (Prob. 3-9.) Part of television modulator circuit.

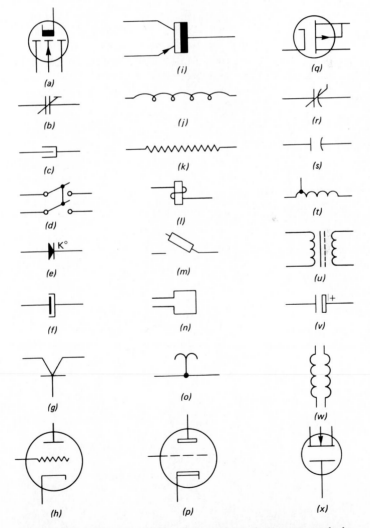

FIG. 3-41 (Prob. 3-10.) Symbol exercise. Incorrect symbols.

3-8. Complete the circuit shown in Fig. 3-39 by adding the symbols listed below at the places where the numbers appear. Rectifiers CR_4 through CR_7 may be redrawn as the bridge in Fig. 3-18e.

Locking switch (make): 1
Multicell battery: 2
Polarized capacitor: 3
Solid-state rectifier: 4
Resistor: 5
Diode: 6
Resistor: 7

The completed drawing should have all lines the same weight. Use $8\frac{1}{2} \times 11$ paper.

3-9. Redraw the circuit shown in Fig. 3-40 to about $2\frac{1}{2}$ times as large as it appears here, on $8\frac{1}{2} \times 11$ paper. Then place the correct standard symbols in the spaces where the standard letter-number designations appear (R = resistor, C = capacitor, L = inductor, G = ground, D = diode). All symbols are "general" except C_3, which is variable. The symbol at R_2 may have to be smaller than the other Rs because of space limitations. Circuit U_1 is an integrated circuit. Pin numbers and identification may be required by your instructor.

3-10. Each symbol in Fig. 3-41 is incorrect or not an American National Standard. First, identify the symbol that is being depicted. Then draw the correct symbol and label it, using the same general layout as in Fig. 3-35. Use $8\frac{1}{2} \times 11$ or 11×17 paper.

3-11. Identify some symbols assigned by your instructor. Use the format assigned.

4

Wiring, Cabling, and Chassis Drawings

Before a machine or system using electronic parts is made or built, instructions must be given to the persons who will be involved in the fabrication or assembly. Such persons are likely to be engineers, machinists, electrical installers, assembly workers, time-and-motion engineers, computer and numerical control (N/C) programmers, printed-circuit (PC) board designers, robotics technicians, and electrical and electronic drafters and detailers. Some of these people may be giving the instructions, some may be on the receiving end of the instructions, and some may be doing a little of both. Drawings convey these instructions very well, although they sometimes must be accompanied by other information. There are several different kinds of drawings used for this purpose. Correct or recommended practice for most of these drawings is set down in certain ANSI, IEEE, and Military Standards. Some of the more relevant standards are listed below:

ANSI Y14 "Drafting Manual," which includes

14.1 "Drawing Size and Format"

14.2 "Line Conventions, Sections and Lettering"

14.3 "Multi and Sectional Drawing Views"

14.4 "Dimensioning and Tolerancing for Engineering Drawings"

14.6 "Screw Threads"

14.15 "Electrical and Electronics Diagrams"

Mil Std 242 "Electronic Equipment Parts (Selected Standards)"

Mil Std 196 "Joint Electronics Type Designation System"

Mil Std 429 "Printed Wiring and Printed Circuits Terms and Designation"

Mil Std 681 "Identification, Coding and Application of Hook Up and Lead Wire"

Company standards, if any, that may be more specific or restrictive than ANSI, IEEE, or Military Standards

In this chapter we will discuss the following types of drawings that are commonly used for production:

Wiring or connection diagrams
Cabling diagrams
Harness diagrams
Sheet-metal layouts
Assembly drawings

Connection Drawings

4-1 Connection, or Wiring, Diagrams

In order to connect various parts (motors, resistors, capacitors, connectors, displays, and circuit boards) it has been the custom to use or follow a *connection diagram.* The older name for this type of drawing is *wiring diagram,* and, because it is so widely used, we shall use these two terms interchangeably.

In a manufacturing company these drawings are used by the time-and-motion, quality-control, and estimating sections, as well as by individual workers. Wiring diagrams are also used in the maintenance area by persons doing checking, troubleshooting, and modification. Sometimes they are used with schematic diagrams (explained in Chap. 7) to provide additional information.

In low-volume production, a connection diagram may be used at first and then put away after the worker has memorized the steps and connections. In some instances companies with active methods sections are able to produce without such a drawing.

In high-volume production (ICs, PC boards, consumer products, etc.) the connection diagram alone is not used for connecting parts, but rather is just a step in the design-manufacture process.

4-2 Types of Connection Diagram

These diagrams may be classified according to their general layout as follows:

1. Point-to-point
2. Highway, or trunkline
3. Baseline, or airline
4. Straight-line

There are also other ways of labeling wiring diagrams, as the reader will discover. Such titles might include *cabling, harness,* and *interconnection* diagrams.

4-3 Point-to-Point Diagrams

This is the oldest type of wiring drawing. Figure 4-1 shows a pictorial form of the diagram in which each part and terminal are drawn about as they appear to the viewer. This type of drawing has met with great success in the do-it-yourself kit industry and other places.

Figure 4-2 is another type of pictorial point-to-point diagram, but it is not as close to what the actual object looks like as is the first drawing. Many of the symbols are pictorial, but their physical relationship to each other is not accurate. Furthermore, the lines representing the wires (conductors) are drawn either horizontally or vertically, not as they actually are placed in the chassis or frame of the automobile. A distinctive feature of this drawing is the color code and identification system required when there are as many wires as are presently utilized in motor vehicles. The wire in the upper right corner, for example, is L7-18BK. This means that this is conductor L7, which is of size 18 wire (see p. 117) and is black in color.

In the two drawings discussed so far, the component parts have been drawn pictorially. In most wiring diagrams, these parts are shown by means of symbols or by simple geometrical figures such as squares, rectangles, and circles. This will be the case with the connection diagrams that follow, including the aircraft wiring diagram in Fig. 4-3.

Figure 4-3 shows only a small part of the original drawing, but it contains enough details to give a fair picture of what such a drawing looks like. Note the switch symbols* S_8, S_9, etc., the battery symbol, BT_1, and the circuit-breaker symbols, CB_1, CB_2, etc. Note, also, the uniform spacing of wire conductors, drawn about $\frac{1}{4}$ in. apart, with their rounded "corners." Actually, most of the wires are routed in bundles (harnesses), and the physical routing of wires probably does not bear close resemblance to the way they are drawn here. Item J_9 below center and right, is a female receptacle into which male connector D_7 is plugged. The letters represent the pins in the connector. Readers will have a difficult time tracing the paths of many of the wires in this drawing because only part of the drawing is here. They can trace line H_{27} from switch S_8 to CB_5 and line H_{37} from S_8 to J_9, however. In Figs. 4-2 and 4-3, medium-weight lines have been used to represent wires. This is common practice, although some manufacturers draw all or certain conductors more heavily.

4-4 Interconnection (external- wiring) Diagrams

Figure 4-4 is another example of a poir t-to-point wiring diagram. It consists of several PC boards and motors of a rob ot and individual wires or cables that go from terminals in one board or motor to terminals in another board. The large rectangle called I/O / CPU has the microprocessor and related circuitry as well

---

* In many diagrams in this book (and elsewhere) letters and numbers (subscripts) are made the same height.

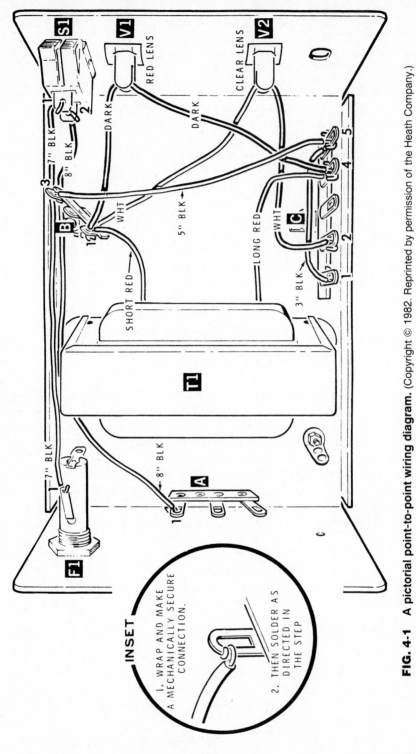

FIG. 4-1 A pictorial point-to-point wiring diagram. (Copyright © 1982. Reprinted by permission of the Heath Company.)

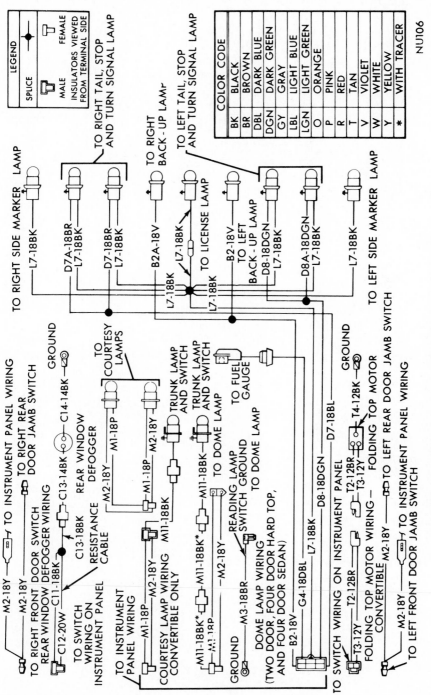

FIG. 4-2 Pictorial wiring diagram for an automobile. (Chrysler.)

117

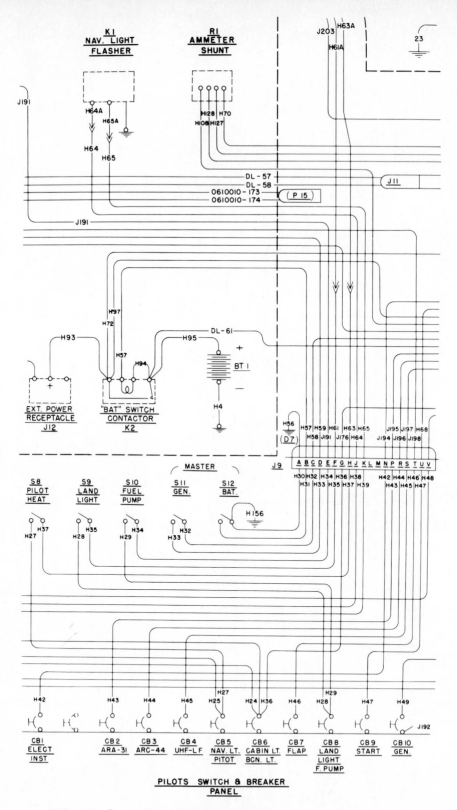

FIG. 4-3 Connection (wiring) diagram for a small aircraft. (Cessna.)

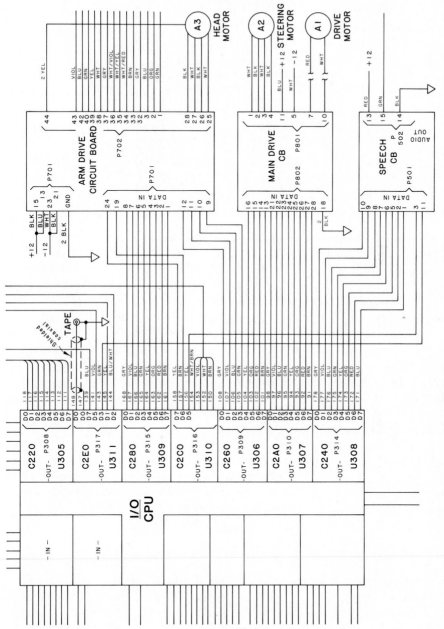

FIG. 4-4 Part of the interconnection diagram for a robot. "Ports" and wiring at left are incomplete. (Copyright © 1982. Reprinted by permission of the Heath Company.)

119

TABLE 4-1   Component Number Assignments by Location*

0–099	Parts found on robot chassis
101–199	Parts found on sonar transmit board
201–299	Parts found on power supply board
301–399	Parts found on I/O board
401–499	Parts found on CPU board
501–599	Parts found on speech board
601–699	Parts found on sense board
701–799	Parts found on arm drive board
801–899	Parts found on main drive board
901–999	Parts found on remote controller

* Last five lines omitted. This list was printed in the lower right-hand corner of the block-interconnect diagram shown in Fig. 4-4. (Reprinted by permission of the Heath Company.)

as input ports on the left side and output ports on the right. Each port has one or more pin connectors, labeled P_{308}, P_{317}, and so on. Internal connections on the boards are not shown, only the external connections. Therefore this type of diagram is called an *interconnection* (or *interconnect*) diagram.

As in the previous example, all conductor paths are drawn horizontally or vertically, but with sharp corners. Each path has a number and color and is assigned to a pin location on the circuit board. A three-letter color abbreviation system is used, with BLU for *blue,* GRN for *green,* and ORG for *orange.* A shielded coaxial conductor is shown as No. 148 leading from C2EO to the *tape-out* jack and is attached to a common chassis connection. Figure 4-4 is not quite a pure point-to-point diagram because the wires leading from I/O port C_{220} feed into a *highway.* This aspect is discussed in Sec. 4-6. Only part of the diagram has been shown, and some lettering has been deleted to facilitate reading of the drawing.

Another body of information appears on the same sheet as the interconnect diagram in Fig. 4-4. This is the number assignment for components (parts) mounted on the various PCBs of the robot. Table 4-1 shows most of this information. We have explained, with examples in Chap. 3, how components are identified by number and sometimes by function.

4-5 An Example for Types of Wiring Diagram

A stereo system is shown in Fig. 4-5 with the wires connecting the various components. A typical point-to-point diagram of this system is shown in standard form below the pictorial view. Each part has been shown as either a rectangle or a circle, which is generally standard practice. And each part has been assigned a number, pretty much in an arbitrary manner, although as the reader will notice, we did start with the item (record changer) that was at the top and go downward, 1, 2, 3, etc. The rectangles and circles representing the major components need not be arranged to correspond to the actual location of these

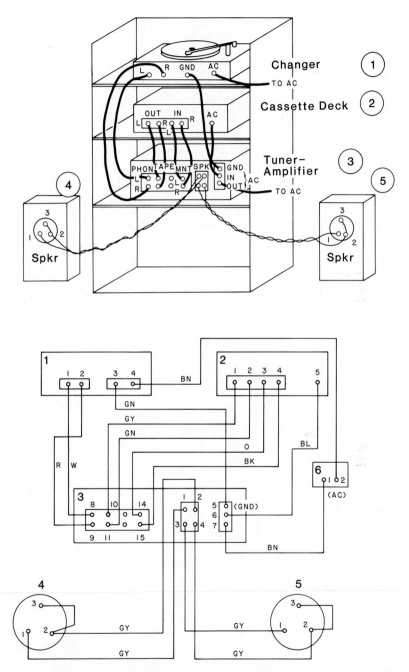

FIG. 4-5 Pictorial diagram and point-to-point diagram for stereo equipment wiring.

items with regard to each other, although an attempt to do so will make the diagrams easier to draw and read. The large boldface numbers representing each item should follow some sequence, left to right or up and down, if possible. As was the case in Figs. 4-3 and 4-4, neat arrangements of parallel lines that include equal spacing where several lines are in a particular area is obtained in the point-to-point diagram. The lettering code is different from the preceding example in that one or two letters are used. One-letter designations are adequate for such colors as red, white, yellow, and orange, but two letters are necessary for certain others such as black (BK), blue (BL), brown (BR), green (GN), gray (GY), and slate (SL). The wires joining terminals 2 and 3 of the speakers are called *straps* or *pigtail* leads and do not require color designations.

4-6 Highway, or Trunkline, Type of Wiring Diagram

Practically any electrical assembly or system can be shown wired or connected graphically by the point-to-point method. But the same assembly or system can also be connected graphically by other methods. Sometimes the point-to-point diagram gives the clearest picture and fastest reading, yet, at other times, one of the other methods might give the best picture and provide easier and quicker reading. One of these other methods is the highway type of connection diagram. (We mentioned that a *highway* was shown in the interconnection diagram in Fig. 4-4.) This type differs from the point-to-point style in that *conductors* (*conductor paths* might be better) are merged into long lines called *highways* (or *trunklines*) instead of being drawn as separate, complete lines from terminal to terminal. The short lines leading from the terminals to the highways are called *feed,* or *feeder, lines,* and must have some sort of identification near the highway so that the conductor path can be followed by the reader.

One or more highways may be shown, depending on the wire routing necessitated by the physical (or graphical) arrangement of components. These trunklines do not have to conform to actual bundles or harnesses, if such are used. Figure 4-6 shows the connection of the stereo equipment of the example (Fig. 4-5) in the form of a highway diagram. The components have been drawn and numbered exactly as they were in Fig. 4-5. The highways were located about 1.25 in. (32 mm) from the rectangles; thus the feeder lines from the terminals were comparatively short, yet long enough for some identification to be placed close by. We were able to prevent any highway from crossing another and feeders from crossing each other or a highway. (This may not be possible in certain diagrams, but crossings should be kept to a minimum.)

Each conductor (feeder) has identification in the form of (1) component or part destination, (2) terminal at the component destination, and (3) wire color. Thus the feeder leading from terminal 1 of component 1 has the following identification: 3/8-W, which means that it goes to part 3, terminal number 8, and the wire is white. Going to component 3, terminal 8, we find the other end of this connector, which has the "reverse" designation, 1/1-W. The highway is usually drawn the same weight as the feeders, but it may be drawn heavier.

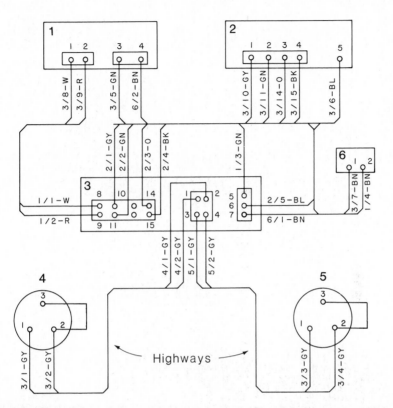

FIG. 4-6 **The wiring of the equipment shown in Fig. 4-5 drawn as a highway diagram.**

Where the feeder joins the highway, an arc or 45° line is drawn, and this line or arc should be in the direction the wire is headed.

A different method of identifying circuit paths is in wide use. This ANSI-recommended method suggests that each feed line have this designation near the point where it joins the trunkline: (1) component destination, (2) terminal or component part, (3) wire size or type (if necessary), and (4) wire color. In Fig. 4-7, the top feeder line at terminal board 12 carries the following identification: *TB*14/4-*B*2. Broken down, this would be: *TB*14 — destination; 4 — terminal at destination; *B* — size of wire (say, No. 22 hookup wire by a prearranged code); and 2 — color (red, by Mil Std 122 color code shown on the drawing and in Appendix C). Going to *TB*14 and terminal 4, we find the other end of this connector, and here it has the reverse identification, *TB*12/1-*B*2.

The choice of location and number of highways may depend on several factors. Separate highways for different cable groupings may be desirable. If a group of wires needs to be segregated or shielded, a separate highway for that group may be desirable. Such a highway might be drawn close to, and parallel to, another highway. Occasionally situations may arise where certain wires,

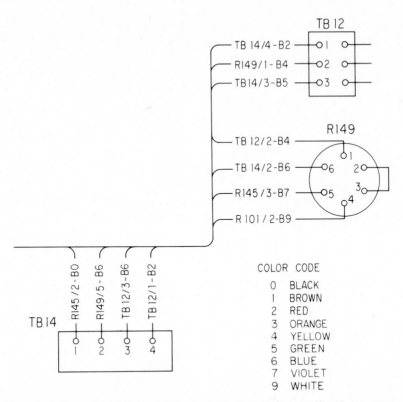

FIG. 4-7 Part of a highway diagram using the ANSI identification system.

called *critical wires,* should be drawn from point to point and not merged into a highway. Special identification for these wires may be necessary.

The 45° lines where the feeder joins the highway are about 0.20 in. (4 mm) long, and the arcs, if used, should have radii that are about 0.20 in. (4 mm).

The highway type of connection diagram is particularly suited for showing the wiring of large panels where there are many terminals and conductors. Figure 4-7 is part of a panel-wiring diagram. A panel-wiring diagram generally shows the back, or wiring, side of the panel. Terminal strips, switches, breakers, etc., are usually shown in exact or approximate position with regard to each other. Sometimes the panel is drawn as a scale drawing showing all terminals, lights, lettering, and switches, and the wiring is shown diagrammatically usually on the same sheet.

4-7 Airline, or Baseline, Diagrams

These wiring drawings, with a rather misleading name, are similar in some ways to highway diagrams. The *airline,* or *baseline,* as it is sometimes called, is an imaginary, usually horizontal or vertical, line conveniently located so that short feed lines may be drawn from component terminals to it.

The stereo system used previously has been drawn in this manner in Fig. 4-8. It has six horizontal baselines and one vertical one. (In many such diagrams, all baselines are horizontal.) In our example these lines have been drawn heavier than the others, but in many such diagrams the baselines and feeder lines are of the same weight. The feed lines are drawn at right angles to the baseline but without the curves or 45° lines that are in highway diagrams. Good identification of conductors is a necessity. In Fig. 4-8, the same identification system is used as in Fig. 4-6. The major difference between airlines and highways is that a highway must go from one component to the destination component, sometimes making several bends and turns to get there, but an airline may stop at any convenient place and another airline may be drawn near the destination component. In airline drawings having many component parts there are many airlines spotted at different places close to groups of parts. In such cases, a feed line may enter one airline and come off another airline. This makes for slower reading and tracing of conductor paths, but for systems or packages containing many wires and terminals, the airline type of diagram may be less cluttered and confusing than point-to-point and highway drawings are.

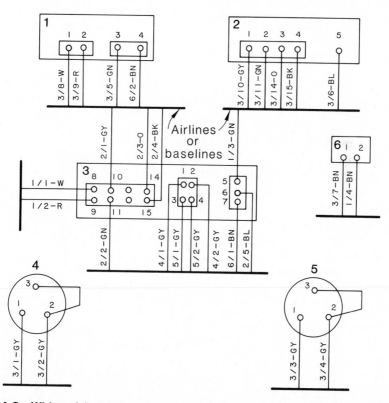

FIG. 4-8 Wiring of the equipment shown in Fig. 4-5 drawn as a baseline, or airline diagram.

Sometimes, so many components are present in a large system that if they were shown as one column or strip, the airline diagram would be too long and narrow for practical purposes. To avoid this, the diagram may have to be laid out in several columns side by side.

4-8 Straight-Line and Hybrid Diagrams

Another type of wiring or interconnection diagram is the straight-line type that is shown in Fig. 4-9. Instead of drawing the components or terminals as rectangles, circles, etc., each component number is listed at the top with a vertical line directly underneath. Conductors that go to each component are shown with a connection dot and the terminal number close to the dot. Wire colors are placed above, often near the center of the "span."

Component numbers usually start with the lowest at the left and increase, with the highest at the right.

We have started with the wires or conductors from part 1 (refer to the lower part of Fig. 4-5) that go to part 3 since there are no connections between 1 and 2. After those three wires are placed on the diagram, and the wire from terminal 4 to part 6 is drawn, we start with the connectors that are on part 2. This type of diagram has similarities with the wiring list shown in the next section. Although

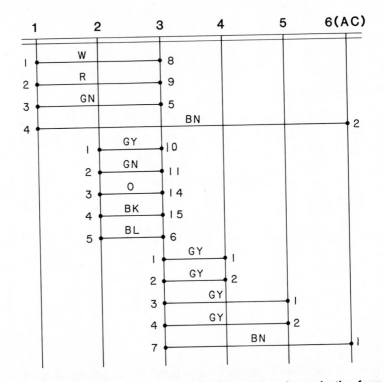

FIG. 4-9 The equipment of the preceding examples drawn in the form of a straight-line diagram.

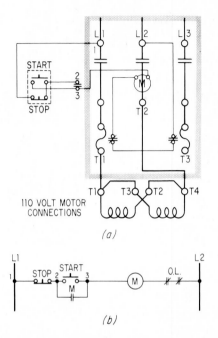

FIG. 4-10 Connection diagram *(a)* and elementary diagram *(b)* of a motor starter.

not as explicit as the previously explained diagrams, such a diagram is very helpful to estimators and installers of very large and complex systems.

Frequently, it is desirable to show the wiring diagram and the schematic (elementary) diagrams on the same drawing. Figure 4-10 shows these two diagrams for a motor controller. In this panel-wiring drawing, two different line widths are used for conductors. Narrow lines, say, $\frac{1}{100}$ in., are used for the control, or pilot, section of the circuit; and wide lines, say, $\frac{1}{50}$ in., represent the line-voltage part.

For very complex installations, it may be necessary to compile *wiring lists* or *to-and-from* diagrams in addition to, or in place of, connection diagrams. A typical heading for such a list might be:

Item. No.	Symbol	Color	Size	From		To	
				Component	Terminal	Component	Terminal

Additional information, such as routing (conduits, ducts, etc.), function, and military type number, may be shown.

4-9 Line Spacing and Arrangement

The reader may or may not have noticed the neat, uniform line spacing in some of the examples—particularly the point-to-point diagrams. This spacing and layout were not achieved accidentally, but rather by careful planning and good

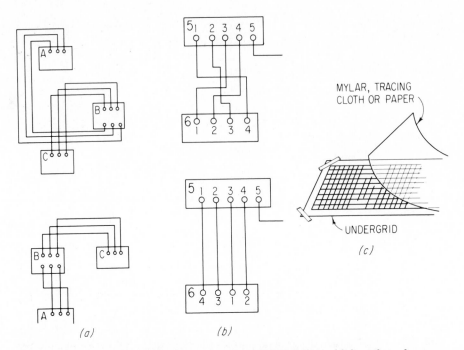

FIG. 4-11 Layout of wiring diagrams: *(a)* experimenting with location of components; *(b)* changing terminal locations; *(c)* using preprinted undergrid.

use of drafting aids. Two especially helpful aids are (1) the preliminary freehand sketch (or sketches) and (2) a commercially produced or office-made undergrid.

When making the original sketch, it is advisable to consider laying out the components so that the simplest, neatest wiring diagram will result. This may include changing the location of the components with regard to each other (if such juggling is permitted) and changing the order of the terminals. Rather elementary graphic examples are shown in Fig. 4-11. The shorter and more direct the lines are, the quicker and easier it is for the reader to trace the paths. Keeping the crossings to a minimum also facilitates reading.

Spacing of parallel lines is most often at $\frac{1}{4}$ in. This permits use of a prepared undersheet having $\frac{1}{4}$-in. grids, which, if a transparent drawing medium is used, greatly facilitates the making of the final drawing. For various reasons (lettering requirements, reduction of drawing) other spacings may be required.

4-10 Wiring Harness or Cabling Drawing

A large portion of the expense in producing complex electronics assemblies can be laid to the wiring. The wiring operation can be optimized (made simpler and cheaper) by study of the chassis and connection drawings, followed by the drawing of a harness diagram. This is generally done in a series of steps:

1. Study of the mechanical assembly, noting the general path which the harness should take and any obstacles which might cause difficulty.

2. Making a drawing of the outline of the harness. This can best be done if a suitable full-scale chassis or assembly drawing exists. A sheet of tracing paper or other translucent drawing medium is then placed over the drawing and the harness outline done on the top sheet.

3. Completion of the harness (sometimes called *local-cabling*) diagram itself, using the assembly drawing and the wiring drawing for the assembly. This may include "breakout" points where individual wires (or several wires) break out of (or enter) the harness and where nails or pegs will be driven into a board or jig on which the cabling will be assembled.

4. Proper identification of each wire on:
    a. The drawing itself.
    b. A table or listing of wires by color, etc.

Figure 4-12 shows a typical local cabling, both by itself and as it later appears connected to the assembly. There is additional cabling and wiring on the same assembly.

Figure 4-13 shows how such a diagram may be begun, and Fig. 4-14 shows the final full-scale drawing of the harness with breakout points and lines showing where the wires are stripped. For example, the wires in the upper left corner extend beyond the line marked 104, 5 but are stripped back to this line, and the bare wire is cut so that $\frac{3}{4}$ in. remains beyond the line. The approximate thickness of the cabling is shown and is based on the number (and thickness) of wires that are in the harness at these places. If 26 wires are in the center part, the approximate thickness there would be about $\sqrt{26} \times \frac{1}{20}$ (thickness of insulated No. 22 wire), or about $\frac{1}{4}$ in. plus, which is only slightly less than what the finished harness measures at this point.

Table 4-2 shows the tabular form of wire identification and routing. It is often placed on the same sheet as the harness drawing. This system uses a sort of "station" method whereby leads numbered in the 100s are at one area of the harness and those numbered in the 300s and 400s are at other areas, or "stations," along the harness. Other numbering systems are sometimes used. Figure 4-14 is one of a set of drawings made for the assembly of the electrical equipment. Other drawings might include a schematic diagram and connection diagrams.

Another harness drawing is shown in Fig. 4-15. This is a pictorial view of a harness and flat cable on a hinged chassis or panel with the door shown in its open position. Breakout points are prominently identified by number as BO #1, BO #2, etc. Wires are identified by color at critical places. In the case of the flat cable, the edge with the brown wire is identified at both ends. Pin connectors and connector shells are shown in their exact locations and identified.

More explicit information on connections may be provided as in Fig. 4-16. These show the connections for each wire at the right in Fig. 4-16*a* and similar treatment at the left end. (An insert not shown here provides more information about how the capacitors and the individual plug-in connectors are attached to

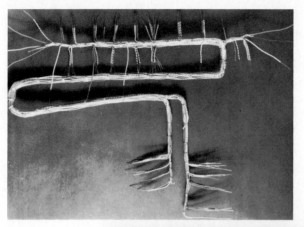

(a)

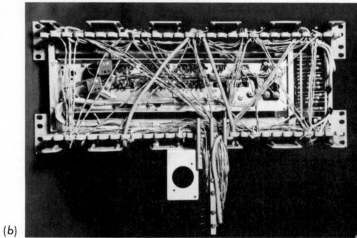

(b)

FIG. 4-12 Photograph of a local cabling harness: *(a)* harness by itself; *(b)* harness attached to chassis. (AT&T Technologies, Inc.)

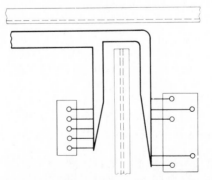

FIG. 4-13 Fitting a harness drawing into the spaces available in a chassis.

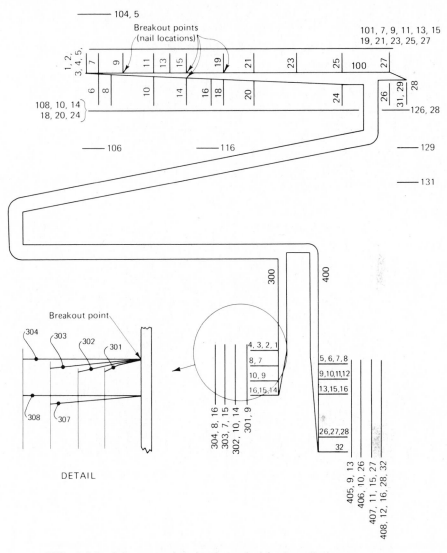

FIG. 4-14 A harness drawing intended for installation purposes.

the wires.) The connections for the flat cable in Fig. 4-16*b* are shown at the right end, and individual plug-in connectors are illustrated at the left end.

Similar drawings are sometimes accompanied by a parts list which might identify each end connector by type and manufacturer part number, type of sleeving, length and color of each wire, solder specification number, etc. The many different types of end connectors and terminals make it almost impossible to cover this subject in a book.

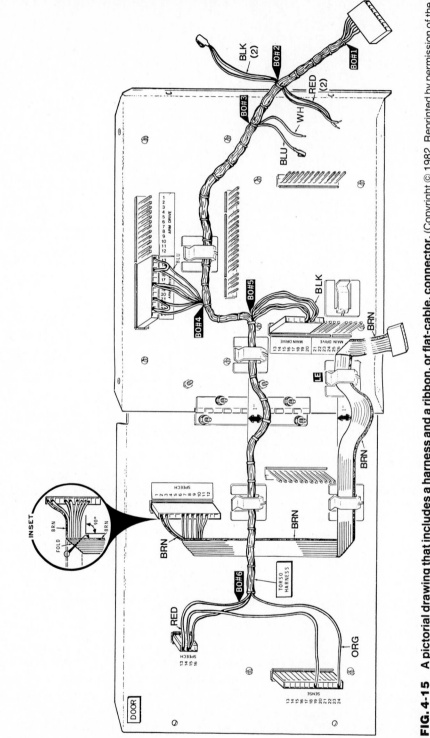

FIG. 4-15 A pictorial drawing that includes a harness and a ribbon, or flat-cable, connector. (Copyright © 1982. Reprinted by permission of the Heath Company.)

TABLE 4-2 Harness-Wire Routing

COLOR	START	BREAK OUT AT	FINISH	TERMINAL DATA
BL	101		119	CAP(*C*34)
BL	120		114	TERMS
BL	111		121	103 1
BL	118		427	104 2
R	408		120	105 INDR
R	109		412	(*L*31)
R	428		114	106 CAP(*C*31)
R	115		111	REL(AL)
R	110		316	TERMS
R	304		128	107 77*R*
R	129		121	108 BOT 3, 4
BK	107		103	109 TOP 1, 2
BK	106	108,118		110 REL (DB)
		120,123	415	

Note: This tabulation is part of the drawing shown in Fig. 4-14.

Construction and Assembly Drawings

4-11 Sheet-Metal Layouts

In order that the frames, chassis, shields, and other such parts of electrical or electronics assemblies can be correctly and economically manufactured, drawings must be made that will tell the builder exactly how the item is to be made. Figure 4-17 shows the pattern of a shield for some electronics equipment of a satellite. It is laid out flat; the lines on which it is to be folded are shown as thin lines with two short dashes in the center. In addition, there is usually some lettering or printing (not shown in this illustration) that tells exactly what kind of material is to be used and how the material (an aluminum alloy) is to be finished. Although it does not tell how to fold the metal, a little study will reveal that this will make a boxlike enclosure if the sides are folded at 90° to each other. Chassis diagrams are treated in the same way, unless they have too many dimensions to make the folding-out graphical concept practical. Figure 4-17 uses the standard two-decimal system of dimensioning, which has become quite popular in U.S. industry. All except the most critical dimensions are rounded to the nearest hundredth of an inch (then usually to the nearest even hundredth). Note the .015 dimension which the designer considered to be more critical. The aligned system (guidelines parallel to dimension lines) is employed.

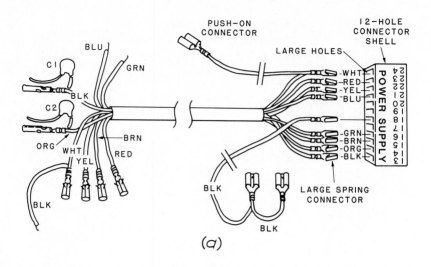

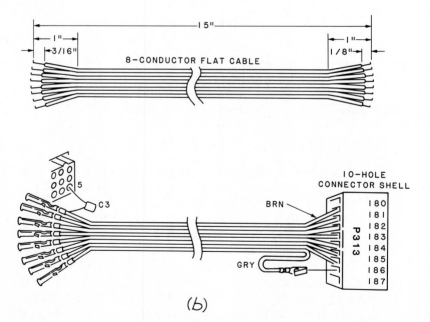

FIG. 4-16 Details of harness and cable connections: *(a)* sleeve harness; *(b)* flat cable. (Copyright © 1982. Reprinted by permission of the Heath Company.)

Figure 4-18 shows the same type of drawing for another piece of satellite equipment, except that a different dimensioning system is used. Here, horizontal location dimensions are given from the left edge because of the critical distances to the square projections, which must fit into mating recesses on another piece. This method of dimensioning to a well-defined datum plane (it

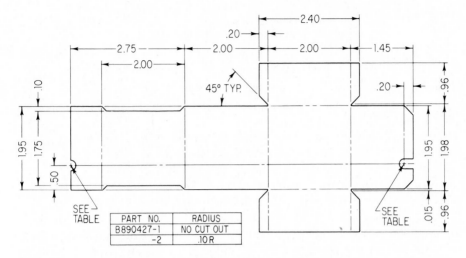

FIG. 4-17 **Pattern layout of a shield for a communications satellite.** (AT&T Technologies, Inc.)

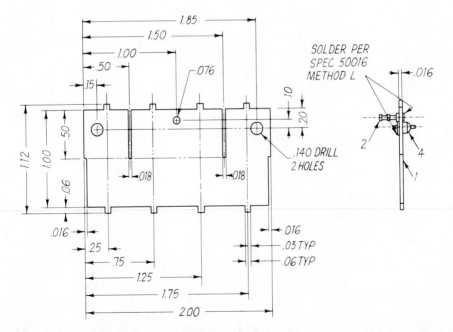

FIG. 4-18 **Manufacturing drawing of a center fin of a chassis.** (AT&T Technologies, Inc.)

should be on a finished surface) is often the best way for numerical-control (N/C) manufacturing. It does require two straight edges, however.

4-12 Chassis Manufacture

Since chassis[1] are produced in different shapes, sizes, and complexity, the number of views required for their manufacture will vary. A rather simple U-shaped chassis with a few holes or openings might be well described in two views. More complex ones with many holes or cuts require several views or drawings. The next example illustrates a large, rather complex chassis.

Figure 4-19 shows only part of a construction drawing for a chassis like that shown in Fig. 4-12, with the wiring harness. The complete drawing has four views, plus supplemental drawings showing the exact shapes of some of the holes and other cutouts. Because of size limitations, we have shown only the left quarter of two adjacent views and some of the cutout details, which are placed around the edge of the sheet. The cutout dimensions are given in separate details to avoid cluttering the main views with too much information and to provide a practical way to give the correct tolerances for mounting holes and their components in many situations (Figs. A, C, and D in Fig. 4-19).

For complete manufacture of this chassis, a separate set of instructions, entitled "Manufacturing Layout and Time Rate," is issued to the manufacturing section. Some of the 22 steps included in this publication are shown in Table 4-3. Note that even the tools are specified.

This chassis drawing is a typical engineering drawing made to full scale (1 in. = 1 in.). It can be used later on as a basis for an assembly drawing (for putting the rest of the chassis together) and for the harness, or local-cabling, drawing. The advantages of making it to full size are now fairly evident. However, chassis for some miniaturized packages cannot be drawn to full scale because they are too small. They must be drawn larger than actual size in order that details and dimensions can be appropriately shown.

More and more layout drawings will use metric dimensioning as the United States slowly converts to SI (Système International d'Unités) in which the base unit of length is the meter. Figure 4-20a shows a bracket that is dimensioned in SI and the unidirectional placement of figures, in which all guidelines are drawn horizontally and all numbers and letters are read from the same direction. The unidirectional system has long been approved by ANSI Y14-4, as has the aligned system exhibited in the previous three examples. Many companies are employing dual dimensioning in the manner shown in Fig. 4-20b, with millimeters added above the inches. Dimensioning to the nearest millimeter is accurate enough for most manufacturing facilities and equipment. Sometimes more accuracy is needed, in which case dimensions are given to the nearest 0.1 mm.

---

[1] The plural of *chassis* is spelled the same way as the singular, which is somewhat confusing. In the above sentence we mean the plural. In the following sentence we are writing about a single chassis. A chassis is a frame, on which something is built or assembled.

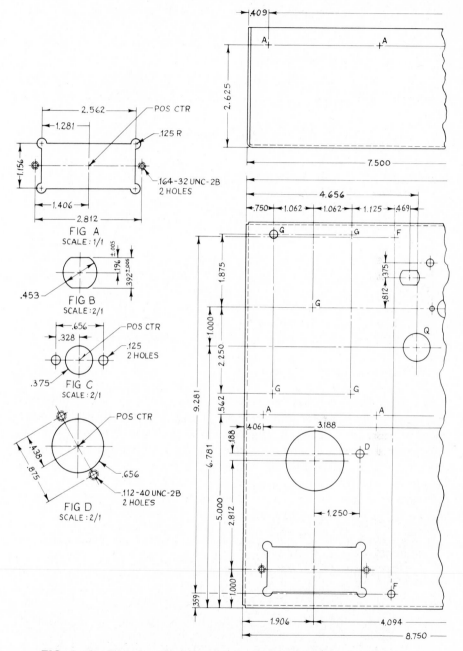

FIG. 4-19 Part of a detail drawing for the manufacture of a chassis.

TABLE 4-3 "Manufacturing Layout and Time Rate"*

RM 512261	Brass sheet, 0.063- × 24- × 84-in. Grade A
(Stock No.)	(5) parts per sheet
	(20) sheets per 100 parts

OPERATION DESCRIPTION	MACHINES, TOOLS, GAGES, TEST SETS, ETC.
Metal parts	
1. Shear strips to 15.923 ± 0.010 in. × 24 in.	Niagara shear (11005)
a. Gage part from front measuring scales on shear (backup)	Use front gaging.
	18-in. vernier caliper
b. Place parts on pallet	
2. Trim parts to 22.674 ± 0.010 in. × 15.923	Niagara shear (11005)
a. Gage part from front measuring shear (backup)	Use front gaging
	24-in. vernier caliper
3. Perforate blank complete except that (4) corner notches (cutouts) are not to be blanked at this time	No. 5 Minster punch press (62221)
(1) Setup	P&D C-729515
(2) Handlings	Bolster C-673918-9
4. Omitted	
5. Deburr part	Bench
a. Place part on pallet	Pneumatic sander
6. Omitted	
7. Tap (2) holes 0.164–32 for "B" cluster	Snow tapper (79003)
(1) Setup	(1) H.S.S. tap, 0.164–32
(2) Strokes	(2) Flute plain pt.

* Five of twenty-two steps are described.

The detailer or designer must be certain that the manufacturer has metric drills. Otherwise it will be necessary to specify standard U.S. twist drills. The nearest U.S. drill to a 5.00-mm (0.1968-in.) hole is No. 9 (0.1960 in.), and the closest drill to a 9.00-mm (0.3543-in.) hole is size T (0.358 in.). Both the metric and the U.S. twist-drill tables are listed in Appendix D.

4-13 Hole and Terminal Data

Chassis and wiring boards may be dimensioned according to standard drafting practice. There are several methods in use. Two systems that would be quite suitable for parts containing many holes are shown in Fig. 4-21. These methods can be used for manual operation of standard drill presses or for numerical control (N/C) machine tools. Figure 4-21a illustrates the type of drawing that is ideal for a programmer to use. Various sizes of holes are indicated by different letters, and the holes within a letter group are numbered sequentially. The xy reference point is located close to a corner, in this case at the center of hole A1. The programmer will be able to write the program (which will later—through tape—give instructions to the machine tool about how to do the job) in a very efficient manner. Holes that are the same size will be drilled one after another. After these holes have been drilled, the drill bit will be changed (manually on

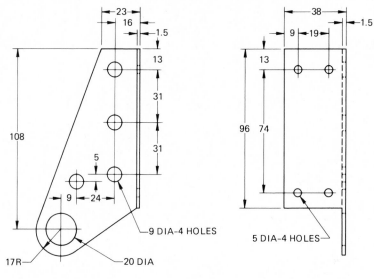

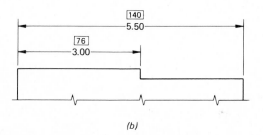

(b)

FIG. 4-20 Metric and dual dimensioning: *(a)* using millimeters as the units; *(b)* dual dimensioning with inches below and millimeters above.

some tools, automatically on others) and then the next series of holes will be made.

The table in Fig. 4-21*a* is so constructed that the *absolute* method of N/C can be easily performed. However, the other method, called *incremental,* may also be used by subtracting *x* and *y* values of one hole from the next. Figure 4-21*b* also lends itself to both absolute and incremental programming approaches. Actually, it is not necessary to draw the holes on this type of drawing. Some companies follow the practice of showing centerlines only. Holes B1 to B4 are, of course, threaded. They are No. 10 taps of the coarse-thread (NC or UNC) series having 24 threads per inch. (See table in Fig. 4-21*a*.)

Standard threads can be identified in tables such as are found in Appendix D. The unified (UNC, UNF, etc.) series includes standard U.S. threads in inches or fractions. In this table, we find the No. 10 coarse (UNC) thread, with 24 threads per inch, and the fine (UNF) thread, with 32 threads per inch. Either

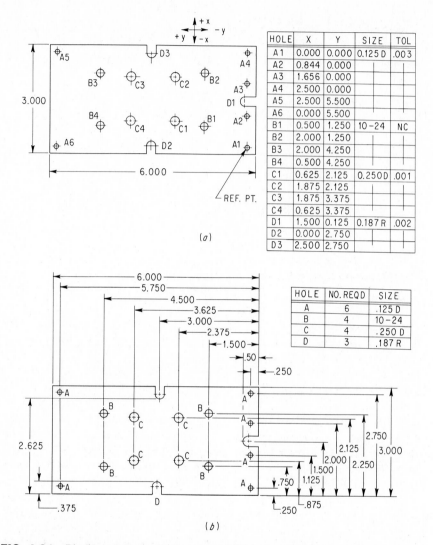

FIG. 4-21 Positional dimensioning on a drilling drawing: *(a)* coordinate system; *(b)* baseline system.

type of thread could be used; the coarse is more common. Another table, which identifies standard coarse and fine metric threads, is also included in Appendix D.

The pitch of a thread is the distance between the corresponding part of two adjacent threads. For U.S.-English (UNC and UNF) threads, the reciprocal of the pitch (e.g., 24 or 32 threads per inch) is given in the table. But for metric screw threads the pitch is given in millimeters. Thus for a 12-mm thread the pitch for a coarse thread is listed as 1.75 mm, and the fine thread as 1.25 mm.

4-14 Assembly Drawings

Figure 4-22c shows the required components drawn in place on a circuit board. Some reasons for locating the components on such an assembly drawing are:

1. To fit the parts into the space available

2. To satisfy wiring requirements, such as accessibility for connections and short wiring paths

3. To achieve a pattern without crossings if printed circuitry is to be used

4. To satisfy electrical requirements such as shielding, built-in capacitances, etc.

5. To satisfy manufacturing requirements such as assembly by N/C equipment.

Such a layout is made by (1) studying the original schematic, or elementary, diagram to determine what components have common connections, where common lines (e.g., ground, B+) are, etc.; (2) drawing one or more point-to-point wiring diagrams freehand in order to get an optimal arrangement; (3) further refinement, if justified or necessary, which might include cutting out outlines of parts on heavy paper and shifting them into different arrangements; and (4) drawing the final assembly drawing, such as is shown, to scale. Each component is given a designation, such as R_1 (resistor No. 1) and C_5 (capacitor No. 5), or the value, such as 100 for a 100-Ω resistor. The assembly in Fig. 4-22c was derived from the schematic diagram in Fig. 4-22a and then by experimentation in the location of components. Figure 4-22b represents one start in the layout of components, with the switch and volume control at the front, wires to the speaker and automobile radio at the back. Because the volume control is so bulky and the circuit board and chassis so small (less than 4 in. long), the final layout (Fig. 4-22c) includes a recessed portion for R_7. As will be explained in Chap. 5, the layout was made with a 0.1-in. (2.54-mm) undergrid. Note that all holes (for connection to printed circuitry on the other side) are on the intersections of the grid lines. A photograph of the PC-board assembly is shown in Fig. 4-23.

This drawing can be used for two or more purposes: (1) to give a parts-location layout, along with a list of parts and catalog numbers, to production and service personnel and (2) to provide the location of holes for a drilling drawing such as Fig. 4-21.

Another type of assembly drawing is that used for the assembly of a large chassis itself. Figure 4-24 is one such view of such an assembly. The drawing also lists the various parts as follows: (1) chassis (also shown partially in Fig. 4-19), (2) panel, (3) and (4) angles, (5) brackets, (6) and (7) rivets, (8) grommet, (9) Penn fastener, and (10) self-clinching fastener. (Numbers 7 and 8 are shown on another view.)

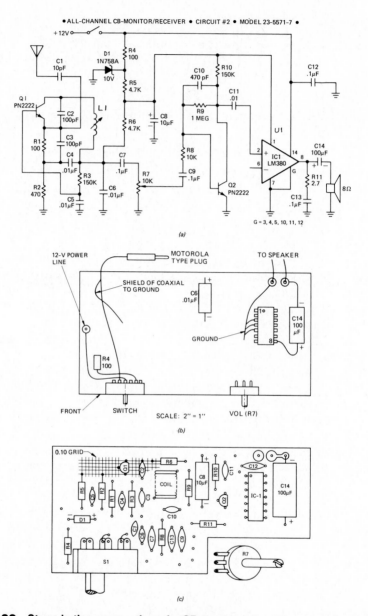

FIG. 4-22 Steps in the preparation of a CB receiver circuit board: *(a)* schematic diagram; *(b)* trial location of components (incomplete); *(c)* final component assembly, or components location, drawing. (Kantronics, Inc.)

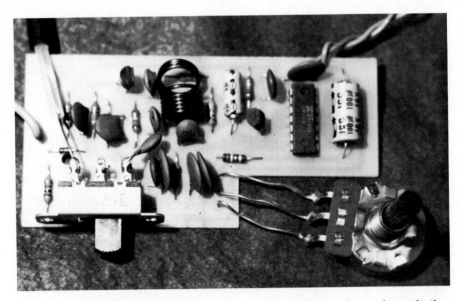

FIG. 4-23 Photograph of the CB-receiver board which has been drawn in the previous figure. (Kantronics, Inc.)

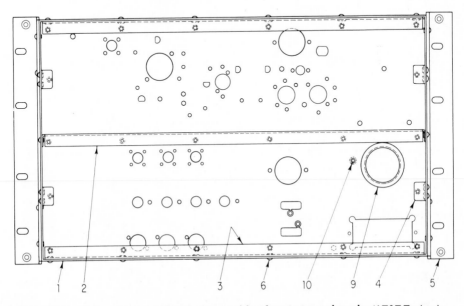

FIG. 4-24 A drawing used for the assembly of parts on a chassis. (AT&T Technologies, Inc.)

TABLE 4-4 Parts List

LIST OF COMPONENTS FOR CB RECEIVER

R_1	100	C_1	10 pF	D_1	IN758A
R_2	470	C_2	100 pF	Q_1	PN2222
R_3	150 kΩ	C_3	100 pF	Q_2	PN2222
R_4	100	C_4	0.01 μF	U_1	LM380
R_5	4.7 kΩ	C_5	0.01 μF	L_1	650 μH
R_6	4.7 kΩ	C_6	0.01 μF	ANT	ROD
R_7	10 kΩ	C_7	0.10 μF	SPKR	8 Ω
R_8	10 kΩ	C_8	10 μF		
R_9	1 MEG	C_9	0.10 μF		
R_{10}	150 kΩ	C_{10}	470 pF		
R_{11}	2.7 kΩ	C_{11}	0.01 μF		
		C_{12}	0.10 μF		
		C_{13}	0.10 μF		
		C_{14}	100 μF		

As in the case of Fig. 4-19, this drawing is accompanied by a "Manufacturing Layout and Time Rate" sheet which lists each assembly step and the tools with which the operations are to be performed. Chassis are available in various sizes and shapes. Their main purpose is to hold, or support, component parts, but they also often furnish shielding and rigidity and even act as parts of electric circuits.

A parts assembly drawing is usually accompanied by a list of components. Table 4-4 shows the list of all the components of the CB receiver illustrated in Figs. 4-22 and 4-23.

4-15 Photodrawing

An interesting development in the graphics area is photodrawing. Figure 4-25 shows a photograph of an assembled radio with the accompanying letter-number designation of parts. These photographs perform a function similar to that depicted in Fig. 4-22c. In some ways a photograph is clearer than a drawing. Certain things, like disk capacitors and coils, show up better in a photograph than in a drawing, especially if the assembly is crowded. Electrical installations that have been extensively modified can be shown better by photographs than by well-worn drawings that have had many changes made on them. Numbers for parts identification, alignment symbols, and connection points are drawn directly on the photograph. A complete list of parts usually accompanies a photodrawing.

SUMMARY

Many types of production drawings are used for the construction, wiring, and assembly of electrical and electronic equipment. Examples are connection or

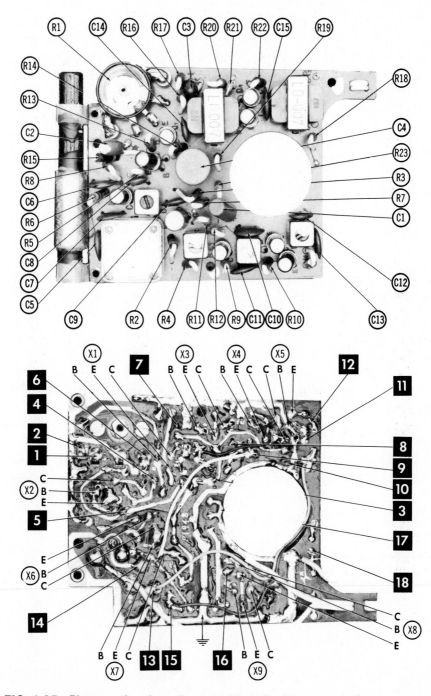

FIG. 4-25 Photographs of a radio assembly, both sides. (A Howard W. Sams Circui-trace photograph.)

wiring diagrams, panel diagrams, chassis layouts, cabling diagrams, and assembly drawings. Connection diagrams are classified as point-to-point, straight-line, highway, or baseline (airline). Careful identification of conductors must usually be made; this may include such items as color, wire size, component and terminal destination, shielding, and function.

The objects being connected are sometimes shown in pictorial form, symbol form, or elemental (rectangular or circular) form. Interconnection diagrams and cabling diagrams are variations of the connection diagram. Rather detailed drawings may be required to show how the individual wires of a cable or harness are attached to a terminal or connector.

Chassis are necessary to hold the components of an electronic system together and to provide cover, rigidity, and sometimes electrical properties to a circuit. Drawings are used for the manufacture of the assembly itself and are often used as the basis for assembly of the electrical components on the chassis and wiring that is within the chassis area. These drawings usually provide all the information to make the chassis and may be supplemented by complete parts lists and sets of manufacturing instructions. The method of manufacture often dictates how such a drawing is to be dimensioned.

QUESTIONS

4-1. What are the differences between a point-to-point and a straight-line wiring diagram?

4-2. What are the similarities between a highway diagram and a baseline diagram?

4-3. When is it desirable to make a wiring diagram that is pictorial?

4-4. If a connection diagram is to show such items as switches and antennas, to what source would you refer in order to portray those items?

4-5. What is the difference between a highway diagram and an airline, or baseline, diagram?

4-6. Are connection diagrams sometimes accompanied by other types of diagram? If so, what might the other drawing be?

4-7. Is it true that highways are generally drawn horizontally in a highway wiring diagram?

4-8. What abbreviations would you use for the colors red, white, orange, slate, gray, green, brown, and blue?

4-9. What is the sequence, or order, recommended by American National Standards Institute (ANSI) for the identification of feed lines in a highway-type connection diagram?

4-10. For what reasons would you use different line widths in a connection diagram?

4-11. Show, by examples, how resistors on each of three different circuit boards belonging to an extensive electrical system would be identified. (Let each resistor be no. 2 on its board.)

4-12. What is meant by a break-out point in a harness? How would you identify such a point on a drawing?

4-13. How are fold lines drawn on a flat layout drawing of a chassis?

4-14. What is the closest metric drill to a No. 16 U.S.-English system drill?

4-15. What is the diameter in inches of an E-size U.S. drill?

4-16. What is the diameter of a No. 20 wire? Of a No. 14 wire?

4-17. What is the number of threads per inch of length of a No. 8 thread or tap for a (a) coarse thread and (b) fine thread?

4-18. How would you decide how many views to draw for a complete chassis drawing?

4-19. Before drawing the final components location drawing, what type (or types) of diagram is necessary?

4-20. If the holes in a chassis, or cover, are to be drilled by an automatic drill (N/C), a table listing the locations and sizes of the holes is made. In what sequence, or order, are the holes arranged in such a table?

PROBLEMS

4-1. A stereo system is shown pictorially and as a point-to-point diagram in Fig. 4-26. Redraw this as a highway type of connection diagram with at least three highways. For each lead show the following identification: (1) component, (2) terminal, and (3) color of wire. Sheet size: 9 × 12 minimum.

4-2. Redraw the stereo system shown in Fig. 4-26 as a baseline (airline) diagram. Use the same identification as specified in Prob. 4-1. Sheet size: $8\frac{1}{2}$ × 11 or larger.

4-3. Redraw the stereo system shown in Fig. 4-26 as a straight-line diagram.

4-4. Redraw the installation shown in Fig. 4-27 as a highway-type connection drawing, with at least three highways. For each lead use standard identification as follows: (1) component destination, (2) terminal, and (3) color of wire. For example, the lead from terminal 7 of component 3 would read D1/1/Y and would be placed near component 3. Drawing sheet size: 11 × 17 or 12 × 18.

4-5. Redraw the installation shown in Fig. 4-27 as a straight-line connection diagram. Use standard abbreviations for wire colors. Use 11 × 17 or 12 × 18 paper.

4-6. Figure 4-28 is a baseline diagram of a large-scale amplifier. Redraw this

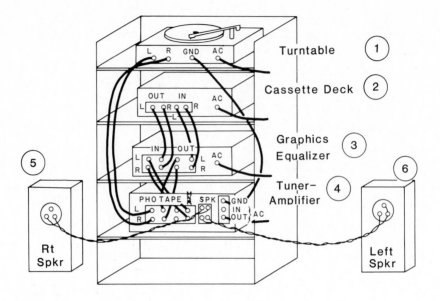

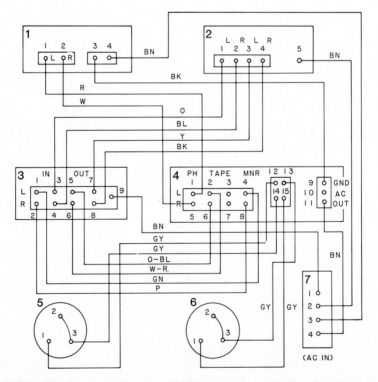

FIG. 4-26 (Probs. 4-1, 4-2, and 4-3.) Pictorial view of the rear of a stereo sound system and a point-to-point wiring diagram.

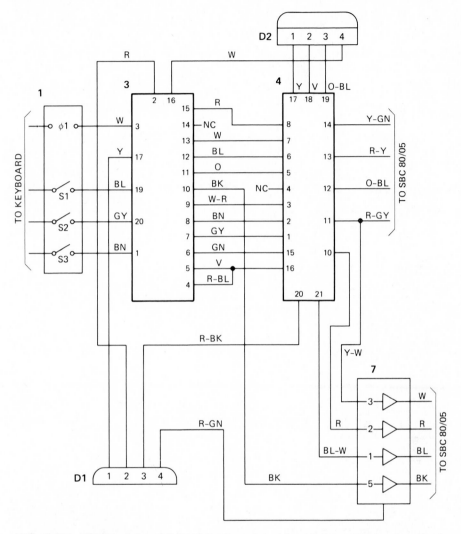

FIG. 4-27 (Probs. 4-4 and 4-5.) Point-to-point wiring diagram of a keyboard-microprocessor interface.

as a point-to-point diagram using the wire colors shown. Use 9×12 (minimum) paper.

4-7. Redraw the apparatus shown in Fig. 4-28 as a highway diagram. Use 11×17 or 12×18 paper.

4-8. Draw the wiring diagram of the Honda CB 750 as it appears in Fig. 4-29. This may be improved by using standard symbols for contacts, diodes, etc. and abbreviations for wire colors, such as GN for green and BL for blue. Switching schedules may be included or omitted, as your instructor indicates. Use 11×17 or 12×18 paper.

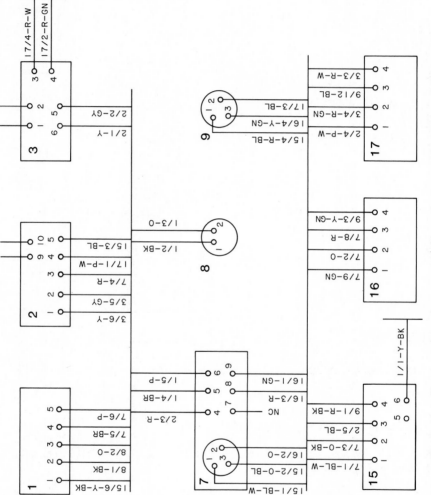

FIG. 4-28 (Probs. 4-6 and 4-7.) An airline or baseline diagram of an amplifier.

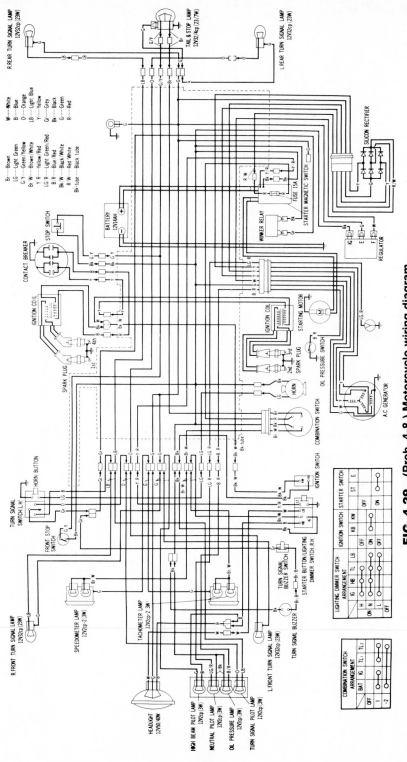

FIG. 4-29 (Prob. 4-8.) Motorcycle wiring diagram.

151

4-9.   The rudimentary diagram of a local-cabling harness appears in Fig. 4-30. Complete the diagram using the following information:

START	FINISH	AREA	START	FINISH	AREA
101, 3, 5	f	400 (call 401, 3, 5)	301, 5, 9	n	200
102, 4, 6	e	200	311, 12	i	400
105, 7, 9	d	200	313	m	200
110, 111	c	200	321, 22	h	400
115, 116	b	200	323	l	200
119, 120	a	200	330	g	400

Since starting line 101 terminates at f in the 400 area, call the termination 401; call the termination of 103, 403; etc. Then add a wire-routing diagram to your drawing. At different places the harness should have different thicknesses, which you should estimate. (Use a 9 × 12 or 11 × 17 sheet.)

4-10.   Figure 4-31 shows four views of a subchassis for an amateur-radio

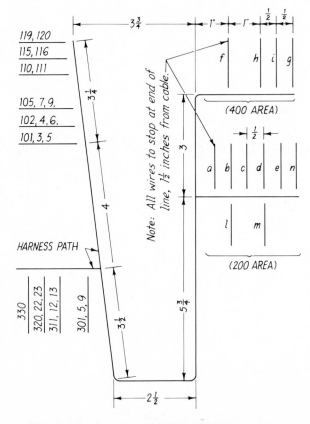

FIG. 4-30 (Prob. 4-9.) Cabling harness problem.

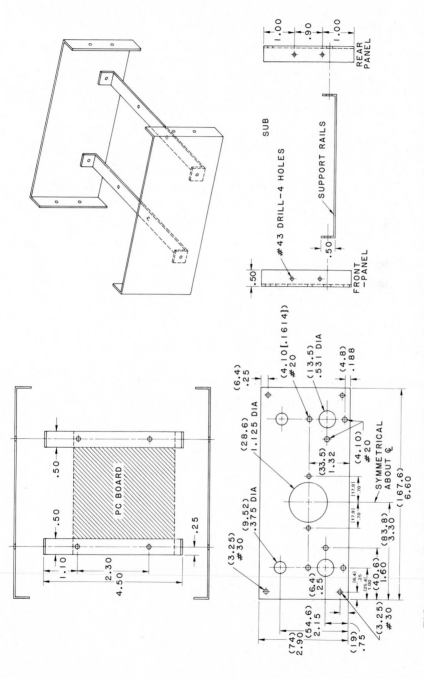

FIG. 4-31 (Prob. 4-10.) Orthographic and pictorial views of a subchassis for an amateur-radio receiver.

receiver. Not shown are a cover and an anodized outside front panel. The front view has dual English-metric dimensioning, whereas the top and side assembly views are dimensioned in inches. Draw to full scale three or four views of this chassis. Use metric, dual, or English system of dimensioning, as your instructor directs. Without dimensioning this will fit on 11 × 17 paper; with dimensioning, on a C-size (17 × 22) sheet, unless scaled down.

4-11. Figure 4-32 shows a pictorial view and partial views of a bracket, plus a schedule for hole sizes. This bracket is part of the assembly shown in Fig. 4-24. Make a complete drawing for the construction of this part. The bracket is to be made of steel sheet, cold-rolled and commercial quality (CRCQ). The holes and ovals are to be punched out, and "C" holes countersunk for a 0.190 – 32 flathead machine screw. The part is to be degreased per Spec. 51606 and zinc-plated for 289A finish. Tolerance is ±0.016 in. Dimensions may be converted to the metric system. Use 11 × 17 or 12 × 18 paper.

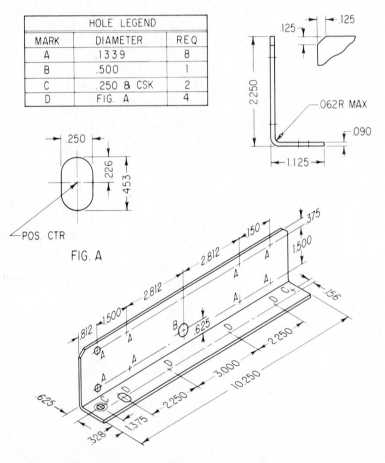

MARK	DIAMETER	REQ
A	.1339	8
B	.500	1
C	.250 & CSK	2
D	FIG. A	4

HOLE LEGEND

FIG. 4-32 (Prob. 4-11.) Bracket for a telecom relay chassis.

4-12. Make a three-view drawing of the chassis shown in Fig. 4-33. Flanges A and C are identical. The chassis is to be made of 20-gage (0.95-mm) CR steel and degreased. Finish is to be gray enamel, baked 525A, 0.025 mm thick. Dimension it according to one of the methods prescribed in this chapter. One of the views may be a flat foldout, or development. This would permit dimensioning for N/C drilling of the chassis. Dimensioning may be done in the dual or metric system. Dimensions shown are in millimeters. Use 11 × 17 paper.

4-13. A chassis and cover are shown pictorially in Fig. 4-34, and the front face is dimensioned below in millimeters. Make a flat layout (development) of the chassis including fold lines (see Fig. 4-17). Put all dimensions on the figure. Dimension the holes with a table such as that shown in Fig. 4-21, lower part. Make a set of notes to include the following: material 20-gage 1020 steel; all dimensions in millimeters, tolerance 0.50; deburr all edges; degrease per Spec. 51606. Use 8 × 11 or 9 × 12 sheet if drawn full size. (Dimensions may be converted to inches and the chassis drawn in the English system if requested.) Other views of the chassis may be drawn if your instructor so directs. Diameters of holes are: A, 7 mm; B, 3 mm.

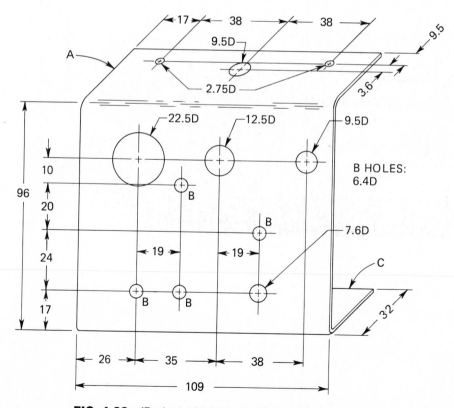

FIG. 4-33 (Prob. 4-12.) Pictorial view of a preamplifier.

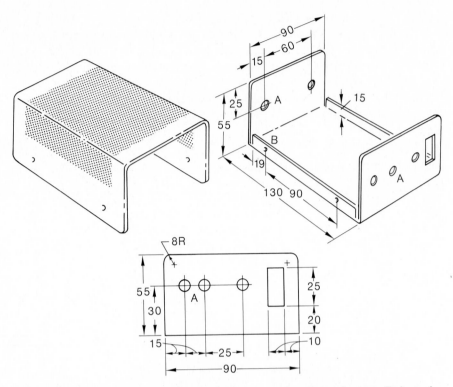

FIG. 4-34 (Prob. 4-13.) Above: pictorial view of chassis and cover. Below: detailed drawing of front face of chassis. Dimensions are in millimeters.

4-14. Figure 4-35 shows a pictorial view and several other partial views of a chassis. The hole sizes are as given in the table below. Make a complete set of drawings for the construction of this chassis, which has one face labeled A, two panels on each side marked C (each having two holes), and one panel in front labeled B. There is no panel in the rear. Panel B is swung out in the pictorial view to show how C is folded. Panel B is actually vertical. Dimensions are in inches.

HOLE	LETTER OR NUMBER DRILL	DIAMETER (IN IN.)
D	N	0.302
E	See detail	See detail
F	#38	0.1015
G	K	0.281
H	#27	0.144
J	#20	0.161

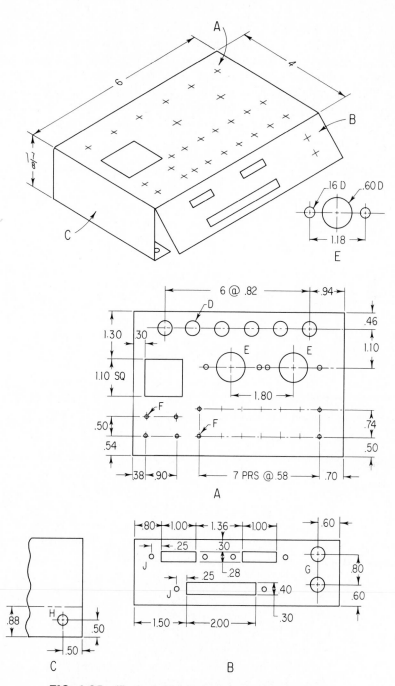

FIG. 4-35 (Prob. 4-14.) A chassis drawing problem.

The chassis may be drawn in two ways. Several orthographic views, with dimensions, of the chassis in its completed folded form would be satisfactory. Showing the single piece of metal laid out flat, with fold lines and a small pictorial view of the folded chassis, would also be satisfactory.

All burrs should be removed. The 20-gage steel should be degreased per Spec. 5160 and coated with clear varnish per Spec. 5160. Tolerance is ±0.02 in. Dimensions may be converted to the metric system. Sheet size: 11 × 17, 12 × 18, or larger.

5

Printed-Circuit Boards

Printed-circuit (PC)-board development and improvement have been slow but continuous since the switch from the wired connection of all components to the printing of circuits, which began in the 1950s. This development is still continuing, possibly at a more rapid pace with the introduction of surface-mounted parts.

However, the drawing or layout of printed circuits has not changed much in recent years. More of this drawing is being done by computer-aided equipment, of course. But every day, somewhere, hundreds of persons are making drawings on drafting tables of printed circuits for new products.

In general, two systems are employed to place a printed or plated circuit on a circuit board. One method is to use a copper-clad laminate (board) and etch away the unwanted copper foil, leaving the desired circuit pattern. In this system a plating resist is applied in the image of the desired pattern. The board is subjected to the etching process, wherein that portion of the copper that is not covered by the resist is removed. One of several methods is employed to get the resist pattern onto the board.

The other process is the *additive* one, in which the conductor pattern is deposited on the board. Usually this is copper. Tin lead is then deposited on the copper by the electroplating process. After large mounting holes, if any, are made and gold connection tips, if any, are added, the circuit board undergoes a *reflow* process. During this process the copper and tin paths become solder, and the component connections are soldered to their respective conductor paths at their pads or plated-through holes.

Often a solder-resistant plastic is printed over the side of the board that has the solder and covers those areas that are to have no solder. This *solder mask* prevents solder from bridging closely spaced conductors and splashing on other unwanted places.

Regardless of how the PC board is to be manufactured, the layout of the final artwork is pretty much standard. Considerable thought and drafting go into the layout. A study of the calculator board (Fig. 5-1) should indicate to the reader that careful layout of components and conductor paths is required.

The most recent version of what is essentially the same calculator (Fig. 5-2) shows a less complex board but the same orderly drawing and printing of

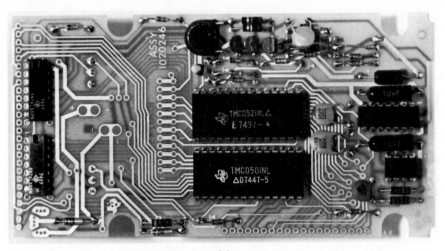

(a)

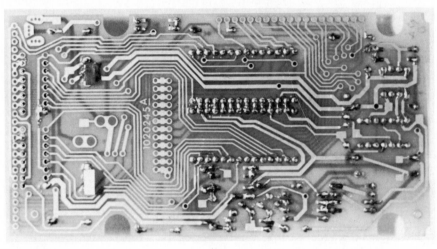

(b)

FIG. 5-1 Printed-circuit board for a scientific calculator: *(a)* component side; *(b)* foil side. (Texas Instruments, Inc. and Precision Arts, Ltd.)

conductor paths. This flexible laminated "board" is transparent: You are look-ing at the component side and seeing the circuit paths which are printed on the same side and covered with a protective coating. The board shown in Fig. 5-1 has printed wiring on both sides. The latest model of calculator costs a little more than *half* of what the earlier (by 7 to 8 years) model cost.

5-1 Making Drawings for a Printed Circuit

The first step in making a set of drawings for a printed circuit is to study the elementary diagram. Such a study will provide such information as:

1. What groups of components have common connections
2. What the peak potential differences will be
3. Grounds, voltage supply lines, etc., that are required
4. Ground paths, large and heavy, and heat sinks that will be necessary
5. Other pertinent information

The next step is to make a *preliminary sketch* from the schematic diagram, such as that shown in Fig. 4-22a. The components involved in this circuit are mostly capacitors* (C_1, C_2, etc.), resistors (R_1, R_2, etc.), two transistors

* In the figure, letters and subscripts appear in the same height (for example, C1).

FIG. 5-2 Printed-circuit board for the most recent model of the scientific calculator. The two ICs have replaced six ICs and about 18 discrete components that were in the older model. (Texas Instruments, Inc.)

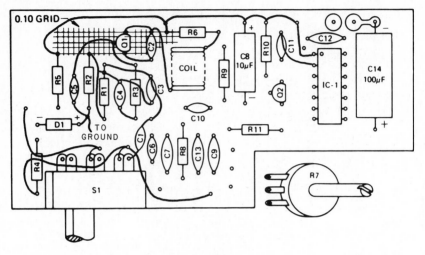

FIG. 5-3 A preliminary sketch for the layout (artwork) of a PC board.

(PN2222) which we shall call Q_1 and Q_2, and an integrated amplifier circuit IC 1. In the sketch (or sketches, because there may be several trials before the most likely pattern results), all components are located and circuit paths are sketched in. This is in reality a rough wiring diagram. Such a diagram was made before the component layout in Fig. 4-22c could be made. A preliminary sketch, in which the final component layout was used and freehand sketching of circuit paths was begun in the left portion, is shown in Fig. 5-3. (Since we are looking at the component side and because the printing of circuit paths is usually done on the other side, the final drawing should be made for the other or "foil" side.) Now this sketch can be checked to determine whether:

1. Crossovers are zero or a minimum.

2. Bypass and grid lines are as short as possible.

3. Longer lines, such as ground and high-current, are placed near or around the edge of the board.

4. The design is neat and compact.

5. Other specifications (see Sec. 5-4) are met.

Figures 4-22 and 5-3 show the final component layout as seen looking at the component side of the CB-receiver board. The components, as mentioned before, have been drawn so that their leads will pierce the board at the intersection of the 0.10-in. (2.54-mm) grid lines. Figure 5-5 shows two grids that are frequently used. Not shown is a 0.05-in. (1.27-mm) grid system. The 0.05 (or 50-mil) grid is becoming quite popular because of its compatibility with surface-mounted devices to be explained later. The 50-mil grid is similar to the 100-mil (0.1-in.) grid, except that the holes are "centered" 0.05 in. apart. Through connections for grid systems are shown in Fig. 5-5c and d.

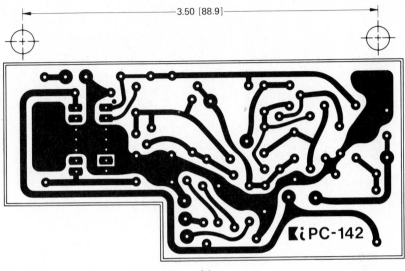

(a)

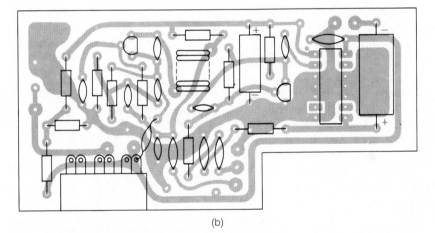

(b)

FIG. 5-4 *(a)* **Final layout (final artwork) for a CB receiver, foil side.** *(b)* **Component layout of CB PC board.** (Kantronics, Inc. and Bishop Graphics, Inc.)

In order to position components properly it is necessary to know their size, how far apart their leads must be placed, and how close together the components may be positioned on a board. Table 5-1 and Fig. 5-6 supply this information for many of the more frequently used components.

The final artwork for the CB circuit board is shown in Fig. 5-4*a*. It was done by a company that specializes in PCB layout with adhesive aids. The ground path is shown extending from the upper right corner, going down in the right center part, and then going up to the integrated circuit and capacitor, upper left. It is extra wide and heavy at each end and thus acts as a heat sink in those areas. (A heat sink removes heat by radiating it away from heat-producing elements.)

TABLE 5-1 Component Spacing on PC Boards

	DIM B	DIM C	DIM A
Resistors			
130 Family			
$\frac{1}{8}$ W	0.140	0.300 or 0.400	0.100 (2.54)
$\frac{1}{4}$ W	0.250	0.400 (10.2)	0.100
$\frac{1}{2}$ W	0.375	0.550 (14.0)	0.150 (3.81)
1 W	0.562	0.750 (19.1)	0.225 (5.72)
2 W	0.687	0.850 (21.6)	0.325 (8.25)
136 Family			
13-xxx-2	0.281	0.425 (10.8)	0.100
13-xxx-22	0.343	0.500 (12.7)	0.100
13-xxx-12	0.560	0.725 (18.4)	0.150
Diodes			
D-41	0.200	0.300 or 0.400	0.100
D-26	0.450	0.500	0.300 (7.62)
D-14	0.300	0.400	0.100
D-13	0.350	0.500	0.300
Tubular capacitors			
0.01 μf			
Small	0.625	0.800 (20.3)	0.300
Large	1.000	1.300 (33.0)	0.450 (11.4)
0.10 μf			
Small	0.875	1.100 (28.0)	0.500
Large	1.625	1.950 (49.5)	0.650 (16.5)
Ceramic capacitors			
Small		0.250 (6.35)	0.200 (5.08)
Medium		0.350 (8.89) or 0.400 (10.2)	0.200
Large		0.400 or 0.500 (12.7)	0.200 or 0.300 (7.62)

Dual-In-Line Integrated-Circuit Packages Ceramic and Plastic

NUMBER OF PINS	DISTANCE BETWEEN ADJACENT PINS	DISTANCE BETWEEN ROWS OF PINS
8	0.100 (2.54)	0.300 (7.62)
14	0.100	0.300
16	0.100	0.300
18	0.100	0.300
20	0.100	0.300
22	0.100	0.400 (10.2)
24	0.100	0.600 (15.2)
28	0.100	0.600
40	0.100	0.600

Note: Dimensions at left are in inches. Those in parens are in millimeters.
Resistor data from King Radio Corp.
DIP data from Texas Instruments, Inc.

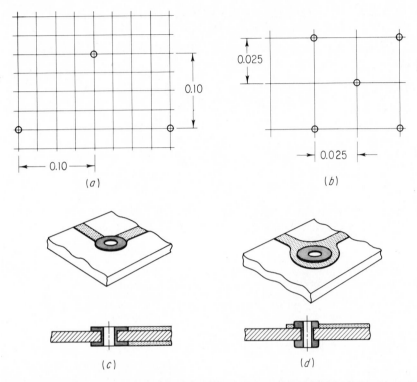

FIG. 5-5 Grid systems and through connections: *(a)* 0.10-in. (2.54-mm) grid; *(b)* 0.025-in. (0.635-mm) grid; *(c)* plated hole connection; *(d)* plated eyelet.

This PC board is a single-sided circuit board with components on one side and printed wiring on another. A photograph of the wiring side of the manufactured CB receiver PC board is shown in Fig. 5-9. As in the case of many similar products, the circuit paths were made by the etching process, in which a thin layer of conducting material (called a foil) is bonded to one side of the board and then etched away so that only the conductor paths remain.

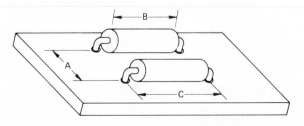

FIG. 5-6 Axial lead spacing for resistors and tubular-shaped components. (See Table 5-1.) (King Radio Corp.)

TABLE 5-2   Weight and Thickness of Copper Foils for PC Boards

FOIL WEIGHT		FOIL THICKNESS	
oz/ft²	mg/cm²	in.	mm
0.5	16.74	0.00068	0.0173
1.0*	33.5	0.00135	0.0343
2.0†	66.95	0.0027	0.0686
3.0	100.4	0.0042	0.107
5.0	167.4	0.0068	0.173

* Often referred to as 1-oz foil.
† Often referred to as 2-oz foil.

TABLE 5-3   Width of Conductor Paths in Relation to Current for a 40°C Temperature Rise above Ambient

CURRENT, A	1-oz 0.00135 (0.0343) COPPER	2-oz 0.0027 (0.0686) COPPER
1.5	0.015 (0.381) wide	0.008 (0.203) wide
2.5	0.031 (0.787)	0.015 (0.381)
3.5	0.062 (1.58)	0.031 (0.787)
4.5	0.125 (3.18)	0.062 (1.58)

Note: Minimum spacing between conductors is 0.031 in. (0.787 mm) for voltages up to 150 V.

The final artwork or master layout of the CB-receiver circuit board of Fig. 5-4 was made by placing Fig. 5-4*b* upside down and then putting a standard acetate medium over it and applying commercially available tape and PC patterns. This single-sided PC board has served as a typical example. Later, a double-sided board layout will be presented.

The great bulk of PC boards are made from a paper-base phenolic or a glassy epoxy laminate, called G-10, which is superior to the phenolic in that it has a higher resistance to warping. They are of several standard thicknesses ranging from $\frac{1}{32}$ (0.79) to $\frac{1}{4}$ in. (6.35 mm). Copper foil comes in several thicknesses or weights as shown in Table 5-2. The cross section of the conductor should be large enough to carry the required current with a certain temperature rise. Table 5-3 is a partial table that provides some of this information for a 40°C rise above the ambient temperature (that which exists in the immediate environment).

5-2 Use of Tape and Other Contact-Adhesive Aids

Making a PC-board layout with tape is not quite as easy as it looks to the beginner. The preprinted appliqués (contact-adhesive patterns) are usually easy enough to separate from their protective backing. However, getting them on the exact desired location and oriented correctly is difficult for the person who has no experience or correct tools. (The latter should include a very sharp X-acto knife or equivalent and a thin, flat spatula-type tool for manipulating the pattern onto the drawing medium.) Getting a straight line with tape also requires experience. One way is to hold the tape down at its starting point (usually on a pad which has already been emplaced), pull it slightly at the other end, and then lay it down and cut it to the desired length. One must try to allow for the stretching of the tape. Unfortunately, the tape is not supposed to be stretched, and the manufacturer's instructions so indicate. The key to this approach is the word *slightly,* for as the authors and many others have discovered, it is extremely difficult to get a straight line if the tape is not stretched.

Pads and registration marks are usually put down first, then conductor paths, grounds, and heat sinks (if any) are added. Mylar for this artwork is available with grid lines spaced 0.050 (1.27), 0.100 (2.54), or 0.125 in. (3.18 mm) apart. Such a grid may be placed under a clear sheet of Mylar, which is to be the final artwork (master layout), or a grid with blue or brown "dropout" lines[1] may itself be used for the master layout. When the tape is laid down, it should overlap onto the adjacent part of the lands, elbows, etc. (see Fig. 5-7).

It is very convenient to have 90° elbows available commercially, but not all corners are 90°. Other bends can be made with tape if one crimps it on the inside radius; or they can be done with ink, in which case one makes a line the same width as the tape. It is also possible to obtain 45° elbows commercially.

---

[1] These lines will not appear on the photographic reproduction.

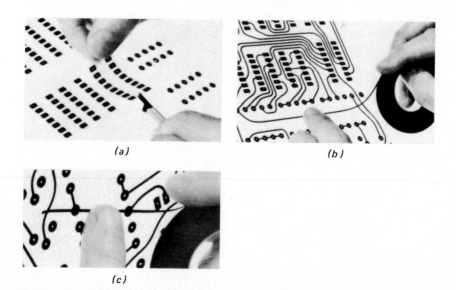

(a)

(b)

(c)

FIG. 5-7 *(a)* Putting a "stick-on" pattern on a PC-board layout; *(b)* and *(c)* applying tape for long circuit path and short path distances. (Bishop Graphics, Inc.)

The contact-adhesive aids shown in Fig. 5-8 deserve some comment. First, the dimensions listed are full, or actual, size. (The original drawing was reduced to about half size to fit in this book). Second, the aids come in a number of different sizes (widths, in the case of lines). Conductor tapes are available 0.015 (0.38), 0.020 (0.51), 0.026 (0.66), 0.031 (0.787), 0.040 (1.02), 0.046 (1.17), 0.050 (1.27), 0.062 (1.57), 0.090 (2.29), 0.100 (2.54), and up to 2 in. (50.8 mm) wide. Some patterns are such that they can be cut to desired lengths (Fig. 5-8*d*), and some are in fixed sizes (Fig. 5-8*e* and *f*). Figure 5-8*i* represents a set of lands which have the same center-to-center spacing as Fig. 5-8*e*. But the pads are small enough that a small conductor can be placed between them as shown — a situation that is sometimes unavoidable. Otherwise the larger pads are preferred because they provide better soldering and are more stable if a lead or component has to be removed and replaced. The IC connector pattern (Fig. 5-8*e*) is available in different sizes and shapes. An optional arrangement is to buy flatpack patterns which have long "fingers" resembling Fig. 5-8*d*. The outline of the IC package can be drawn over the pattern, which can be trimmed to the exact size and shape with a small knife.

Target patterns (Fig. 5-8*g*) are designed so that when superimposed any slight mismatching can be seen and corrected. Land or donut patterns (Fig. 5-8*h*) are available in many sizes and combinations of inner and outer diameters. Careful attention must be given to inner diameters because different wire sizes are often present in any single printed circuit. Thus, wires to speakers, potentiometers, and outside power, for example, are usually larger than the leads to typical components on the board (Fig. 5-9).

Our master layout was drawn to a 2:1 scale (twice size). Scales of 2:1 and 4:1 are common, but 10:1 and even 100:1 have been used where extreme accuracy has been required. Tapes and aids can be obtained for most of these scales. For instance, a 0.062-in. (1.57-mm) tape on a 2:1-scale layout will produce a line that is actually 0.031 in. (0.787 mm), which is large enough to accommodate most currents (Table 5-3).

The checklist that follows later provides other specifications for good PC-board layout. Figure 5-4*a* includes two reduction targets with the required final dimension provided so that the finished product will be the desired size. More accurate layouts can be made with the coordinatograph (Fig. 5-10, which has a plotting accuracy of 0.001 in. (0.0254 mm).

FIG. 5-8 Commercially available contact adhesive aids for PC-board drawings: *(a)* conductor tapes, 0.046 in. (1.17 mm), 0.050 in. (1.27 mm), and 0.10 in (2.54 mm) wide; *(b)* 90° elbows or bends 0.125 in. (3.18 mm) and 0.10 in. (2.54 mm) wide or thick; *(c)* Tee 0.125 in. (3.18 mm); *(d)* connector pattern, 0.20 in. (5.08 mm) center to center; *(e)* 16-lead DIP pattern 0.20 in. (5.08 mm) center to center; *(f)* 12-lead TO-5 pattern 0.74 in. (19 mm) outer diameter, 0.075 in. (1.91 mm) inner diameter; *(g)* target or registration marks; *(h)* pads or lands 0.281 in. (7.14 mm) outer diameter, 0.062 in. (1.58 mm); *(i)* seven-lead pad pattern, 0.20 in. (5.08 mm) center to center, with conductor-path tape between two lands; *(j)* older appliqué examples.

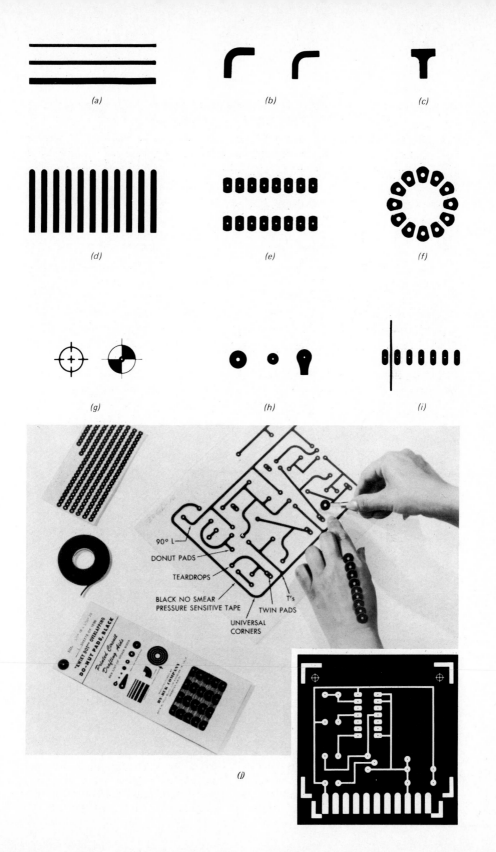

(a)

(b)

(c)

(d)

(e)

(f)

(g)

(h)

(i)

90° L

DONUT PADS

TEARDROPS

BLACK NO SMEAR
PRESSURE SENSITIVE TAPE

T's

TWIN PADS

UNIVERSAL
CORNERS

(j)

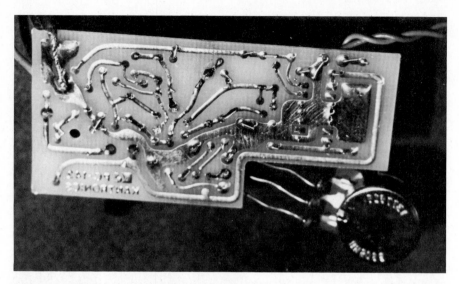

FIG. 5-9 Foil side of the CB receiver PC board. The shield of the coaxial cable can be seen soldered to the ground at upper left. Note the larger wires to the potentio-meter at bottom right. These require larger pads and holes than most of the other component leads. (Photograph: Precision Arts, Ltd.)

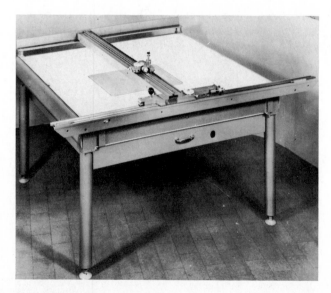

FIG. 5-10 The coordinatograph is often used for precision drafting of printed circuits. (Keuffel & Esser Co.)

5-3 Other Aspects of PC-board Layout and Construction

Orientation of components is important from the production standpoint. Figure 5-11 shows in diagrammatic form the PC board moving along the waves of the wave-soldering equipment. The solder just touches the locations on the foil (solder) side of the board where leads from the components go through the board by way of the holes which have been drilled through their pads or lands. This makes small, neat, strong solder connections. In order to minimize unavoidable "bridging" of solder, dual-in-line packages, flatpacks, etc., are positioned with their long dimensions parallel to the direction in which the board moves across the solder waves. This is often the long dimension of the PC board, but that depends on the production equipment and methods in use. (In the additive system of manufacture, the solder bath is not used. Interconnections are formed during the reflow process. If this system is used, the orientation of components to reduce bridging is not important.)

Part of a N/C machine which automatically puts components on a board and clinches them into position is shown in Fig. 5-12c. This method requires that the pattern be limited to the X and Y axes as shown in Fig. 5-12a. The components are first placed on tapes in the order in which they are to be inserted on the board. The tape (Fig. 5-12b) moves through the machine much as an ammunition belt is fed through a machine gun. Components are then placed on the board one after the other in a line. The idea is to minimize the number of passes that the board and components have to make through the machine.

A semiautomatic insertion machine is shown in Fig. 5-13. This machine is very suitable for small companies where circuit boards are not complex, or where the number of boards produced is not high. Larger, more expensive, automatic insertion machines are used for PC boards where the density of components is high and production quantities large.

Transistors and capacitors are available with preformed "standoff" leads that facilitate their assembly on the board. Dual-in-line package elements can be mounted directly on the board with their leads projecting through, or they can be separated with a thin spacer to eliminate possible short-circuiting. They

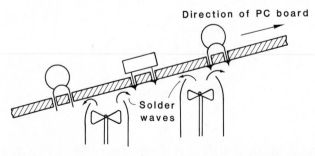

Direction of PC board

Solder
waves

FIG. 5-11 Dual-wave soldering process. The first wave provides good wetting of the component leads to the signal paths. The second, more gentle, wave removes excess solder.

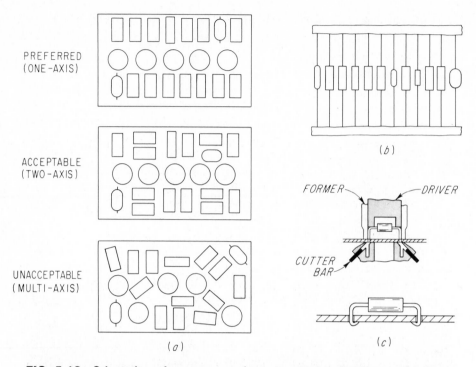

PREFERRED
(ONE-AXIS)

ACCEPTABLE
(TWO-AXIS)

UNACCEPTABLE
(MULTI-AXIS)

(a)

(b)

FORMER — DRIVER

CUTTER
BAR

(c)

FIG. 5-12 Orientation of components for automated assembly of PC boards: *(a)* preferred and unacceptable designs for automatic assembly; *(b)* components assembled on tape; *(c)* section of insertion machine action.

can also be plugged into *sockets* which are fastened to the board as other components are. This facilitates removal and replacement of ICs, microprocessors, etc. Completed boards are often sprayed with a material that inhibits the deleterious effects of humidity, fungus, etc. This is called *conformal coating.*

Over a long period of years many devices have been designed and built for use with PC boards. In addition to the sockets just mentioned and shown in Fig. 5-14a and microprocessor, IC packages, and relays, there are DIP switches (Fig. 5-14b) and various displays, some of which are shown in Fig. 5-14c. The vacuum fluorescent display has gained popularity in automobiles because the characters show up well in bright light or darkness, and also because all letters of the alphabet — as well as numerals — can be formed. [Not all letters can be formed with a seven-segment light-emitting diode (LED) or liquid crystal display (LCD).]

5-4 Some Specifications for Good PC-board Design

The following are some minimum design standards prepared by a company that produces thousands of printed-circuit boards a year:

1. All components will be oriented on the X or Y axis wherever possible. Dual-in-line IC packages will be oriented on one axis and keyed in the same direction wherever possible.

2. Components will be mounted on only one side of the PC board.

3. Maximum board size shall be 5.80 (147) $\times$ 11.70 in. (297 mm), if possible, to allow wave soldering and programmed assembly. (The sizes may vary from

FIG. 5-13 Insertion machine for PC-board components. With this machine the operator can quickly position devices onto a board. (Kansas City Star.)

(a)

(b)

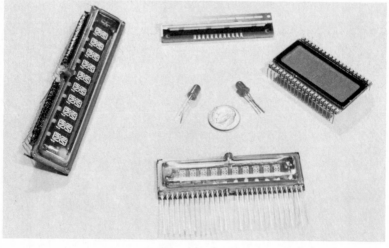

(c)

FIG. 5-14 These items can be mounted on standard 0.10-in.-grid PC boards: *(a)* sockets for ICs, with a chip carrier socket in the center and a programmable memory (PROM) package mounted on a socket at the top; *(b)* DIP switches which have on-off switches 0.10 in. apart; *(c)* displays, with vacuum fluorescent displays at left and bottom, LED bar display at top, LCD display at right, and single LEDs near the dime. (Courtesy Kantronics, Inc.)

company to company, depending on the equipment used. This specification will not apply if the electroplating process is used instead of the solder bath.)

4. Components should be located so any component can be removed from the board without removing any other part.

5. When large areas of ground plane are required, crosshatch or solder mask to avoid excessive buildup and warping during soldering.

6. All boards must have at least two holes referenced to the artwork. These holes, which may be mounting or extractor holes, must be 0.062-in. (1.57-mm) diameter or larger.

7. Component pad centers will conform to standard spacing and should be maintained on even 0.025-in. (0.635-mm) graduations.

8. Minimum spacing shall be maintained, as shown in Table 5-1 and Fig. 5-6.

9. Components that dissipate more than 2 W should not be mounted on a PC board, but should be heat sunk directly to the chassis.

10. Determine pad size as follows:
   a. Maximum misregistration allowed: 0.002 in. (0.051 mm).
   b. Minimum working tolerance: 0.008 in. (0.203 mm).
   c. Plated-through holes require a minimum of 0.006 in. (0.153 mm) for through-plating.
   d. Allow 0.002 in. (0.051 mm) for etching tolerance per ounce of copper.
   e. The above plus the minimum required annular ring multiplied by two for each side of hole.

As an example, we might have a finished hole size of 0.031 in. (0.787 mm), misregistration 2 × 0.002 for 0.004 in. (0.101 mm), working tolerance 0.008 in. (0.203 mm), through-plating 0.006 in. (0.153 mm), etching tolerance 0.002 in. (0.051 mm), minimum land 2 × 0.005 for 0.010 in. (0.254 mm), for a total of 0.061 in. (1.57 mm), which would be the minimum pad size.

11. Component bodies will be at least 0.05 in. (1.27 mm) from the edge of the board.[1]

12. No conductor path shall be closer than 0.025 in. (0.635 mm) to the edge of the board.[1]

13. No component pad perimeter will be closer than 0.05 in. (1.27 mm) to the edge of the board.[1]

14. Conductor paths should be oriented to run parallel to the longest axis on the solder side of the board. (This reduces bridging during wave soldering.)

15. Conductor paths should be oriented to the *XY* coordinate system. Necessary deviations should be at 45°.

16. All final artwork will be done at a 4:1 scale minimum. Any board requiring very critical tolerance may be done at 10:1.

---

[1] This is a product design specification. It may or may not apply to PC boards designed by other firms.

17. Wherever possible the following markings will be added to the component side of the board:

- *a.* Termination numbers or wire colors
- *b.* Polarity for capacitors, diodes, etc.
- *c.* Test-point identification
- *d.* Adjustment identification

5-5 Double-sided and Multilayer Board Layout

Figure 5-15 is the schematic diagram of the display-board circuit for a small radar unit. The components consist of three integrated circuits I_1, I_2, and I_3; three cold cathode display lamps DS_1, DS_2, and DS_3; two transistors Q_1 and Q_2; two resistors R_1 and R_2; and a connector J_1. After some analysis of the schematic, an assembly drawing (or component layout) was made. Figure 5-16 shows a satisfactory arrangement of the devices. Note that the cathode lamps and integrated circuits are positioned with I_1 close to DS_1, etc. The schematic diagram shows seven connections between I_1 and DS_1; therefore, if they are located close to each other, crossovers will probably be minimized. If one allows space for some circuit paths between the devices and lays out to scale (double, or $2:1$, in this case), the size of the board that results is about 3 (76) $\times$ 2¾ in. (70 mm).

Using this layout, one or more rough sketches of possible circuit patterns have been made. One such sketch is shown in Fig. 5-17. This was made on a sheet of tracing paper placed directly over the layout of Fig. 5-16. A logical step is to run ground and power paths where they can conveniently be connected to components. Such a line starts at terminal 9 of DS_3, connects at e of Q_1, and goes to J_{1-16}, ground. Other logical procedures are to sketch paths between components in one direction, say, vertically such as between 7 of DS_2 and 13 of I_2, and between 4 of DS_2 and 14 of I_2. Note that there are few crossovers in the area between DS_2 and I_2 on the left and DS_1 and I_1 on the right. Also note that it is possible to have conductor lines going under the ICs and cathodes. This is possible even on the component side because the bottoms of the devices, when mounted, are raised almost a millimeter off the board.

Careful layout and preliminary sketching have not been able to eliminate quite a few crossovers of the radar circuit. This indicates that a double-sided board is the most logical and economic construction. (It is possible to treat a crossover as a component on a one-sided board by making a "jumper" out of wire. But the economics of modern PC-board manufacture suggests that such crossovers be held to a bare minimum.) Therefore it was necessary to make more sketches for the printed paths on the component side and for the printed paths on the other side. One approach is to put all the vertical paths (as in Fig. 5-17) on one side and the horizontal ones on the opposite side. It usually is not this simple, but this is one way to begin.

The final results of such preliminary sketching are shown in the two master layouts of Fig. 5-18. Each layout was drawn double size. The distance between

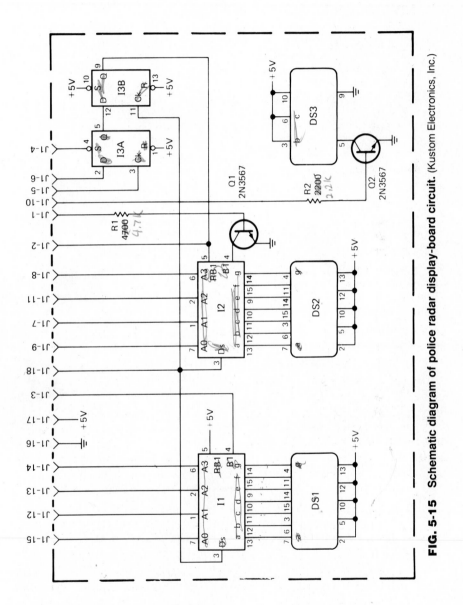

FIG. 5-15 Schematic diagram of police radar display-board circuit. (Kustom Electronics, Inc.)

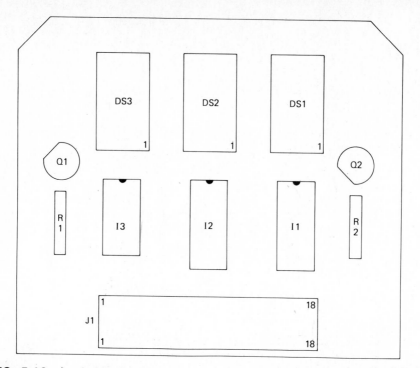

FIG. 5-16 Assembly drawing (component layout) for hand-held radar circuit board.

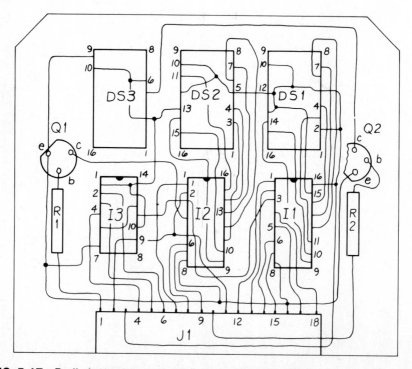

FIG. 5-17 Preliminary sketch of connections for radar circuit board. Such preliminary drawings are a link between the schematic diagram and the master artwork.

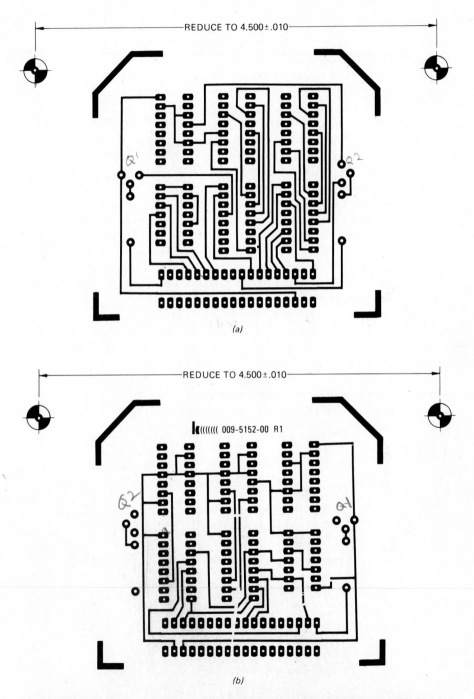

FIG. 5-18 *(a)* Master layout for component (near) side of PC board. *(b)* Master layout (final artwork) for far side of PC board. (Kustom Electronics, Inc.)

pads was 0.2 in. (5 mm). The paths were laid down with 0.026-in. (0.66-mm) tape, and the same figure was used for minimum space between paths. Most paths are horizontal or vertical, but there are some paths in congested areas (between I_2 and I_1, for instance) where it was necessary to use angled lines in order to meet minimum spacing requirements.

The master layout must be made on a dimensionally stable medium to a scale larger than the final product. Registration is important, and two (in some cases three or more) registration points quite far apart are required. The two points used in Fig. 5-18 are such that the small holes in the centers remain white when both sides are exactly aligned with each other. The heavy corner markers establish the board outline on the inside of these markers, as shown in Fig. 5-19. Lines in the master artwork must be solid black, unless a colored taping system is used. If the color system is used, blue tape is used for conductor paths on the component side and red tape is used on the opposite side. These specially made tapes can both be put on one sheet of clear polyester film. With the use of filters, the photographic process can "hold" one color while "dropping" the other color, and vice versa. Thus two circuit patterns, one for each side of the PC board, can be made from the two-color layout.

Once the master artwork is completed, it can be used to make other drawings, such as the drilling drawing (Fig. 5-19), the marking drawing (Fig. 5-20), and solder masks. Marking drawings include such items as polarity markings, serial number, vendor's number, and the name of the company using the board. (Many product manufacturers have their PC boards made by firms that specialize in PC-board production.) Marking drawings are usually made with commercially available adhesive lettering and markings. Care should be

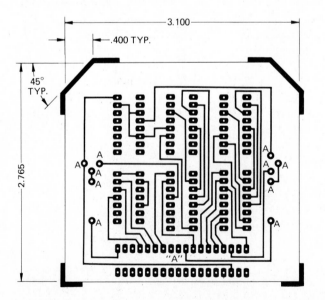

HOLE INDEX		
SYM.	QTY.	DESCRIPTION
A	46	.040 DIA.
B	94	.031 DIA.

FIG. 5-19 Drilling drawing for PC board. All holes not labeled A are B size.

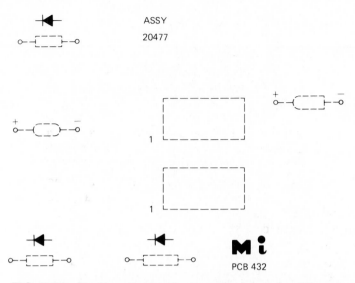

FIG. 5-20 Marking drawing for a small printed circuit board.

taken in locating the markings and lettering so that they are not hidden by the components. Lettering and registration marks should be at least 0.015 in. (0.381 mm) thick, and letters should be high enough to read after reduction to final board size. The number 1 is used to indicate where pin 1 of the IC is to be located. The dashed lines for component outlines are not actually shown.

In some cases the identifying number for each component (e.g., R_1, C_5) is put on this drawing and printed on the board. For small components this identification is located so that the device is placed directly over the printing. (Unfortunately the device usually hides the printing, and locating a part for replacement becomes a chore.)

If the density of components (and wiring) that are to be in a circuit is high, two sides of a single board may not be enough to accommodate all the printed circuitry. In such an instance, more than one board may be necessary. A photograph of a multilayer board is shown in Fig. 5-21. This high-density board with a lot of integrated circuits has four layers of printed circuitry printed on two pieces with a thin layer of what is called *prepreg* material in between. In the lower right-hand sector, just below the last integrated circuit and to the right or left, can be seen three colors (or shades) of lines, which represent circuit paths in various layers. (Unfortunately, only two layers can be seen in the photograph.) Boards with as many as 14 layers have been manufactured. Sometimes a number of circuit boards interconnect with (plug into) such a multilayer board. In these cases such a board is called the "mother" board. Figure 5-22 shows a cross section of a six-layer board. After the board patterns have been made from the artwork, they are aligned with precision and the layers of prepreg material are placed between the boards. Then the assembly is placed in a heated press under high pressure. A six-layer board made of three 0.012-in.-(0.30-mm-) thick

FIG. 5-21 Photograph of a four-layer PC board. At least two layers of conductor paths can be seen in the lower right sector. (Kustom Electronics, Inc.)

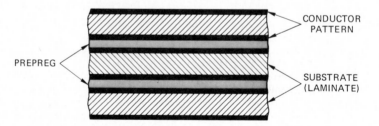

CONDUCTOR
PATTERN

PREPREG

SUBSTRATE
(LAMINATE)

FIG. 5-22 Cross section of a six-layer circuit board. Total thickness is about 0.062 in. (1.57 mm).

laminates might have a total thickness of 0.062 in. (1.57 mm), including substrate, prepreg, and thicknesses of printed wiring.

The design and layout for multilayer boards is beyond the scope of this book. Considerable experience is desirable. For one thing, a decision must be made about which circuits are to be placed in which layers. Sometimes ground planes and voltage buslines are the only items in a single layer. Accuracy of registration and manufacture are very important. Computer-aided layout is often used.

Photographs of another double-sided PC board are shown in Fig. 5-23. Approximately 150 electrical components are on this 5.75×6.75 in. board. One memory device, U_{10}, has been programmed by the manufacturer and has a

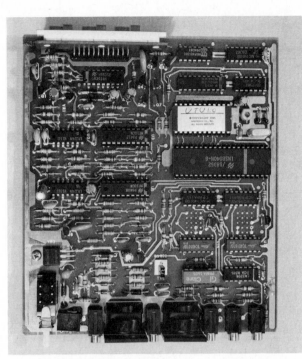

(a)

(b)

FIG. 5-23 Photographs of a PC board for a computer-transceiver interface unit: *(a)* looking down on component side; *(b)* viewing at an angle, with connectors on one edge near bottom. The back side of a LED bar display is shown at the top.

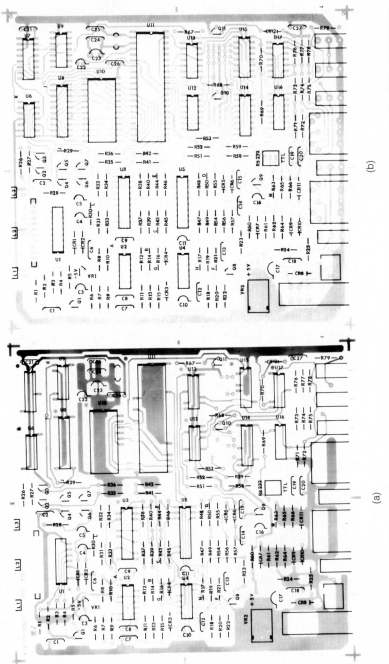

FIG. 5-24 Two of four photographs of the transceiver interface board: (*a*) component side showing markings for small components and outlines for IC packages and a few other devices — circuit paths shown are on the same (near) side; (*b*) x-ray photograph of the component side, showing the PC paths of the far side. (Kantronics, Inc.)

(b)

(a)

184

note to that effect pasted on top. This computer-transceiver interface (for amateur radio operators) has an operator's manual that includes a schematic diagram and component layout (placement diagram). The printing from the latter drawing shows up on the board itself. In Fig. 5-23 such lettering as U_3, U_{15}, U_{17}, and C_{17} is visible. However, identification of most resistors is hidden by the resistors themselves. For this reason, the operator should have the component placement diagram, for troubleshooting.

A recent development in photography has made it possible to make a single photograph of a board that shows the components on one side and the printed wiring on the opposite side. Figure 5-24 shows two photographs of the PC board in Fig. 5-23. Figure 5-24b shows the outlines of the ICs and identification of the smaller devices as seen looking down on the component side of the board. But the printed wiring is that of the other side. This is called an *x-ray photo.* Two other photos were made looking at the other side of this PC board but are not shown. One is an x-ray view in which the printed circuits of the component side show through. These photographs are useful to the designer and production personnel, but are also reproduced and sent to the users of products. They are very helpful to whoever has to replace a part or perform modification.

5-6 Surface-mounted Devices

We have shown a few surface-mounted devices in Figs. 3-28 and 3-35. Surface mounting has already begun to change many aspects of electronic assembly. A whole family of tiny active and passive[1] devices is being developed to meet the demand for lighter, smaller, cheaper, and better boards. These components are soldered to board solder pads which are matched to the "footprints" made by the leads of the package. Through holes are eliminated, and the proper mix of active and passive components cuts the length of leads, thus reducing parasitic capacitance and inductance in the wiring. Such technology is producing smaller, denser, and cheaper boards whose high-speed ICs can operate at optimal efficiency.

In Fig. 5-25a we have shown a section view of a PC board having pin-type and leaded-type components mounted in through holes and soldered on the bottom (foil) side of the board. In Fig. 5-25b a chip carrier is shown with J-type leads and a small outline package with gull wing leads soldered onto a board. The J and gull wing leads produce a rectangular "footprint," and the PC drawings and final artwork have pads whose rectangular shapes match the footprints. Widths of lines and spaces between lines have become smaller. Line widths now in use are 6 mils (0.006 in. or 0.15 mm) to 12 mils (0.012 in. or 0.30 mm) wide, and minimum line spacing is about equal to line widths. Figure 5-26 shows the lands or pads and 10-mil lines of a small part of the final artwork for a circuit containing small-outline integrated circuits (SOICs) and the outline and end view of a 14-pin SOIC. Standard dimensions are being developed for

---

[1] An example of an active device is a transistor. Examples of passive devices are resistors and capacitors.

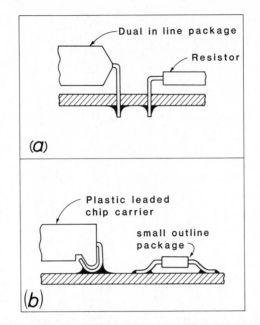

FIG. 5-25 The old and the new. *(a)* Through-the-hole mounting of leaded components. *(b)* Surface mounting of chip carrier with J-shaped leads at left, and small-
outline IC package with gull-wing leads, right. (Reprinted from *Electronics*, February 9,
1984; Copyright © by McGraw-Hill, New York, 1984.)

14-, 24-, and 48-lead SOICs. Standard 3- and 4-lead small-outline packages,
called *SOT 23* and *SOT 143* for transistors and certain other devices, are
$114 \times 51 \times 43$ mils. New equipment has been developed for production of
small-outline PC boards. Two soldering methods are used: (1) *reflow* in which
components are placed over solder that has been screened over the pads and
then is heated and melted and (2) dual-wave soldering in which the devices are
exposed to two soldering procedures for 2 s each.

A circuit board with surface-mounted devices is shown in Fig. 5-27, and in
the center part of Fig. 5-28, with two other boards. This 3.75-in. (95-mm) ×
2.25-in. (57-mm) unit is part of a portable transceiver. It has more than 200
active and passive devices mounted on both sides. These can be identified by
their shapes and colors, not all of which (especially browns and grays) show up
well in the accompanying photographs. The colors and devices are:

White	Resistors
Silver (metallic)	Semiconductors
Black	Integrated circuits and electrolytic capacitors
Brown	Capacitors
Gray	Capacitors

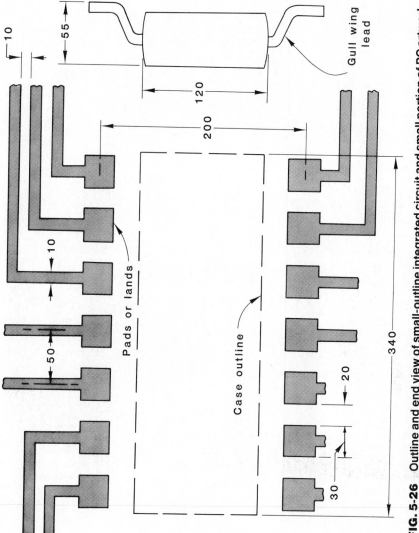

FIG. 5-26 Outline and end view of small-outline integrated circuit and small portion of PC artwork for surface-mounted components. Dimensions are in mils (1 mil = 0.001 in.).

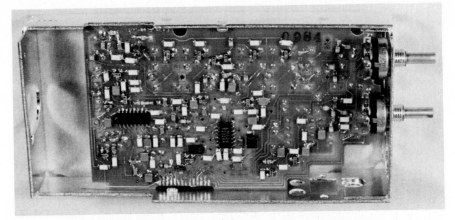

(a)

(b)

FIG. 5-27 *(a)* and *(b)* Two views of a circuit board having surface mounted components. (King Radio Corp.)

FIG. 5-28 A view of a transceiver in which the circuit board of the previous photograph is mounted with two other boards having mostly surface-mounted devices. (King Radio Corp.)

Two SOICs can be seen with their gull-wing leads projecting away on each side. On either side and close to one IC are two black electrolytic capacitors. White resistors are easy to see. Brown and gray capacitors, nearly square in shape, appear gray in the photograph and are not as plainly visible as the black-and-white devices. In Fig. 5-27a a SOT 23 transistor is shown just below the black capacitor, which is to the left of the IC whose long dimension is perpendicular to the long dimension of the board (and chassis). The SOT, which is silver-metallic color, shows up as gray in the photograph and has two leads above and one below. Printed-circuit paths are also visible in several places in Fig. 5-27a.

The general layout and drawings for boards with surface-mounted components are much the same as those for PC boards of recent years. One might expect the following differences:

1. Larger scales
2. Narrower paths
3. Smaller parts

4. Closer spacing

5. Different pads

6. Greater density

7. More components (on both sides if necessary)

8. No, or fewer, through holes

9. 50-mil grid spacing

Surface-mounted technology is one of the waves of the future. However, PC boards using present state-of-the-art components with 100-mil spacing will be manufactured and used for many years to come.

5-7 Mounting of PC Boards

Printed-circuit boards are attached to a chassis in various ways. The method is determined by the size and shape of the board, the size and shape of the chassis, grounding and other electrical requirements, ease of removal if such is desirable, ventilation requirements, and the standard procedures used by the manufacturer.

The three figures starting with Fig. 5-29 show three different chassis. The accompanying printed matter describes briefly how the boards are attached. Figure 5.29a is the case (and chassis) for the transceiver interface unit shown in Figs. 5-23 and 5-24. The 5.75-in. (146-mm)-wide board is simply slipped into the case with two edges sliding along, and in, the grooves on each side. After front and rear panels are put in place, the black end frames are pressed on (with a tight fit), completing the assembly of this 1.90-in (48-mm) $\times$ 5.90-in. (150-mm) $\times$ 7.00-in. (178-mm)-deep unit. The case-chassis is made of extruded aluminum. The finished unit is shown in Fig. 5-29b.

The boards shown in Fig. 5-30 are fastened onto the chassis with 2-56 screws (No. 2 screws having 56 threads per inch). This chassis is made of cast aluminum. Tapped holes can be seen on its top edge. One of several features of this KT 79 transponder is a double-tuned coaxial line filter near the lower left corner. Near the bottom of the picture, U-shaped coupling probes transmit the signal from the circuit board at the right through the filter and out through the small board at the left.

The photograph of the Heathkit robot in Fig. 5-31 shows three of the 13 circuit boards. The large CPU board is fastened on to the square "wrap around" aluminum chassis by means of No. 6-32 screws and nuts and 0.30-in. (8-mm)-long spacers between chassis and board. However, the two smaller boards are "pressed" onto plastic "bayonet"-type fasteners which go through the small holes with a little pressure and hold the boards firmly in place. The spacers for these boards are plastic. Approximately five PC boards are mounted on the other two sides of the chassis, while the rest of the boards are in various places. One is located under the hexadecimal keyboard, which is visible at the top.

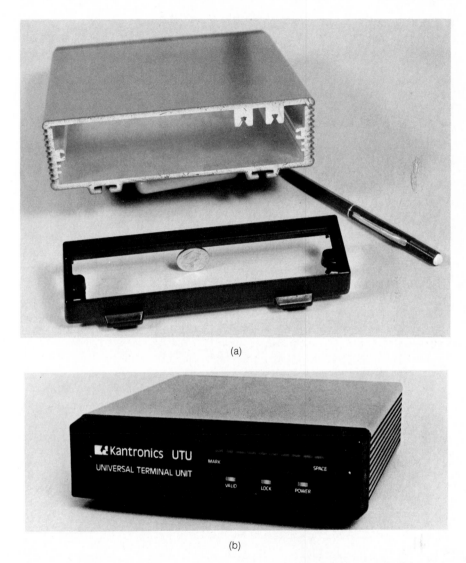

(a)

(b)

FIG. 5-29 **Case and chassis for the UTU transceiver interface unit: *(a)* empty case with the front parts removed; *(b)* the complete unit.** (Kantronics, Inc.)

The flexible board for the calculator shown in Fig. 5-2 has two small holes near each end of the LCD display which is at the right in the photograph. (One hole can be seen on the flexible board above the display and one can be seen "below" the display.) These 0.08-in. (2-mm) holes are pierced by small plastic pins on the bottom part of the calculator case when the circuit board is placed between the top and bottom parts of the calculator, and these parts are pressed

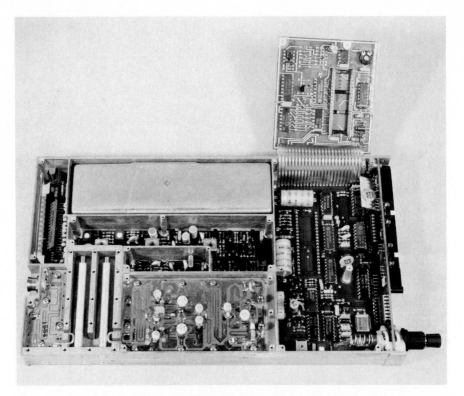

FIG. 5-30 The chassis for a transponder for aircraft. This cast aluminum chassis holds several circuit boards and some other parts. (King Radio Corp.)

together. The flexible board is also clipped to the inside half of the keyboard, which fits below or underneath the keys of the keyboard. Thus the board is held in place by the two pins and the clips and the fact that its shape exactly fills the space that is available.

These are some of the ways that boards are attached or mounted. There are other ways which we haven't covered. Designers and detailers must know how a board is to be attached before the final artwork is done.

SUMMARY

Printed-circuit board technology is a mature one, but changes are still occurring. In the original boards the components were mounted on one side and the printed circuits on the opposite side. Now, in many boards the components are mounted on one side and PC paths on both sides. It is also possible to have components and printed circuits on the same side. In the future, some boards may have surface-mounted devices on one side, present state-of-the-art compo-

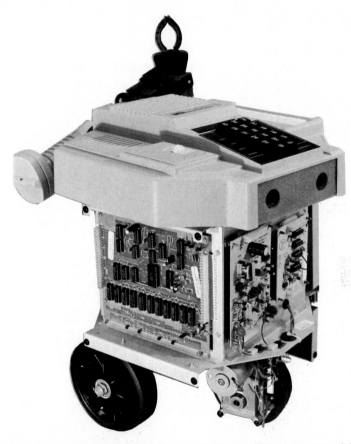

FIG. 5-31 Photograph of the Heathkit robot with side panels removed. Three of its 13 PC boards are visible.

nents on the other side, and printed wiring on both sides. Printed-circuit technology is such that the density of components and circuit paths may be high enough to require multilayer boards.

Artwork for PC boards must be accurate and therefore is often drawn two or four times the size of the finished conductor pattern. It may be done in ink on a stable medium such as Mylar, or with tape and preprinted pads, bends, etc. Companies that make or use PC boards usually have extensive lists of specifications that must be followed by the drafter or designer. Some of the drawings that are often included in the production of a PC board are:

1. Preliminary sketches for component location and conductor paths
2. Component layout
3. Master layout, final artwork
4. Drilling drawing

5. Marking drawing

6. Solder mask

Many scales are in use for PC-board layout. Both 2:1 and 4:1 are common. Some computer-assisted drawings are drawn full size. Scales of 10:1, 20:1, and 100:1 have been used. Printed-circuit paths are often drawn horizontally or vertically. Sometimes electrical requirements, such as resistive pinching and inductive coupling, dictate what directions adjacent lines should go. Production methods and economics (minimum use of materials) also must be considered in circuit-board layout.

QUESTIONS

5-1. What drawings may be required for the production of a rather complex PC board?

5-2. What are some typical widths for printed-conductor paths?

5-3. What are some typical scales used in making master layouts?

5-4. What grid spacing is being used for surface-mounted devices?

5-5. What is the first step in making drawings for a printed circuit?

5-6. Show, by means of sketches, the following: eyelet, board outline, pad, elbow, and registration mark for use in PC-board construction.

5-7. What is the purpose of putting a socket on a PC board, and then inserting an IC into that socket?

5-8. What information is included in a marking drawing?

5-9. Where is a good place to show the number of a resistor on a PC board? Show by means of a sketch.

5-10. Where is the outline of a PC board in relation to the heavy corner markers?

5-11. What is the difference in the pad shape for, say, an IC on a typical PC board and the pad shape for a PC board having surface-mounted components?

5-12. What is the purpose of a solder mask in PC-board production?

PROBLEMS

5-1. The completely dimensioned working drawing of a small PC board appears in Fig. 5-32. Using a scale of 1 in. = 0.30 in., draw this board on an $8\frac{1}{2} \times 11$ or 9×12 sheet. Your instructor may specify a completely dimensioned drawing with or without lettering, a drilling drawing, a marking drawing,

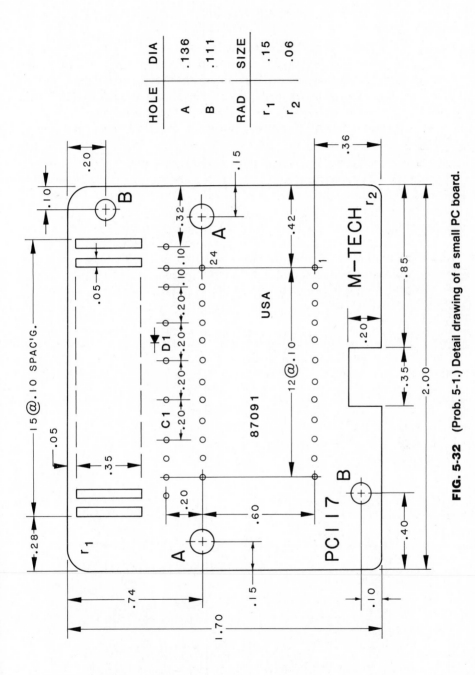

FIG. 5-32 (Prob. 5-1.) Detail drawing of a small PC board.

or a combination drilling-marking drawing. (An optional $\frac{3}{8}$-in. = 0.10-in. scale will require the use of a 9 × 12 or 11 × 17 sheet.)

5-2. The artwork for the foil side of the PC board for Prob. 5-1 is shown in Fig. 5-33. Make a scale drawing to 1 in. = 0.30 in., 4:1, 5:1, or $\frac{3}{8}$ in. = 0.10 in. of the master layout for this board. Use line widths of 0.015 or 0.020 in. for the thin conductor lines, or heavy pencil or ink lines if you do not have tape. The (shaded) ground plane should be about as shown and may be within 0.030 in. from the edge of the board. Add registration marks and any other items your instructor may request.

5-3. The dimensioned drawing of a PC board is shown in Fig. 5-34 with the PC paths on the component side. Do one or more of the following at your instructor's request: (a) make a drilling drawing using 0.03-in. (7.6-mm) holes; (b) make a marking drawing showing all lettering; and (c) make a final artwork drawing for the PC paths for this side using line widths of 0.015, 0.02, or 0.03 for the narrow lines and 0.05 for the wide line. Do not show pads for the ICs unless you have the appropriate appliqués. The addition of pads to those holes that are joined to printed-conductor paths at C_1 and C_2 would be quite appropriate.

This problem will barely fit on an $8\frac{1}{2}$ × 11 sheet if a scale of 1 in. = 0.30 in. is used. A $\frac{3}{8}$-in. = 0.10-in. or 5:1 scale (1 in. = 0.20 in.) will require the use of a 11 × 17 sheet.

5-4. Draw the 2 × 2 in. circuit board shown in Fig. 5-35b to four times actual size. Then complete the location of the components for the schematic diagram shown in Fig. 5-35a. Show the necessary wiring (as hidden lines on the other

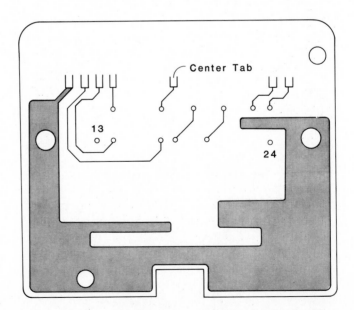

FIG. 5-33 (Prob. 5-2.) Foil side of PC board No. 117 with printed wiring.

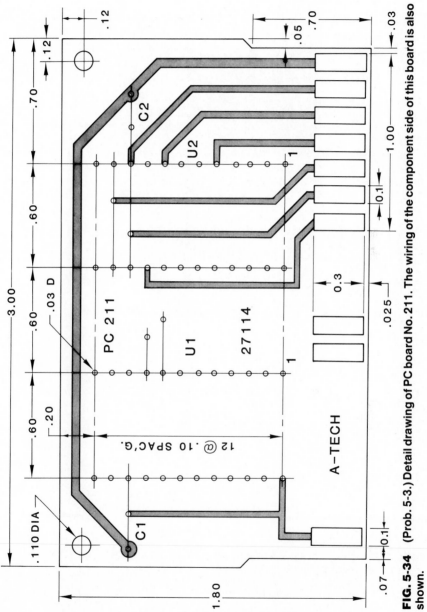

FIG. 5-34 (Prob. 5-3.) Detail drawing of PC board No. 211. The wiring of the component side of this board is also shown.

197

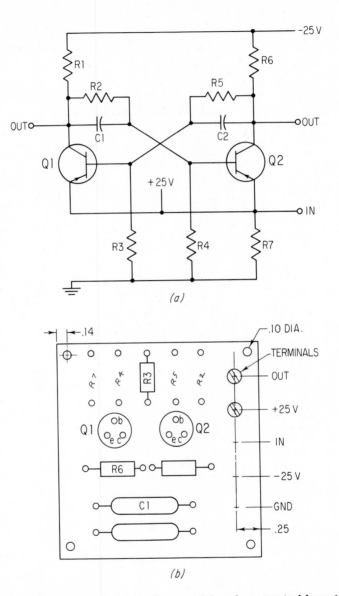

FIG. 5-35 (Prob. 5-4.) Schematic diagram *(a)* and suggested layout *(b)* for a wiring or PC-board layout.

side of the board) to agree with the schematic diagram. Add notes as required by your instructor. An alternative solution would be to draw a mirror image of the board and put the components in as hidden parts and the wiring in as solid. Use 11 × 17 or 12 × 18 paper. Typical component sizes are: resistors, 0.14 × 0.30 in.; capacitors, 0.20 × 0.80 in.; and transistors, 0.37 in. in diameter. This can also be used as a PC-board layout problem.

5-5. Figure 5-36 shows a preliminary layout of a circuit board that is to contain the sweep-generator circuit. Check the layout against the schematic diagram (no dot system) for improvement in layout. Make a PC-board master layout (outside dimensions are 2.20×2.40 in.) to a $4:1$ or $5:1$ scale. Use the 0.10-grid system for locating components whose sizes are: resistors, 0.25×0.09 in., except R_3, which is 0.375×0.09 in.; capacitors, 0.422×0.135 in.; L_1, 0.400×0.15 in.; and diode, 0.275×0.105 in. maximum. Use 0.062-in. conductor paths, with 0.031-in. minimum spacing unless your instructor specifies otherwise. Use $8\frac{1}{2} \times 11$ paper.

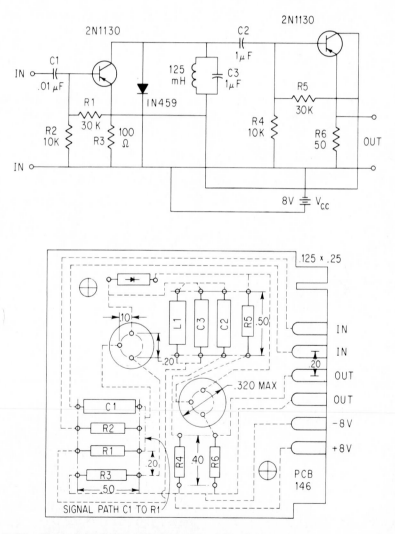

FIG. 5-36 (Prob. 5-5.) Schematic diagram and preliminary component layout of a circuit board for a sweep generator.

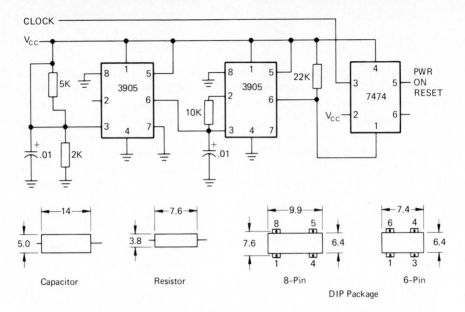

FIG. 5-37 (Prob. 5-6.) Schematic diagram of a timer reset. Major dimensions of the components are shown below.

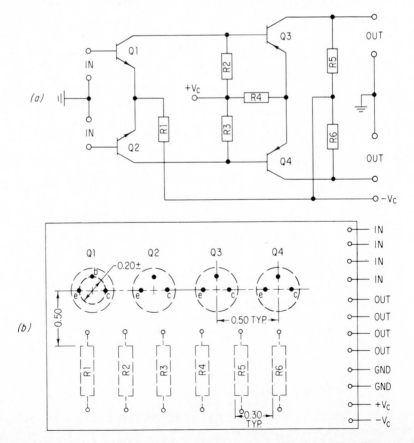

FIG. 5-38 (Prob. 5-7.) A printed wiring problem for a differential-amplifier circuit.

5-6. A digital timer circuit is shown in schematic form in Fig. 5-37. Make a preliminary wiring sketch, a component layout sketch, and a final artwork for a PC board for this circuit. Have all the inputs located at one side with fingers 5.08 mm (0.20 in.) apart. Use a 2.54-mm (0.10-in.) grid. Let conductor paths be 1.57 mm (0.062 in.) wide with a minimum spacing of 0.80 mm (0.031 in.) between paths and outside paths and edge of board. Shape and size of board are to be determined by student or instructor, as is scale. Terminal pads are to be 3.18 mm (0.125 in.) in diameter with holes 1 mm (0.040 in.) in diameter. The DIPs, however, will use standard pin spacing of 2.54 mm (0.10 in.). Dimensions of the components shown are in millimeters. Jumpers are permitted.

5-7. Figure 5-38 includes an elementary diagram and a suggested arrangement of parts for a differential amplifier. Make a scale drawing of the final master drawing for the pattern. A maximum of four jumpers will be permitted. The lower figure shows the wiring side of one arrangement of components. Transistor diameters are 0.360 ± 0.010 in., and resistor dimensions are 0.093 diameter × 0.375 in. Try to get an acceptable pattern on a board that is no larger than 3.00 × 2.00 in. Freehand preliminary sketches are suggested. If you cannot get an acceptable pattern with the suggested arrangement of compo-

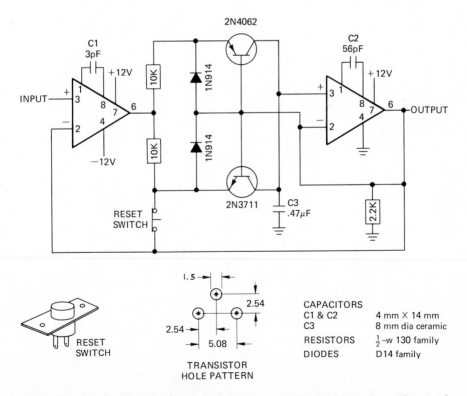

FIG. 5-39 (Prob. 5-8.) A video peak detector circuit. Major dimensions of most of the components are shown below.

nents, make your own arrangement. Hole diameters are 0.032 in. Terminal pads are 0.125 in. across. Use 0.062-in. conductor paths with 0.031-in. minimum spacing. Show registration marks and pin no. 1 (top pin), and label the board PCB 46. If you do not have tape, you may be asked to draw the signal path in a single heavy line or two parallel lines. Use $8\frac{1}{2} \times 11$ paper. Show one critical dimension.

5-8. Figure 5-39 shows the schematic diagram of a video peak-detector circuit. Dimensions of the components are shown in millimeters. Make a sketch of a physical arrangement of components that will accommodate a printed circuit on a minimum-size board. The switch will be attached to the board by wires. It will be mounted elsewhere. Then draw a master layout using conductor path widths of .787 or 1.02 mm, except for a wider $+12$ path. Use a scale of $2:1$ for $8\frac{1}{2} \times 11$ paper or $4:1$ for 11×17 paper. Make pads 1.58 mm in diameter with 0.64 mm-diameter holes, except for leads to switch, which will require larger holes. Refer to Table 5-1 for sizes of components. Jumpers permitted.

6

Flow Diagrams and Logic Diagrams

A primary drawing is one that shows the function of a circuit or a system in a logical manner. It is usually the first drawing sketched by an engineer or technician. The primary drawing in electronics and industrial control used to be the schematic diagram. But the ascendancy of digital electronics has necessitated that the logic diagram become the primary one, and this is also true for some areas of industrial control. As systems become more complex or miniaturized (using LSI — large-scale integration — components, for example), the block and flow diagrams are becoming more popular and necessary for the explanation of the functions of these systems.

6-1 Examples of Block Diagrams for Electric Circuits and Systems

Such a diagram may be used to show the operation of a large electronics system. In such a case, a block would represent a complete and removable chassis, such as a preamplifier, a multivibrator, or even a television camera.

However, in a different situation, a block diagram may be used to facilitate the understanding of a radio receiver or a multistage amplifier, for example. In this case, each block would represent a *stage.* This is the case of the diagram shown in Fig. 6-1. At this time it might be a good idea to define the word *stage.* A stage is considered to be that part of a circuit which includes the main device (e.g., transistor, diode, or tube) and the associated devices that go with it, such as biasing resistors, load resistor, voltage dividers, and capacitors. In other words, a circuit may have several stages, hooked together somehow, so that the signal goes first through one stage, then through the next, and so on. Each block represents a major subsystem. The diagram shows major signal and data paths, inputs, outputs, and control points.

Figure 6-1 shows how easy it is to understand a circuit's operation by means of a block diagram. It is clearly shown that the signal comes through the antenna (usually portrayed by a symbol rather than a block) and then progresses through the mixer circuit, through the intermediate-frequency (IF) stages, and finally to the output stage and speaker. The oscillator, which is an auxiliary circuit, is appended to the main circuit; and, because it is a frequency

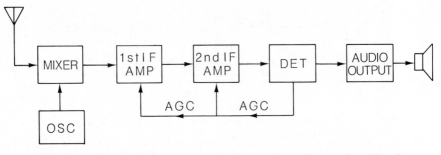

FIG. 6-1 Block diagram of a typical transistor radio receiver circuit.

generator, its output is fed into the signal train as shown by the arrow. A feedback circuit, labeled AGC for automatic gain control, is correctly drawn below the main circuit. If it were desirable to emphasize the AGC circuit, however, it would be appropriate to place it above the main circuit. The purpose of automatic gain control is to prevent fluctuation in speaker volume when the radio signal at the antenna is fading in and out.

6-2 Preparation of Block Diagrams

A block diagram may use standard symbols for certain elements, but it is predominantly one of blocks (usually, but not always, squares or rectangles). The layout can be facilitated by (1) using freehand sketches in the initial stages, (2) using cutout cardboard blocks of appropriate size and experimental arrangement until the best pattern is achieved, or (3) using cross-ruled paper for an undergrid, thus facilitating the construction of blocks of equal size and uniform spacing between blocks. The size of the rectangles is usually determined by the lettering that goes in them; and, since they are usually drawn about the same size, the block with the most lettering will often set the size of the blocks in the entire diagram.

From Fig. 6-1 and other diagrams in this chapter, certain facts about block diagrams can be deduced, and the following rules for their construction can be listed:

1. The signal path should be made to go from left to right, if possible. In large, complex drawings, the input should preferably be at the upper left and the output at the lower right, if possible.

2. Blocks are usually drawn in one of three shapes: rectangular, square, or triangular. (The triangle represents different items in different types of diagram. There are also other shapes for certain specialized diagrams, as will be shown later in the chapter.)

3. Once the size and shape of a block are determined, the same size and shape should be used throughout the drawing. The size of a rectangle, for instance, bears no relation to the importance of the component(s) it represents.

4. A single line, preferably heavy, should be used to show the signal train from block to block. In complex circuits or systems, however, more than one line may have to be drawn leading into or away from a block.

5. Arrows should be used to show the direction of signal flow.

6. Some components, usually terminal ones such as antennas and speakers, are shown by means of standard symbols rather than by blocks.

7. Titles, or brief descriptions, of the components or stages represented should be placed within the blocks.

Aside from the above-listed rules, no standardized procedure exists for the preparation of block diagrams. In Fig. 6-1, for instance, either square or rectangular blocks could have been used. The arrows, which are shown touching the blocks, could have been placed midway between the rectangles if desired.

Figure 6-2 shows the control system for an existing coal-fired electric generating plant to which an addition has been made. The addition is a central collector system in which large mirrors (heliostats) track the sun and beam its reflection to a central receiver that converts this heat to steam which goes to the new steam generator or to storage. This diagram differs from Fig. 6-1 in that the blocks are not all the same size (widely different amounts of lettering), most of the signal paths are bidirectional (arrows at both ends), the master control has a border, and the signal interface (a secondary consideration) is indicated in the dashed lines.

Figure 6-3 shows the block diagram for a digital watch. Not shown are the batteries (2) and light-emitting diodes (LEDs). Practically all the circuitry is on a single chip. A three-view photograph of a transparent module without batteries is shown in Fig. 6-4. Laying out a block diagram for a fairly complex arrangement requires considerable planning, and possibly some experimentation, in order to achieve a neat, well-spaced arrangement of blocks and signal paths.

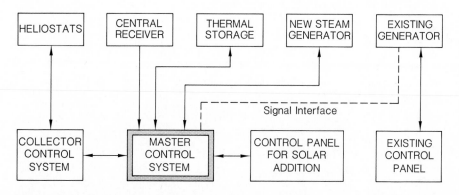

FIG. 6-2 **Flow diagram of the control system of a solar central receiver system.**
(Black & Veatch, Consulting Engineers.)

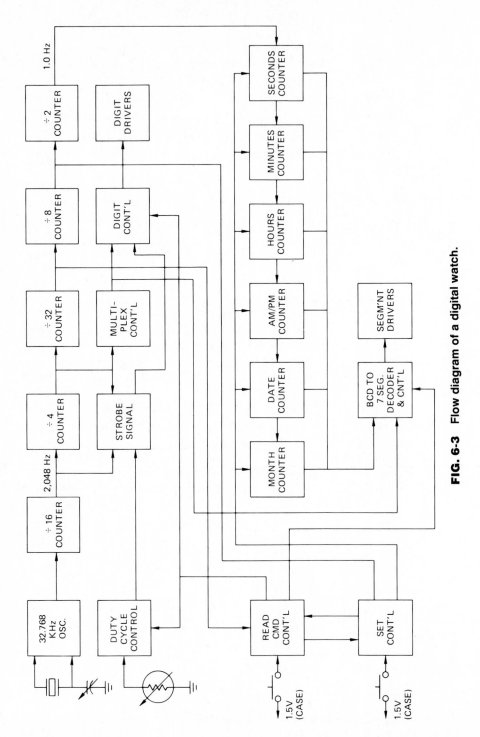

FIG. 6-3 Flow diagram of a digital watch.

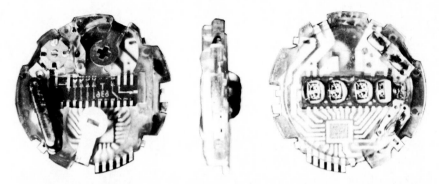

FIG. 6-4 Three views of digital watch module. In view at right can be seen the four LEDs and the IC below them. In view at the left can be seen the contacts for two batteries, the adjustable capacitor (upper left), and the 32,768-Hz quartz crystal oscillator (near left edge, inclined at ≈75°).

6-3 Microprocessor Diagrams

Microprocessors, which are playing and will continue to play an important role in our lives, require flow diagrams that are somewhat different from those shown previously. This is largely because of the many elements which are connected to the internal data bus that is bidirectional. Figure 6-5 is a flow or

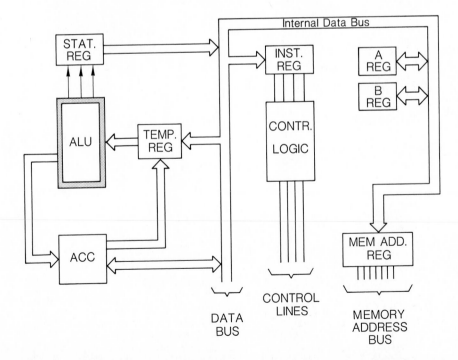

FIG. 6-5 Block diagram of a microprocessor. The data bus, which is bidirectional, moves data words in either direction.

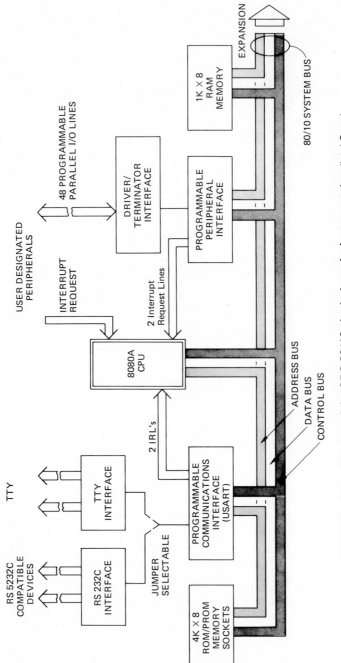

FIG. 6-6 Flow diagram of the SBC 80/10 single-board microcomputer. (Intel Corp.)

block diagram of a generalized microprocessor used for instructional purposes. The *arithmetic logic unit* (ALU) performs the data processing, arithmetic, and similar functions. The *accumulator* (ACC) is a special register that can store one data word. When the ALU adds two data words, one of the two words has been placed in the accumulator. The result of the addition is then placed in the accumulator. Some of the other registers shown in Fig. 6-5 are the *status register,* the *temporary register,* the *instruction register,* the *A* and *B registers*, and the *memory address register.* Our drawing shows the registers and other elements with their relationship to the internal data bus. This bus, which moves data words (usually 8, 16, or 32 bits per word), is connected to all the registers, most of which can either place data on or receive data from the bus.

Before a logic function can place data on the bus, the function must wait for a signal from the control logic. The control signals are not usually considered to be part of the bus. Neither are address signals. For this reason some microprocessor diagrams show control and address buses. Figure 6-6 is such a diagram. This diagram is for a microcomputer [a board having a microprocessor and additional integrated circuit (IC) chips] but is similar to a microprocessor drawing. Here, each of the three buses is shown. Some diagrams show just the address bus and the data bus. Figure 6-7 is the diagram for a learning device consisting of individual elements of a microprocessor. Such learning modules assist programmers in learning microprocessor concepts. The modules consist of the following:

LCM 1001 microprogrammer

LCM 1002 controller

LCM 1003 read-write memory

LCM 1004 input/output interface

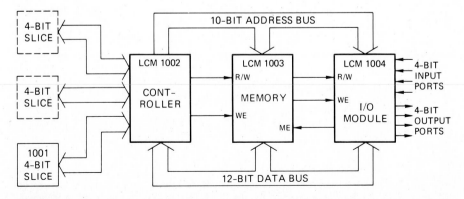

FIG. 6-7 **Flow diagram for the interconnection of microprocessor modules.** (Texas Instruments, Inc.)

6-4 Logic Diagrams

Beginning with computer design in the 1950s, engineers and manufacturers began to work with logic functions which could be performed by basic circuits. Now people working in many areas — transportation, industry, and communications, for example — find it necessary to make and read drawings that contain symbols representing logic functions. We will describe some of the basic logic functions in this chapter. However, examples will also appear in other parts of the book as various circuits and designs are presented.

The symbols that have been used to represent these functions in the past have not been very well coordinated and standardized. For example, Fig. 6-8 shows six symbols that have been used to represent the AND function. Fortunately the most recent standard, ANSI/IEEE Y32E, has included the revised versions of ANS Y32.14 (1973) and the IEEE and Military Standards that relate to logic symbols. Hopefully they will be accepted by everyone who works with these symbols so that only the distinctive and rectangular shapes will be in use. Inasmuch as the distinctive shapes are quicker to follow, they are widely used, and we shall use them for most of our examples. Figures 6-9 and 6-10 show the distinctive shapes for the more commonly used functions. As is the case with other symbols, no exact size has been specified. However, the authors have noted that if the units that are specified in Fig. 6-10 are drawn as millimeters, the symbols will be just about the right size for most large-size drawings.

Now for a description or definition of the functions. AND: If a signal is impressed at A, and at the same instant a signal is impressed at B, there will be a definite output signal at C. The "high" signals are often referred to as the 1-state or the ON-state, and the "low" signals are often referred to as the 0-state[1] or the OFF-state. Hence, the truth (logic) tables in Fig. 6-9 show the numbers 1 and 0.

[1] The number zero.

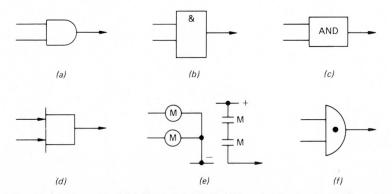

(a)   *(b)*   *(c)*

(d)   *(e)*   *(f)*

FIG. 6-8 Graphical representation of the AND function or gate: *(a)* ANSI-IEEE-approved distinctive shape symbol; *(b)* ANSI-, AIEE-, and IEC-approved rectangular symbol; *(c)* old symbol; *(d)* NEMA symbol; *(e)* JIC symbol; *(f)* old distinctive-shape symbol. Symbols *a* and *b* are recommended.

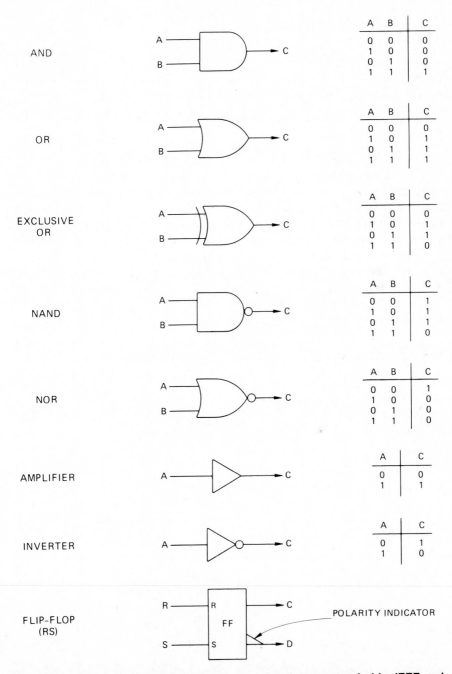

	A	B	C
AND	0	0	0
	1	0	0
	0	1	0
	1	1	1

	A	B	C
OR	0	0	0
	1	0	1
	0	1	1
	1	1	1

	A	B	C
EXCLUSIVE OR	0	0	0
	1	0	1
	0	1	1
	1	1	0

	A	B	C
NAND	0	0	1
	1	0	1
	0	1	1
	1	1	0

	A	B	C
NOR	0	0	1
	1	0	0
	0	1	0
	1	1	0

	A	C
AMPLIFIER	0	0
	1	1

	A	C
INVERTER	0	1
	1	0

FLIP-FLOP (RS)

POLARITY INDICATOR

FIG. 6-9 Distinctive-shape symbols approved and recommended by IEEE and ANSI. Truth, or function, tables for each symbol are shown at the right. The number 1, for 1-state, signifies an input or output that is high or "true." Zero indicates that the input or output pulse is zero, low, or "fake." (IEEE/ANSI Y32E.)

The table for AND indicates that there is a high (sometimes referred to as HI) signal at the output only when there are simultaneous high signals at A and B. The OR table shows that a HI signal or pulse at A or B or both will produce a HI output signal. We believe the other tables in Fig. 6-9 are self-explanatory, except for the RS flip-flop, which will be described as follows:

> *The outputs assume their indicated 1-states when only the S input assumes its indicated 1-state. The outputs assume their 0-states when only the R input assumes its indicated 1-state.*

A polarity indicator symbol is shown on the D output of the flip-flop symbol. This denotes that the 1-state of that output is the less positive level. This symbol is placed at the junction of the input or output line and the function symbol and points in the direction of signal flow.

There are other types of flip-flop circuits, but space does not permit their treatment in this text. And there are numerous other functions that have not been described. These are rather well documented in ANS Y32.14, which is now Sec. 14 of ANSI/IEEE Y32.E.

6-5 Drawing the Symbols

Logic symbols, either the distinctive-shape form or the rectangular form, can be drawn with instruments or special templates that are made for this purpose. The standard proportions are shown in Figs. 6-9 and 6-10. (Dimensions in Fig. 6-10

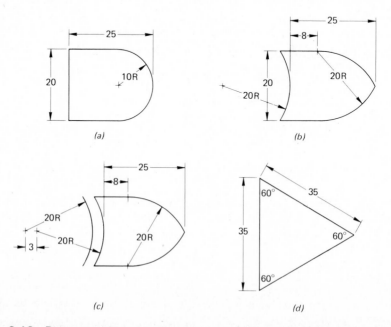

(a)   *(b)*

(c)   *(d)*

FIG. 6-10 Recommended symbol outline proportions: *(a)* AND symbol; *(b)* OR symbol; *(c)* Exclusive OR symbol; *(d)* amplifier symbol. (IEEE/ANSI Y32E.)

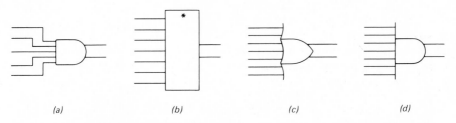

(a)   (b)   (c)   (d)

FIG. 6-11 Accommodation for additional inputs: *(a)* sometimes used for a limited number of inputs; *(b)*, *(c)*, and *(d)* approved by ANSI/IEEE Y32E. The asterisk is to be replaced by a qualifying symbol or letter, such as & for AND, ≧1 for OR.

are in millimeters.) In order to provide for situations where many inputs are to be drawn to a symbol, extensions as shown in Fig. 6-11*c* and *d* may be added.

Sometimes it is desirable to have logic symbols of two or more different sizes on a single diagram. Some of the symbol templates have two sizes of each symbol. One very suitable template is MIL STD 80C.

6-6 Negative and Mixed Logic

The NAND (NOT AND) function symbol shown in Fig. 6-9 has a small circle at the junction of the symbol and the output line. This circle is a negative indicator symbol. Its presence provides for the representation of the output in terms independent of its physical value. That is, the 0-state of the output is the 1-state of the symbol. If a circle is placed on the input side of a logic gate, the state or polarity of the input signal is reversed before entering the gate or function.

The result of the negative circle on the output side of an AND gate is that it now performs the NAND function. Figure 6-12 shows the effect of having negative logic on all leads of an AND function. The truth table is as follows:

Negative Truth Table for AND Function

A	B	C
1	1	1
1	0	1
0	1	1
0	0	0

A glance at the truth table for the OR circuit (Fig. 6-6) will reveal that these two truth tables are the same. That is, the AND circuit performs the same function in negative logic as the OR circuit performs in positive logic.

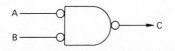

FIG. 6-12 An AND function having negative indicators at inputs and outputs. The result of using this negative logic is that the circuit performs as an OR function.

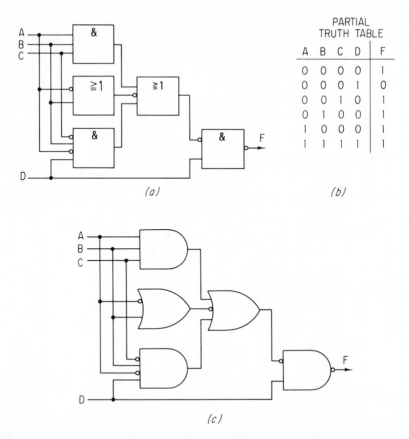

A	B	C	D	F
0	0	0	0	1
0	0	0	1	0
0	0	1	0	1
0	1	0	0	1
1	0	0	0	1
1	1	1	1	1

(a) *(b)*

(c)

FIG. 6-13 Example of a logic diagram in which uniform shapes have been used at *a* and distinctive shapes at *c*. The & symbol is for an AND function, and ≧1 is for the OR function.

Instead of the letter C for output, some truth tables show the letters AB (the output for the AND function), A + B (the output for the OR function), $\overline{A}$ (the output for a converter or NOT gate), $\overline{A}B + A\overline{B}$ (the output for an Exclusive OR function), $\overline{AB}$ (the output for a NAND function), and $\overline{A + B}$ (the output for a NOR, not OR, function). There may be, and often are, more than two inputs into a function, in which case the letters will be changed (see Fig. 6-13) or omitted entirely.

Figure 6-14 has logic symbols that were made by one of the authors on a personal computer. The symbols were developed on the monitor display using *basic* language. They appear at the right side of the CRT display as a *menu*. The remainder of the screen is available for drawing a diagram in which these symbols can be used. The figure is the actual printout from a dot matrix printer. The lines are composed of *pixels*, which are the smallest picture elements that can be made on the tube.

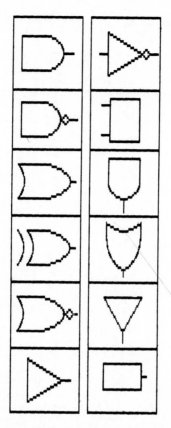

FIG. 6-14 A menu of logic symbols developed on a PC and printed by a dot matrix printer.

6-7 The Decision Table

Somewhat similar to the truth table is the decision table, which is essentially a tabulation of logical relationships consisting of conditions, actions, and rules. Conditions are the variables that influence any decision, while actions are the things to be done once a decision has been made. In the case of BART [San Francisco's rapid-mass-transit system, Bay Area Rapid Transit (Fig. 6-15 and Table 6-1)], a computer makes the decision and one of five actions (0 to 4 listed at the bottom of the table) is taken. A CP is a critical point (control location) at which it is especially desirable for trains to be on time.

6-8 IEC Logic Language and Dependence Notation

The system that is introduced here has been developed by the International Electrotechnical Commission (IEC) to show the relationship between each input of a digital logic circuit to each output without explicitly showing the

TABLE 6-1 Decision Table for an Automated Rapid-Transit System

Rules	Where Event Occurred*	Is Train Approaching CP at ≤ Minimum Tolerance?	Will First-Come–First-Served Give Correct Sequence?	Will Train Behind Be Delayed?	Train Performance Index†	Can Trains Ahead Be Slowed?	Actions‡
1	0	Yes	Yes	Yes			2
2	0	Yes	Yes	No			0
3	0	Yes	No				3
4	0	No		Yes			2
5	0	No		No			0
6	1			Yes			2
7	1			No			0
8	2				0		0
9	2				1		0
10	2			Yes	2		2
11	2			No	2		0
12	2			Yes	3		2
13	2			No	3		0
14	2				4	Yes	1
15	2				4	No	4

* State 0, between a station and a "merge"; state 1, between a station and a CP that is not a merge; state 2, at least one station before a CP.

† State 0, < 10 s late; state 1, 10 to 30 s late; state 2, 30 to 60 s late; state 3, 60 to 120 s late; state 4, > 120 s late.

‡ Action 0, continue with existing schedule; action 1, revise schedules ahead of delayed train to reduce extended gap; action 2, revise schedules behind delayed train to extend reduced gap; action 3, recommend revised sequence at interlocking; action 4, recommend station run-through.

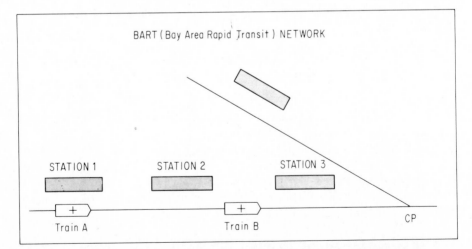

FIG. 6-15 Simplified drawing of a rapid-transit network illustrates how the decision table logic matches system conditions with a rule (in this case rule 14). Rule 14 specifies action 1, which is to reduce the gap ahead of the delayed train by revising schedules of trains ahead of it.

internal logic. It consists of one or more outlines (mostly rectangles) and one or more qualifying symbols. Some of the qualifying symbols are shown in Tables 6-2, 6-3, and 6-4.

Input lines are placed on the left and output lines on the right. When an exception to this rule exists, an arrow pointing to the left, upward or downward, is placed along the line. Outlines of elements may be abutted or embedded. If elements are joined by a (common) line that is parallel to the signal flow, there is no connection. If a common line is perpendicular to the signal flow, there is at least one logic connection between the elements.

Figure 6-16*a* shows three digital logic circuits joined together. The AND and OR symbols are shown in each rectangle. In the upper part, the rectangles are joined, as is customary with this system. The lower part of Fig. 6-16*a* shows the same arrangement, but as we have previously drawn logic circuits. (See Fig. 6-13*a*, where the OR symbols are slightly different.) Figure 6-16*b* illustrates another arrangement in which the lower rectangle is a common output element usually separated by two closely spaced parallel lines. In the lower part of Fig. 6-16*b* the outputs of the upper two OR functions can be seen going to the lower AND function, using the older system.

TABLE 6-2   General Qualifying Symbols

SYMBOL	DESCRIPTION
&	AND gate or function
>1	OR gate or function
$=1$	Exclusive OR gate or function
1	The one input must be active
π	Multiplier

TABLE 6-3   Qualifying Symbols for Inputs and Outputs

—◁	External 0 produces internal 1
▷—	Internal 1 produces external 0
⟶	Active low input; equivalent to —◁ in positive logic
⟶▷	Dynamic input; controlled by clock

TABLE 6-4   Symbols Inside the Outline

D	Dynamic input (clocking)
—⊢ EN	Enable input
J,K,R,S,T	Usual meanings associated with flip flops (R = reset, T = toggle, etc.)
—⊦- m	Shift right inputs, m-1,2, etc.
A	Address dependence; in 1-state it permits action
M	Mode dependence

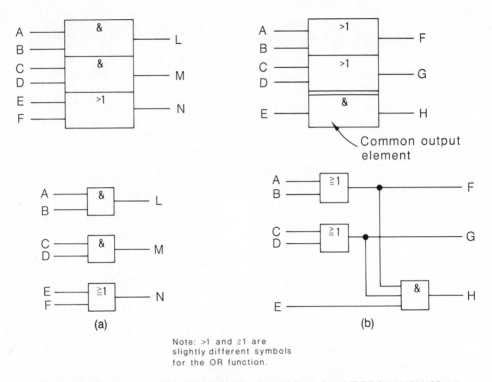

Note: >1 and ≥1 are
slightly different symbols
for the OR function.

FIG. 6-16 **Symbols used for digital logic circuits in revised IEEE 91 ANSI Y32.14.**

While the two lower diagrams in Fig. 6-16 may be about as simple to draw as the upper ones, and easier to read and follow, there is a major reason for the development of the IEC symbols. Many digital circuits now contain hundreds and more functions which cannot be drawn in the more traditional diagram form because of space limitations. Also, certain types of circuit, such as shift registers, coders, and latches, require the addition of more information. The example shown in Fig. 6-17 may help to demonstrate the value of this powerful tool.

Figure 6-17*a* shows the relationship of a common control block symbol. Here we see that the two lower blocks receive the output of the CC block. Figure 6-17*b* makes use of what is called M (mode) dependency. Inputs *b* and *c* control one of four modes (0 through3) that will exist at any one time. Inputs *d, e,* and *f* are D inputs subject to dynamic control (clocking) by the *a* input. The numbers *1* and *2* have been indicated at inputs *d, e,* and *f* so that input *d* is enabled only in mode *2* (shifting down and serial loading) and *e* and *f* are enabled only in mode *1* (parallel loading). The number *4* (at *a, d, e,* and *f*) is an arbitrarily chosen identifying number that labels each input that is affected by the affecting input, which is *a* in this case. Refer to Table 6-4.

Input *a* has three functions. It is the clock for entering data. In mode *2* it causes right-shifting of data, and in mode *3* it causes the contents of a register to

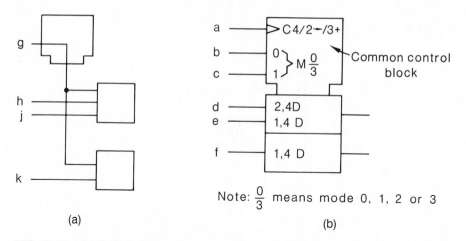

Note: $\frac{0}{3}$ means mode 0, 1, 2 or 3

(a)   (b)

FIG. 6-17 *(a)* Relationship of a control block to other logic elements. *(b)* Logic symbols using M (mode) dependence which affects inputs. Mode 0 (b=0, c=0). Mode 1 (b=1, c=0). Mode 2 (b=0, c=1). Mode 3 (b=c=1).

be incremented by one. This is what the arrow after the *2* and the plus sign after the *3* signify.

Dependence notation is used mostly by persons working with digital design. As can be seen in the situation described above, it can become very complicated. It is beyond the scope of this book to explain the system in more detail. However the reader should be aware that it exists and that he or she might be exposed to it in the future. More detailed information may be found in IEEE/ANSI Y32.14.

SUMMARY

Flow (or block) diagrams are used for different purposes in different situations. They may be used for depicting an electric circuit or system, for the preliminary design work in computers and other electrical installations, and for programming problems to be solved by computers. A left-to-right direction is sought in planning most block diagrams because this is the normal way in which people read. However, this sequence cannot always be followed. In most block diagrams there is a certain amount of lettering within the blocks. This lettering may determine what the sizes of the blocks will be. Liberal use of arrowheads is made in block-diagram construction. Although the rectangle is widely used in flow and block diagrams, other shapes—and sometimes electrical symbols–are used as dictated by standard practice. Blocks and flow lines should be evenly spaced if a pleasing drawing is desired.

In recent years, the use of distinctive shapes for logic diagrams has become popular. These shapes facilitate the reading of diagrams for complex systems.

QUESTIONS

6-1. Where are auxiliary circuits, such as feedback, usually placed with regard to the main signal path of a flow diagram?

6-2. What determines how large the rectangles in a block diagram should be? Show by means of a freehand sketch.

6-3. What standard covers the preparation of block or flow diagrams?

6-4. In making a block diagram of an electrical system, what shapes would you use for the blocks? Why?

6-5. What direction of flow would you attempt to show in planning a block diagram of an electronic circuit?

6-6. What are three devices that may be shown by means of standard symbols, rather than by blocks, according to customary practice?

6-7. What features make the block diagram for a microcomputer distinctive?

6-8. When might a dashed line be used instead of a solid line in a diagram?

6-9. Show by means of a freehand sketch how you would lay out a flow diagram with, say, 10 blocks, too many to fit on a single line on an $8\frac{1}{2} \times 11$ sheet.

6-10. Define the word *stage*.

6-11. Sketch six different shapes that may be used at one time or another in logic diagrams. Label each shape.

6-12. What are five different functions that might be shown with the rectangle in logic diagrams?

6-13. With negative logic (circles) used at all outputs, what does a NAND circuit perform? What does a NOR circuit perform?

6-14. What is the difference between an Exclusive OR circuit and an OR circuit? (Show by means of truth tables.)

6-15. Add the missing lines of the partial truth table of Fig. 6-13b for A and B only, in the 1-state; A and D only, in the 1-state.

6-16. What are the advantages of using distinctive shapes, rather than uniform shapes, in a logic diagram? What is one disadvantage?

6-17. What are the similarities between a *decision table* and a *truth table*? The differences?

PROBLEMS

6-1. Make a block (flow) diagram for an AM radio receiver as follows:
  a. External antenna
  b. RF amplifier

c. Mixer
d. IF amplifier
e. Detector
f. AF amplifier
g. Speaker
h. Oscillator (feeds into mixer)

Show an AGC feedback from the detector to the RF amplifier and IF amplifiers. Make neat uppercase lettering in the blocks using appropriate abbreviations. Use $8\frac{1}{2} \times 11$ paper (tight fit) or 11×17 or 12×18 paper.

6-2.   Make a simplified block diagram for a digital watch as follows. Use a left-to-right sequence for the first four blocks, as follows: (1) quartz crystal, (2) oscillator, (3) frequency divider, (4) wave shaper, (5) battery (with lines flowing to oscillator, frequency divider, and wave shaper), (6) decoder (with lines flowing from battery and wave shaper), and (7) digital readout (line flowing from decoder). Make neat uppercase lettering in the blocks, using appropriate abbreviations.

6-3.   Make a block diagram for the following (TR 9-10 BC-SW) radio receiver (follow the instructions given in Prob. 6-1):
   a. External antenna
   b. Mixer
   c. First IF amplifier
   d. Second IF amplifier
   e. AF amplifier
   f. Driver
   g. Output stage
   h. Earphone or speaker jack
   i. Oscillator to feed into the mixer; *and* AGC feedback around the IF stages

Use 11×17 or 12×18 paper unless drawn as two lines, in which case it might fit on $8\frac{1}{2} \times 11$ paper.

6-4.   Make a block diagram for the following FM radio receiver (see instructions given in Prob. 6-1):
   a. Antenna
   b. FM RF amplifier
   c. FM converter
   d. First FM IF amplifier
   e. Second FM IF amplifier
   f. Third FM IF amplifier
   g. AF amplifier
   h. Output stage
   i. Jack

Use 11×17 or 12×18 paper.

6-5.   Figure 6-18 shows the partially completed flow diagram for the control system for Solar One, the central receiver electric generating plant in California.

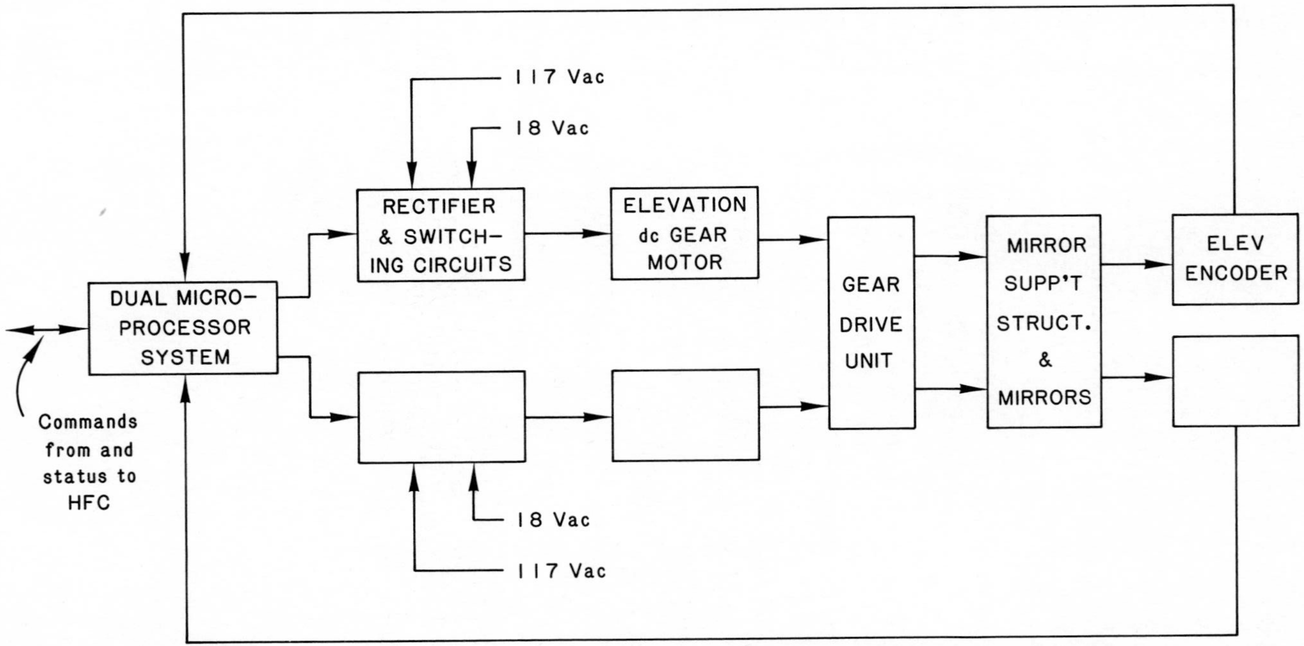
117 Vac
18 Vac
RECTIFIER & SWITCH-ING CIRCUITS
ELEVATION dc GEAR MOTOR
DUAL MICRO-PROCESSOR SYSTEM
GEAR DRIVE UNIT
MIRROR SUPP'T STRUCT. & MIRRORS
ELEV ENCODER
Commands from and status to HFC
18 Vac
117 Vac

FIG. 6-18 (Prob. 6-5.) Flow diagram for solar central receiver electric generating plant.

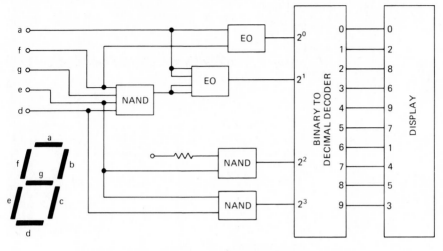

FIG. 6-19 (Prob. 6-6.) Code converter circuit.

Each heliostat reflector has a gearmotor for changing elevation and another for changing azimuth (the horizontal angle). Draw the complete diagram for this control system, showing rectifier, switching, gearmotor, and encoders for the azimuth movement below the corresponding blocks for elevation. Make neat uppercase lettering in the blocks using appropriate abbreviations. Use $8\frac{1}{2} \times 11$ (tight fit) or 11×17 or 12×18 paper. (The abbreviation HFC means heliostat field controller.)

6-6. Complete the drawing of the code converter in Fig. 6-19 by showing distinctive shapes for the NAND and Exclusive OR gates. Draw a border around the entire diagram, including the number 8. Dots are optional. Use $8\frac{1}{2} \times 11$ paper.

6-7. Construct a block diagram for a basic regulating system that has the

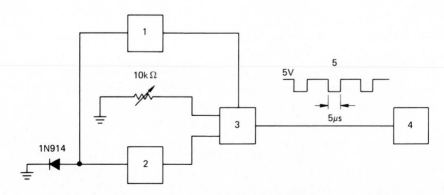

FIG. 6-20 (Prob. 6-8.) Automobile temperature-measuring circuit.

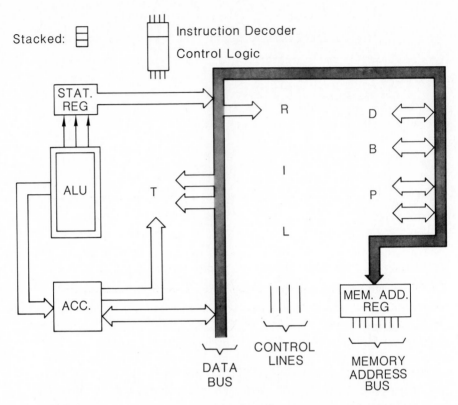

Stacked:

Instruction Decoder

Control Logic

STAT. REG

ALU

T

ACC.

R

I

L

D

B

P

MEM. ADD. REG

DATA BUS

CONTROL LINES

MEMORY ADDRESS BUS

FIG. 6-21 (Prob. 6-9.) Incomplete flow diagram of a microprocessor.

following steps or devices:
   a. Power source
   b. Regulated quantity
   c. Signal-sensing device
   d. Error-sensing device
   e. Reference
   f. Amplifier with feedback
   g. Regulator power source

Such a regulating system could be used for the speed control of a motor. In such a case, the motor would be the regulated quantity and a tachometer would be the signal-sensing device. Use $8\frac{1}{2} \times 11$ paper.

6-8. Make a block diagram of the automobile temperature-measuring and readout circuit shown in Fig. 6-20. Block 1 is the forward-feed compensator, 2 is the capacitor reset, 3 is the operational amplifier, and 4 is the TTL converter. Above the waveform at 5, label "1-Hz square wave." The diode is a temperature-sensing diode. Do not show digits 1 through 5.

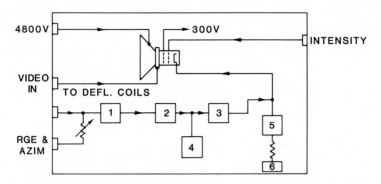

FIG. 6-22 (Prob. 6-10.) Indicator block diagram for airborne radar.

6-9. Figure 6-21 shows the incomplete flow diagram of a microprocessor. Redraw the diagram, and add the following elements where their location is indicated by capital letters:

T—storage registers, two stacked; show connected to the ALU with output arrows and to the accumulator and data bus with input arrows
R—instruction register below which and connected to it is I, the instruction decoder, below which and touching it is L, the control logic (see sketch at top)
D—D register
B—B register
P—Program counters, two stacked

Put the letters or appropriate abbreviation for each element in its block. Label the bus: "Internal data bus." This will fit on a $8\frac{1}{2} \times 11$ sheet.

6-10. Complete the flow diagram displayed in Fig. 6-22 by placing the following titles in the boxes indicated by the numbers: 1, video mixer; 2, video

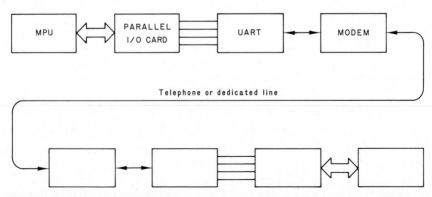

Telephone or dedicated line

FIG. 6-23 (Prob. 6-11.) Diagram showing use of modems for communicating over long distances. Modems (modulator-demodulators) send and receive digital signals by using different tone signals for logic "1"s and "0"s.

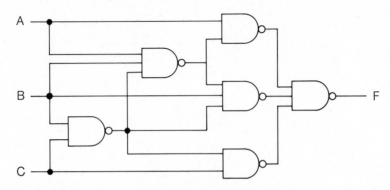

FIG. 6-24 (Prob. 6-12.) A NAND network in a gate array. (Texas Instruments, Inc.)

amplifier; 3, video amplifier; 4, DC restorer; and 5, DC restorer. At 6 replace the block with a ground symbol.

6-11. Refer to Fig. 6-23. The diagram shown is for a system that communicates digital signals over long distances. The UART is a *universal asynchronous receiver-transmitter* that converts parallel data to serial data which can be transmitted by the modem and vice versa. Draw the flow diagram as it is shown and complete the diagram by lettering the names or abbreviations of the elements in the lower blocks. These are the reverse of the ones above; i.e., the modem is at the left and the microprocessor (MPU) is at the right. Use $8\frac{1}{2} \times 11$ paper.

6-12. Refer to Fig. 6-24. The NAND gate is part of a gate-array package. Redraw this gate, using distinctive shape or rectangular symbols with appropri-

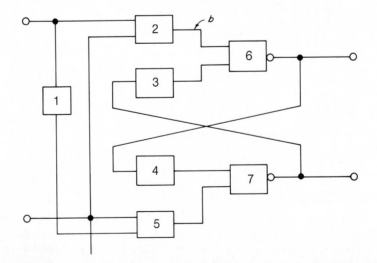

FIG. 6-25 (Prob. 6-13.) A data storage latch. (Texas Instruments, Inc.)

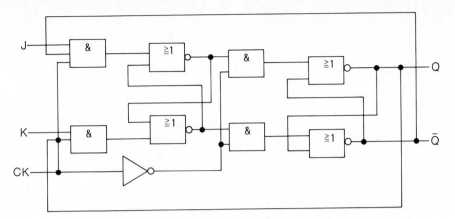

FIG. 6-26   (Prob. 6-14.) A J-K flip-flop circuit.

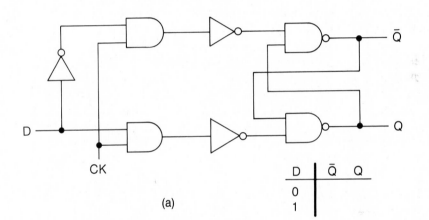

(a)

D	$\bar{Q}$	Q
0		
1		

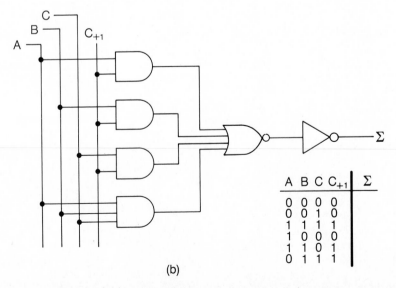

(b)

A	B	C	C_{+1}	Σ
0	0	0	0	0
0	0	1	0	
1	1	1	1	
1	0	0	0	
1	1	0	1	
0	1	1	1	

FIG. 6-27   (Prob. 6-15.) Logic diagrams for *(a)* a flip flop and *(b)* an adder circuit.

ate abbreviations. On the same $8\frac{1}{2} \times 11$ sheet construct a truth table if so requested by your instructor.

6-13. Refer to Fig. 6-25. This logic diagram of a storage latch contains an inverter at 1; AND functions at 2, 3, 4, and 5; and NOR functions at 6 and 7. Draw the logic diagram, using rectangular shapes with suitable notation or distinctive shape symbols, whichever your instructor indicates. The diagram would look neater if symbol 6 were raised so that line b would be a single straight line. Use $8\frac{1}{2} \times 11$ paper.

6-14. Refer to Fig. 6-26. The diagram of the J-K flip-flop circuit consists of rectangular symbols. Redraw this circuit on $8\frac{1}{2} \times 11$ paper, using distinctive shapes for the logic functions.

6-15. Refer to Fig. 6-27. Complete the truth tables for the flip-flop and adder circuits shown. Assume that the clock (CK) circuit is "on" in part a. Show the output at Σ for the six cases of inputs listed for A, B, C, and C_{+1} in part b. Display your results neatly on whatever paper your instructor desires. Draw the logic diagrams if your instructor requests you to do so.

7

The Schematic Diagram

The *schematic diagram* is often the primary drawing of the electronics and communications industry. It is a diagram that shows the functions and relationships of a circuit by means of graphical symbols. It does not show the physical layout of those components, however.

The schematic diagram usually includes electronic and certain passive components. A similar type diagram, involving electromechanical devices, is called the *elementary diagram.* It is discussed in Chapters 9 and 10.

The schematic diagram makes it possible for a person with an electronics background to trace and understand a circuit with comparative ease. For this reason it is used for design and analysis of circuits, instructional purposes, and maintenance troubleshooting of circuits.

7-1 Examples of Transistors in Circuit Drawings

In order to give the student a "feel" for the layout of schematic diagrams, we shall present several figures showing popular formats now in use. Then the problem of laying out such a diagram will be discussed. (Without much background in electronics it may not be possible to understand all the authors' comments about the following circuits, but it should be possible to visualize the different drawing patterns that are shown.)

Figure 7-1 shows the three methods of connecting a bipolar transistor in a circuit: *common base* (or grounded base), *common collector,* and *common emitter.* In the common-base circuit (Fig. 7-1*a*), for example, the signal (shown by the ∼ in the circle, which indicates an ac input) is introduced into the emitter-base circuit and extracted from the collector-base circuit. The base is thus common to both the input and output circuits. The direction of the arrows shows the electron flow. The voltage or power gain may be in the order of 1500, and the phase of the signal is not changed.

In the common-collector arrangement (Fig. 7-1*b*) the signal is introduced into the base-collector circuit and extracted from the emitter-collector circuit. The power gain is lower than in the other two configurations, and there is no phase reversal. This arrangement is used primarily as an impedance-matching device.

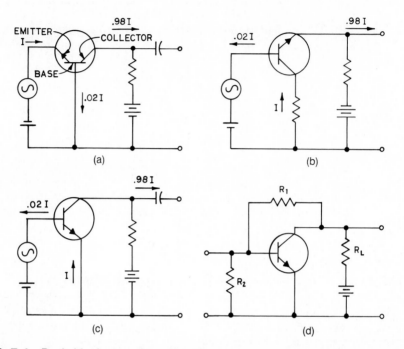

FIG. 7-1 Basic bipolar transistor circuits: *(a)* common base; *(b)* common collector; *(c)* common emitter; *(d)* bias network for common emitter circuit.

The common-emitter circuit (Fig. 7-1c) can provide power gains of 10,000. The input signal is introduced to the base-emitter circuit, and the output is taken from the collector-emitter circuit. However, this arrangement is the most widely used when more than one stage is required.

Figure 7-1d shows a popular biasing arrangement for a common-emitter circuit, which does away with one of the batteries used previously. (Bias is the difference in potential between, say, the collector and the base.) A voltage-divider network composed of R_1 and R_2 provides the required forward bias across the base-emitter junction.

In the examples shown, NPN transistors have been used; PNP transistors could be used instead, in which case the battery polarities and the direction of the arrow on the emitter should be reversed.

7-2 The Basic Amplifier

Three major functions of transistors and electron tubes are amplification, oscillation, and switching.

Figure 7-2a shows a common emitter amplifier circuit using a PNP transistor. Such amplifiers are used in audio circuits. The input is applied across R_1,* and the output is taken between collector and ground. Figure 7-2b shows a

---

* In the figure, R_1 is shown as R1. In many schematic drawings the letter and number are the same height.

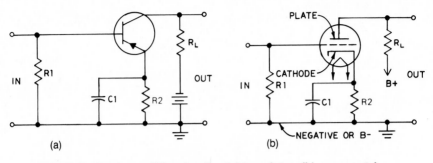

FIG. 7-2 Basic amplifier circuits: *(a)* transistor; *(b)* vacuum tube.

similar amplifier using a triode electron vacuum tube. Again, the input is applied across R_1 and the output is taken between the plate and ground.

7-3 Interstage Coupling

In the main, there are three methods by which two or more stages (amplifier stages, for example) may be hooked together. These three methods are (1) *RC* (resistance-capacitance), (2) transformer, and (3) DC (direct coupling). Each has its advantages and disadvantages. Figure 7-3 shows examples of these methods.

Figure 7-3a shows part of a resistance-capacitance-coupled amplifier. Capacitor C_1 is called the *coupling capacitor.* This method is widely used because of its low cost and the large range of frequencies that can be handled.

Figure 7-3b shows a transformer-coupled network in which one side can be "tuned." (Note that in Fig. 7-3b and c the transistor symbols have not been enclosed by an envelope circle. The reader may recall that the circle is optional. The authors have omitted the circle from these two circuit drawings on purpose to remind the reader that it is not always used. The authors prefer the circle, however, and use it in most examples in the book.) Transformer coupling may also be employed in which both sides or neither side is tuned. Tuning makes possible frequency selection.

Figure 7-3c shows a direct-coupled circuit, which regulates the circuit so that the output voltage is maintained nearly constant. Typical applications include output stages of series-type and shunt-type regulating circuits, chopper circuits, and differential and pulse amplifiers. Obviously, the cost is low for this type of coupling, and it is, therefore, widely used.

7-4 Patterns for Transistor Circuits

The nature of the circuitry, including coupling circuits, often indicates a pattern to the engineer or drafter who is making the final sketch or layout drawing of an electronics circuit. The drawing should be planned to allow the pattern to develop in as auspicious a manner as possible.

Figure 7-4a shows a flasher circuit for emergency vehicles, barricades, boats, and aircraft. It will be noted how the transistors are aligned on a horizon-

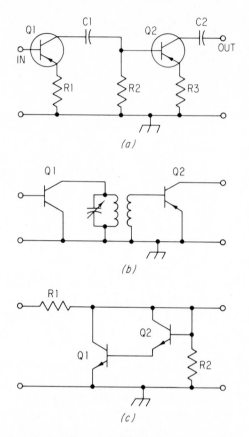

FIG. 7-3 Methods of coupling amplifier stages together: *(a)* RC coupling; *(b)* transformer coupling; *(c)* direct coupling.

tal line (invisible), as are the flasher lamps and some other components of the circuit. This horizontal transistor alignment is possible in many circuits, including radio and television.

Figure 7-4*b* shows an incline pattern which develops because of the direct coupling of the transistors. The second and third stages are biased by the preceding stages.

Figure 7-4*c* shows a staggered pattern which develops when, for instance, PNP and NPN transistors are used in a DC arrangement.

A vertical arrangement of transistors is shown in Fig. 7-5. In a push-pull circuit, such as that shown in Fig. 7-5*a*, each transistor amplifies half the signal, and these half-signals are then combined in the output (collector) circuit to restore the original waveform in an amplified state. This circuit requires transformer coupling, while that of Fig. 7-5*b* does not. In this circuit, showing electron current flow by means of arrows, essentially no dc flows through resistor R_L. Therefore the voice coil of a speaker can be connected in place of R_L without excessive speaker-cone distortion.

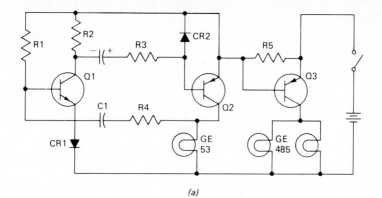

(a)

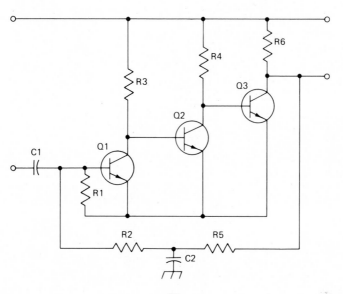

(b)

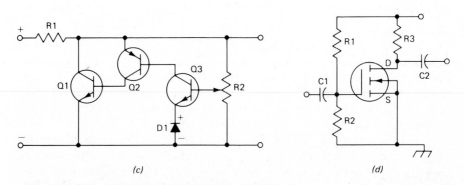

(c)   (d)

FIG. 7-4 Different arrangement patterns of transistors in circuits: *(a)* horizontal alignment as in this flasher circuit; *(b)* incline as in this DC amplifier circuit; *(c)* staggered as in this shunt regulator circuit; *(d)* circuit with MOSFET transistor.

233

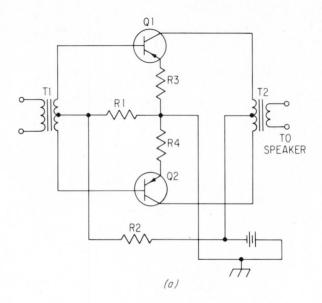

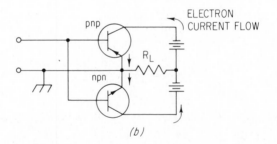

(b)

FIG. 7-5 Vertical alignment pattern of transistors: *(a)* a push-pull amplifier; *(b)* basic complimentary-symmetry circuit.

7-5 Examples of Computer Circuits

Figure 7-6*a* shows a saturated flip-flop circuit having one input, called a *trigger,* at *T* and outputs at A and B. A flip-flop is a memory, or storage, device that may have one of two states, ON or OFF. It always has two complementary level outputs and may have one, two, or three inputs.

A common feature of flip-flop circuit drawings is the crossing, angled, signal-path lines near the center of the circuit. In most other electrical drawings, the signal paths are drawn as horizontal or vertical lines. It would be possible to draw a flip-flop in this manner, too, but most companies now use the angled lines.

Another type of flip-flop is the set-reset, which is shown by means of a block diagram in Fig. 7-6*c*. In this circuit, a pulse at S (set) input causes the

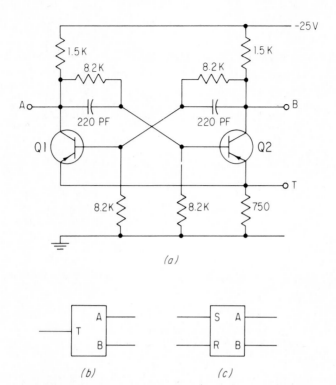

FIG. 7-6 Horizontal alignment of transistor in flip-flop circuits: *(a)* schematic diagram of T type; *(b)* block diagram of T type; *(c)* RS type.

flip-flop to turn on or stay on, depending on the original state. An input at R (reset) causes the circuit to turn off or stay off.

Figure 7-7 shows several typical logic circuits that are manufactured as integrated circuits (ICs). There is only one type of logic, namely Boolean. However, today it is common to hear the component configuration referred to as a type of logic, i.e., transistor-transistor logic (TTL), CMOS logic, etc. These circuits are found in IC manufacturers' catalogs; however, because of the steadily increasing circuit complexity, there is a trend to show only the logic symbols for ICs.

7-6 Reference Designations

Symbols of all replaceable parts should be referenced. A reference may be placed above, below, or on either side of its part. Practice recommended by ANSI includes giving each part a number, such as resistor R_5, and its capacity, such as 100 Ω. Thus we have two lines of designations (combinations of letters and numbers) for most elements in a circuit. Table 7-1 gives letters and examples that are typical of standard practice. Figures 7-10 and 7-15 are good examples.

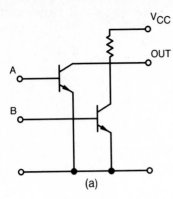

(a)

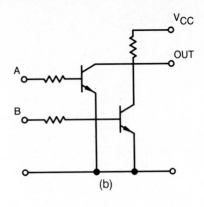

(b)

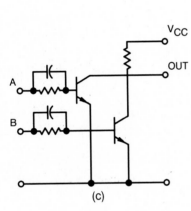

(c)

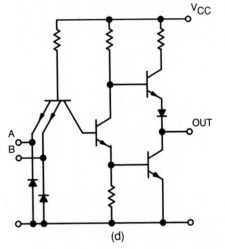

(d)

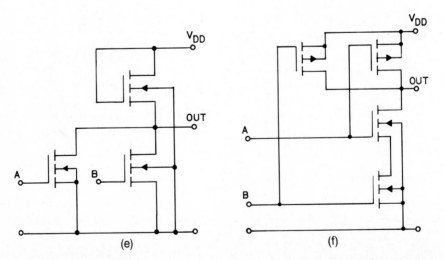

(e)

(f)

FIG. 7-7 Digital ICs: *(a)* direct-coupled transistor logic (DCTL) (archaic) two-input NOR gate; *(b)* resistor-to-transistor logic (RTC) (archaic) two-input NOR gate;

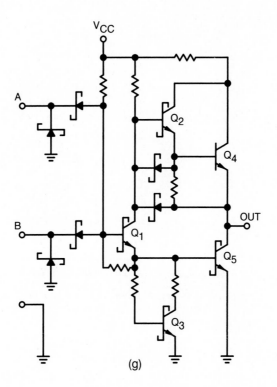

(g)

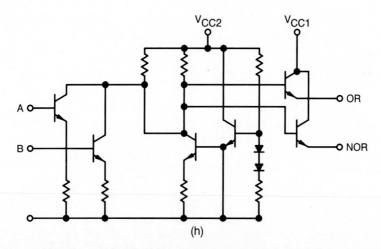

(h)

(c) resistor-capacitor-transist logic (RCTC) (archaic) two-input NOR gate; *(d)* transistor-to-transistor logic (TTL) two-input NAND gate; *(e)* negative channel metal-oxide semiconductor (NMOS) two-input NOR gate; *(f)* complimentary metal-oxide semiconductor (CMOS) two-input NAND gate; *(g)* Schottky TTL (STTL) two-input NAND gate; *(h)* electron-coupled logic (ECL) two-input NOR/OR gate with vertical bias network.

TABLE 7-1

ELEMENT	LETTER	EXAMPLE
Capacitor	C	C5
		10 pF*
Inductor	L	L1
		23 mH*
Rectifier (metallic or crystal)	D or CR	D2 or CR2
Resistor	R	R201
		270
Transformer	T	T2
Transistor	Q	Q5
		2N482
		DETECTOR
Tube	V	V3
		6AU6
		1ST IF AMP

* The abbreviation mH (often MH in drawings) stands for millihenry, a thousandth of a henry. The abbreviation pF (often PF or UUF in drawings) stands for picofarad or micromicrofarad (a millionth of a millionth of a farad).

The letter X is sometimes used for transistors, although Q is recommended by both the Military and American National Standards. The symbol R_{201} does not mean the 201st resistor in the circuit. It means that it is the first resistor in the second (200 series) subassembly. Both the transistor and the tube contain a third line — their function. This is optional information insofar as the drawing is concerned.*

7-7 Laying Out a Schematic Diagram

Figure 7-8 shows a freehand sketch of a light-flasher circuit. This is the type of sketch that might be made by a design engineer or project chief. The purpose of the sketch is to furnish all the information necessary to make a finished drawing of this circuit and to construct it. If the sketch is very rough and there are many places where improvement is indicated, it may be desirable to make a new sketch.

A drafter should produce a well-balanced work that is pleasing to the eye. In order to do this, it may be necessary to change the configuration of the drawing (reorient some symbols and change some lines), but the circuit must still maintain its original technical significance. One must also strive for simplicity and clarity.

A glance at Fig. 7-8 reveals that, among other things, the transistors could be lined up better, it is a little crowded in the shaded region, and the sketcher has

* In a schematic drawing each line of letters or numbers is centered, that is, 270 is centered under R201 and 2N482 and DETECTOR are centered under Q5.

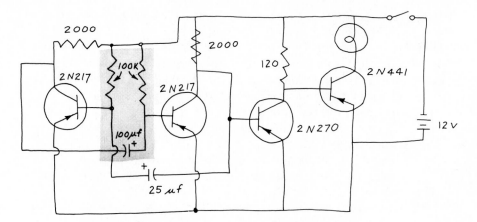

FIG. 7-8 Freehand sketch of a light-flasher circuit.

used loops for crossovers—a nonstandard procedure. Also, the components do not have reference numbers.

Figure 7-9 shows how one can go about correcting some of these deficiencies and starting the layout of the diagram. Figure 7-9*a* shows how the spacing between vertical lines in the crowded area can be determined by lettering the

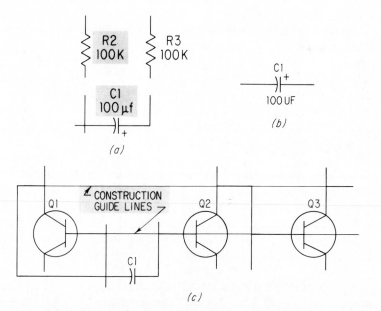

FIG. 7-9 Details in laying out a schematic diagram: *(a)* allowing space for reference designations (identification); *(b)* alternative arrangement for referencing capacitor; *(c)* using horizontal guidelines for transistors and other line work.

appropriate reference designations, possibly on a trial basis on a sheet of scratch paper, or on the new sketch if such is being made.

It is obvious that three, and probably all four, transistors can be aligned on one horizontal line to make for a more pleasing and professional-appearing drawing. Therefore, a light construction centerline can be drawn horizontally at about midheight through the drawing area. At this time, or possibly later, after more components have been drawn, other horizontal guidelines can be drawn in. One other such line is shown in Fig. 7-9c for the collector outputs.

The transistor envelope circles can be drawn in at this time if the spacing between them can conveniently be determined. Figure 7-9a has provided the spacing around C_1,* and, by allowing a little more spacing on each side of the vertical lines (on which R_2 and R_3 are located), we can locate Q_1 and Q_2 without much fear of having to redraw the diagram. Because there are no components between the other transistors, we can locate them at this time without much difficulty. [Note that this is DC (direct-coupled) circuitry. If it were transformer-coupled, much more work and planning would be necessary.]

After putting in the bases, collectors, and emitters, one can draw more lines, such as appear in Fig. 7-9c, and be well on the way toward completing the circuit diagram. One logical step to perform next would be to draw resistor R_4 or R_1 (reoriented from the sketch).[1] Then all five resistors could be located at the same level, and the uppermost horizontal line could be located.

Figure 7-10 shows the finished schematic diagram of the flasher circuit (ANSI referencing has been used). Transistor identification has been placed close to each envelope with larger-than-average letters. The authors have used this arrangement (putting the designations above and in line with the individual transistors) rather than putting the designations at the top of the diagram, which is also approved practice. Note that the entire drawing appears to be symmetrical and that a fairly uniform density is apparent. There are often notes of a general or specific nature at the bottom of an elementary diagram, and this one is no exception. Instead of locating transistor Q_4 as shown, we might have drawn a horizontal line from the output of Q_3 to Q_4 and thus would have placed Q_4 "above" the other transistors.

Another popular way to lay out a schematic diagram is to use standard grids. A preliminary sketch can be made on sketching paper having grids, or a grid sheet can be placed under the tracing paper or film on which the drawing is to be made. Figure 7-11 shows a layout of the flip-flop circuit on a $\frac{1}{4}$-in. (6.4-mm) grid. The grid lines have been used not only as guides for the circuit paths but also as guidelines for making the resistor symbols. The $\frac{1}{4}$-in. (6.4-mm) resistors and 1-in. (25.4-mm) transistor envelopes were larger than those that are drawn on most diagrams before they were reduced for printing.

Figure 7-12 shows a null-detector circuit made on a $\frac{1}{10}$-in. (2.54-mm) grid with heavy lines 1 in. (25.4 mm) apart. All signal paths have been drawn right

---

* We will use numbers and capital letters of the same height in the final schematic drawing.

[1] The reader will have to look ahead to Fig. 7-10 to see which resistors have been assigned designations R_1 and R_4.

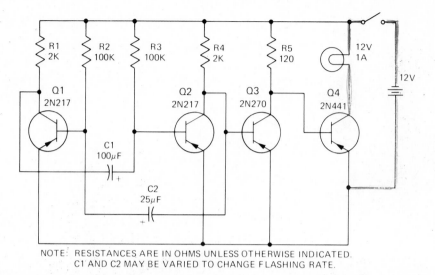

NOTE: RESISTANCES ARE IN OHMS UNLESS OTHERWISE INDICATED.
C1 AND C2 MAY BE VARIED TO CHANGE FLASHING RATE.

FIG. 7-10 **The completed schematic diagram of a flasher circuit.** (From a circuit in the RCA Transistor Manual.)

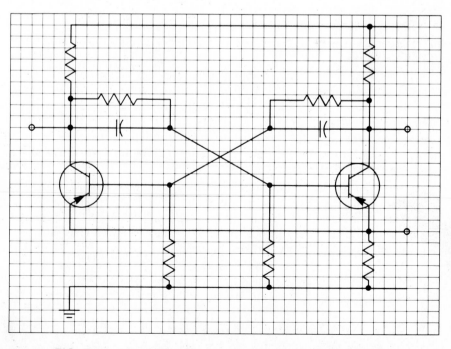

FIG. 7-11 Layout of a flip-flop circuit on a $\frac{1}{4}$-in. (6.4-mm) grid.

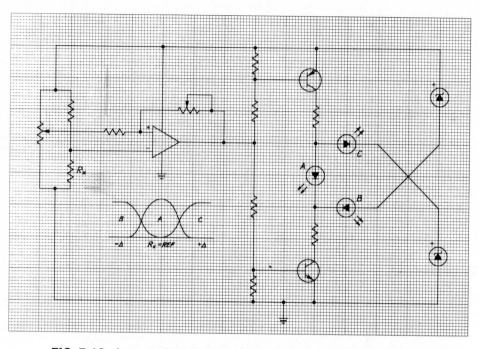

FIG. 7-12   Layout of a null detector circuit on a $\frac{1}{10}$-in. (2.54-mm) grid.

over grid lines, most on the heavy lines. Small grid lines have been used as guidelines for the resistors and to establish the LED envelopes at 0.6-in. (15-mm) diameter. A graph has been added to show that if a resistor is matched against a reference, a diode A will light up when the resistor R_x is exactly or closely matched. If the resistor is less than the reference, diode B comes on, etc. The two transistor envelopes are 0.75 in. (19 mm) in diameter, exactly the size shown in ANSI/IEEE Y32E 1976.[1] Although most signal paths are drawn horizontally or vertically, in Figs. 7-11 and 7-12 there are lines drawn at other angles. These are two types of circuit, along with multivibrators and bridges, in which it is customary to show certain lines in this manner. In most drawings the paths are horizontal or vertical.

Drawings for manuals and journals are made in ink and often are produced using Leroy equipment.[2] Working from a rough sketch, the drafter lays out a neat pencil drawing with or without a grid, as above. The final drawing is made on Mylar or similar film, which is placed over the pencil drawing. Then the symbols are made with the template and scriber as shown in Fig. 7-13a. The next step is to draw the signal paths between the symbols with a technical fountain pen such as the Rapidograph or Staedtler-Mars "700," as in Fig. 7-13b. Lettering is then put on the drawing. This is done with a Leroy lettering

[1] This is somewhat larger than in the previous edition of the standard.

[2] Leroy is a registered trademark of the Keuffel & Esser Company.

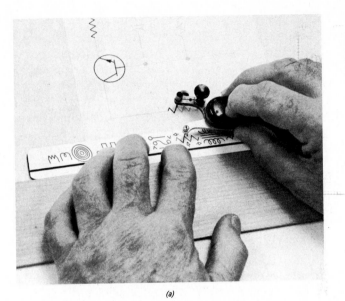

(a)

(b)

FIG. 7-13 The use of mechanical devices (scribes and pens) to lay out a schematic diagram: *(a)* using a Leroy scriber and template to make symbols; *(b)* using a technical pen to draw signal paths between symbols.

template and scriber, or it is accomplished by using an IBM typewriter and cutting out small rectangles of typed lettering which can be pasted on the drawing at the desired locations. Leroy electrical templates are available with three sizes of symbols and thus provide the layout person with considerable flexibility. Both Leroy lettering templates and typed-lettering "stick-ons" also have much flexibility in the size of letters. Typed letters provide a little more uniformity and a cleaner appearance and save a little time if there is considerable lettering on a drawing. (See also Fig. 1-21.)

Still another approach is to use preprinted device symbols in the form of appliqués or stick-ons. Signal paths are made with pressure-sensitive tapes, pencil, or technical fountain pen. As in the method described immediately above, the detailer probably works from a rough sketch. A neat pencil layout of the diagram is then made either on a grid or on plain tracing paper. Symbols may be sketched in, omitted, or partially drawn on this layout if the final layout will be made on another sheet (overlay), which will be placed over the pencil layout.

In the location indicated on the overlay sheet, each pressure-sensitive symbol is oriented in the precise position by permanently and firmly affixing it to the surface of the artwork. A smaller symbol can be separated from the backing material by inserting a knife (with an X-acto No. 11 or No. 16 blade) in between. The symbol is carried to the desired diagram location on the knife, which is also used as a holding and positioning tool. After the symbols are in place, signal paths are put on the diagram. If inked lines are to be used, they can be drawn as shown in Fig. 7-13b with a technical or a ruling pen. Pencil lines will not look as good as ink lines because the printed stick-on symbols will be darker. Lettering is usually put on the drawing at the last. Figure 7-14c shows a sample sheet of pressure-sensitive symbols.

Many students will not have access to appliqués and Leroy equipment. They will probably have to make all lines and symbols with a pencil. The following fundamentals of layout should be observed:

1. Use medium-weight lines for all symbols and signal paths. When emphasis on a component is desired, use a heavier line.

2. Arrange the diagram so that it reads functionally from left to right, if possible.

3. Strive for uniform density so that there are no congested areas and a minimum of large white areas.

4. Draw signal paths vertically or horizontally, except in special cases such as flip-flop circuits.

5. Keep turns or bends to a minimum. Route all connecting lines as directly as possible.

6. Keep the device symbols in proportion to each other. Use only one size of symbol for each device, if possible.

7. Allow adequate space near each symbol for proper identification.

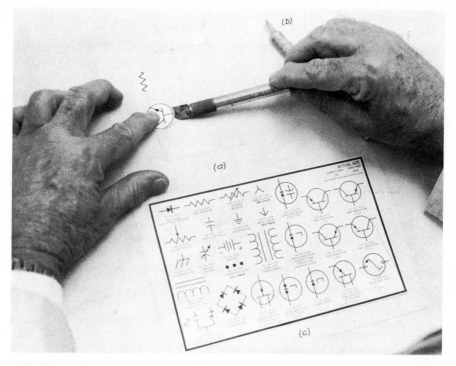

FIG. 7-14 The use of appliqués for construction of schematics: *(a)* applying a symbol; *(b)* a knife with X-acto No. 16 blade; *(c)* typical symbols that are available as pressure-sensitive adhesives. (Bishop Graphics, Inc.)

8. Make integrated-circuit and microprocessor outlines large enough for pertinent lettering to be placed within the outline.

9. Center each symbol along its lead line (signal path).

10. Place the complete identification (and values if necessary) beside or within each symbol, whichever is appropriate. Identify each input and output.

11. Check the final diagram for the correctness of symbols, electrical accuracy, and the presence or absence of connecting dots.

Some aids that are not very expensive might facilitate the drawing of schematic diagrams for students. Grid paper, with or without drop-out lines, is one such aid. A template that makes different sizes of circles or that includes electronic device symbols is another aid. The use of templates is discussed in Chap. 1.

7-8 Line Arrangements According to Voltages

Although the circuit shown in Figs. 7-8 to 7-10 does not become involved in this problem, some circuit diagrams will have lines representing several different voltages. A good practice is to draw the schematic diagrams so that the highest voltage is on the uppermost line, the next highest is on the line below, and, finally, the most negative voltage is on the lowest line. This makes for easier reading of the elementary diagram in many cases. Typical voltage sequences might be as follows:

$$
\begin{array}{rl}
235 & V \\
150 & V \\
6.3 & V \\
0 & V \\
-45 & V
\end{array}
$$

The top line is what is known as the $B+$ line; the zero line is the $B-$, or common, line and may be joined to the ground or chassis. Transistor circuits, as a rule, do not have as many lines as this, but tube circuits may have this many, depending on the circuit and types of tube.

7-9 Examples of Schematic Diagrams

Figure 7-15 shows the complete schematic diagram of a transistor radio. This particular circuit has a separate mixer and oscillator, two IF stages, a detector and audio amplifier, and a class B push-pull output. The transformer coils are drawn as a series of complete loops. This loop, or helix, is one of two symbols that are approved in the standard. Most drawings now have the simpler symbol, as shown in the power-supply transformer (Fig. 7-16). The authors prefer the simpler transformer symbol because it does not clutter up the drawing as much as the old symbol did and is easier to draw. In Fig. 7-15, L_1 has a metallic core and L_2 has an air core. The authors would be tempted to call L_1 and L_2 (transformers) T_1 and T_2.

Figure 7-16 shows the schematic diagram of a video distribution amplifier that takes the output of a video (TV) camera and provides four independent outputs. The two integrated circuits are an operational amplifier and a current driver in a feedback loop. Inputs and outputs are for coaxial cables. The power supply is placed below the distribution amplifier circuit. This is common practice, although the power circuit could be drawn jointly with the other by joining the 6.8-V lines and (probably) rearranging the whole drawing somewhat. Diodes D_3 and D_4 are zener diodes used for regulating. A nonstandard symbol in the original drawing, shown at b, was replaced by the standard connectors at the left of the power supply.

Figure 7-17 shows the schematic diagrams of two versions of scientific calculators. The calculators are manufactured by the same manufacturer, and the new version (Fig. 7-17b) is a replacement for the old version (Fig. 7-17a).

There are some very interesting differences in comparing the two versions and their associated schematics. For example, the old version has six ICs, while

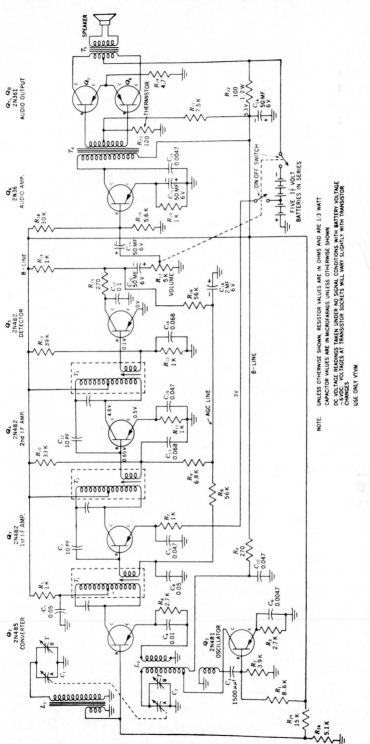

FIG. 7-15 Schematic diagram of a transistor radio receiver. (From Milton S. Kiver, *Transistors*, 3d ed., McGraw-Hill Book Company, New York, 1962.)

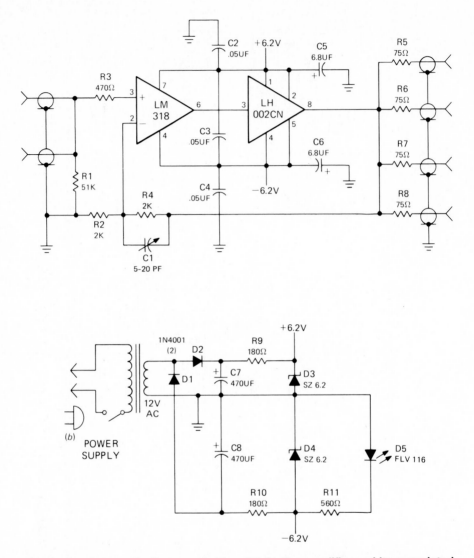

FIG. 7-16 **Schematic diagram of video distribution amplifier and its associated power supply.** (From *Electronics* magazine, October 13, 1977, p. 121; Copyright © 1977 by McGraw-Hill, Inc.)

the new version has combined many functions and requires only two ICs. The old version has a schematic of the keyboard and a partially functioned LED display module, while the new version has a schematic symbol for both the keyboard and the LCD display module. Please note that the ICs in Fig. 7-17*a* contain an identifier beginning with *Z*, while the ones in Fig. 7-17*b* have an identifier beginning with *U*. The authors have seen several other identifiers used.

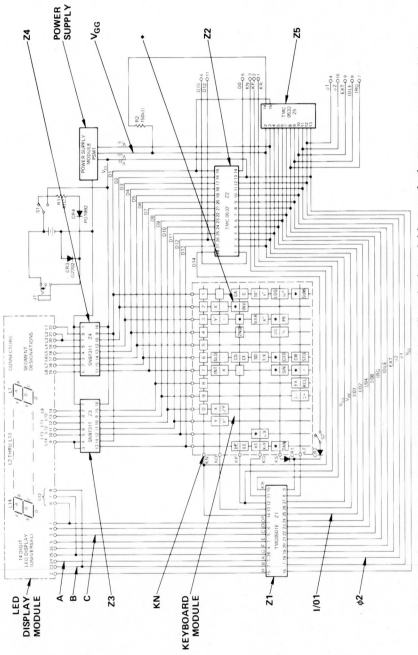

FIG. 7-17a Schematic diagram of 45 key scientific calculator (older version). (Texas Instruments, Inc.)

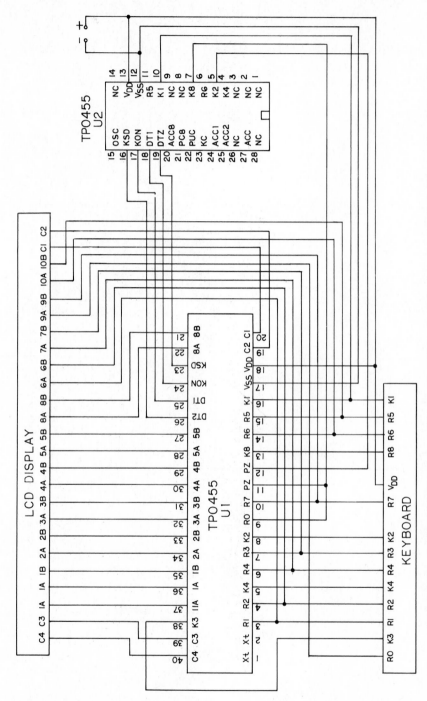

FIG. 7-17b Schematic diagram of 45 key scientific calculator (newer version). (Texas Instruments, Inc.)

These two calculator schematics have many components and circuits identified; please compare these schematics. Overall, the schematic diagram of the new version is simpler and less cluttered than the old version.

7-10 Tube Circuits

Figure 7-18 is the schematic diagram of an audio-amplifier circuit having electron tubes. This older drawing has several features worth mentioning. The tubes, except the rectifier 5Y3-GT, have heaters. The heater circuit XX is not

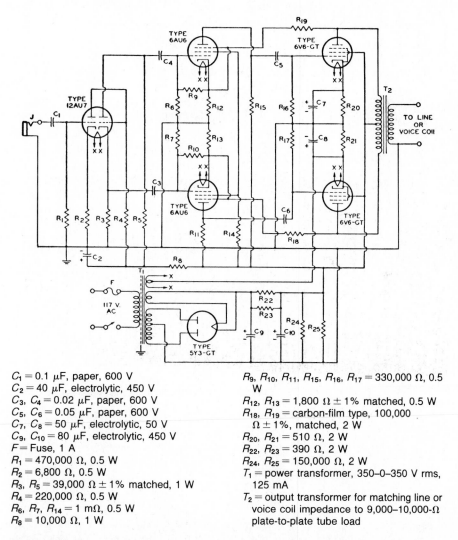

$C_1 = 0.1$ μF, paper, 600 V	R_9, R_{10}, R_{11}, R_{15}, R_{16}, $R_{17} = 330,000$ Ω, 0.5 W
$C_2 = 40$ μF, electrolytic, 450 V	R_{12}, $R_{13} = 1,800$ $\Omega \pm 1\%$ matched, 0.5 W
C_3, $C_4 = 0.02$ μF, paper, 600 V	R_{18}, $R_{19} =$ carbon-film type, 100,000
C_5, $C_6 = 0.05$ μF, paper, 600 V	$\Omega \pm 1\%$ matched, 2 W
C_7, $C_8 = 50$ μF, electrolytic, 50 V	R_{20}, $R_{21} = 510$ Ω, 2 W
C_9, $C_{10} = 80$ μF, electrolytic, 450 V	R_{22}, $R_{23} = 390$ Ω, 2 W
$F =$ Fuse, 1 A	R_{24}, $R_{25} = 150,000$ Ω, 2 W
$R_1 = 470,000$ Ω, 0.5 W	$T_1 =$ power transformer, 350–0–350 V rms,
$R_2 = 6,800$ Ω, 0.5 W	125 mA
R_3, $R_5 = 39,000$ $\Omega \pm 1\%$ matched, 1 W	$T_2 =$ output transformer for matching line or
$R_4 = 220,000$ Ω, 0.5 W	voice coil impedance to 9,000–10,000-Ω
R_6, R_7, $R_{14} = 1$ mΩ, 0.5 W	plate-to-plate tube load
$R_8 = 10,000$ Ω, 1 W	

FIG. 7-18 **Schematic diagram of a high-fidelity audio amplifier.** (From RCA Receiving Tube Manual.)

drawn in order to simplify the schematic. Loops (no connections) and dots are used. Finally, the values for each component are placed below the diagram. Such an arrangement may make for slower reading of the diagram but has the advantage of supplying more information about each part than can be conveniently shown with the other system. It should be stated that tubes are making a comeback in some audio applications.

7-11 Commonly Used Units for Components

In addition to units such as henrys and ohms, certain prefixes (multipliers) are used when components have extremely large or small values. The most often used prefixes are shown in Table 7-2. Suggested units with their prefixes appear in Table 7-3.

Some lettering can be saved by using notes such as the one shown in Fig. 7-10 about resistance values. A similar note might be: *All capacitance values are in microfarads unless otherwise shown.*

Tables 7-2 and 7-3 present what the authors believe to be current practice relative to prefix values for units. We believe this will continue for years to

TABLE 7-2

PREFIX	MULTIPLIER	SYMBOL (COMMON)	SYMBOL (SI PREFERRED)
Giga	10^9	G	G
Mega	10^6	MEG or M	M
Kilo	10^3	K or k	k
Milli	10^{-3}	m	m
Micro	10^{-6}	μ or u	μ
Nano	10^{-9}	n	n
Pico	10^{-12}	uu or p	p

TABLE 7-3

RANGE	UNITS TO BE USED	EXAMPLES
$\leq 999\ \Omega$	Ohms	520 or 210 Ω
$1000-99,999\ \Omega$	Ohms or kilohms	45 kΩ (or K) or 45,000
$100,000-999,999\ \Omega$	Kilohms or megohms	500 kΩ or 0.5 MΩ (or MEG)
$\geq 10^6\ \Omega$	Megohms	2.5 MΩ (or MEG)
$\leq 9,999$ pF ($\mu\mu$ F)	Picofarads	150 pF (or PF) (old $\mu\mu$F)
$\geq 10,000$ pF	Microfarads	0.05 μF (or UF or MFD)
≤ 0.001 H	Microhenrys	5 μH (or UH) (=0.000005 henry)
$0.001-0.099$ H	Millihenrys	3 mH (or MH) (=0.003 henry)
≥ 0.1 H	Henrys	20 H

TABLE 7-4

HIGHEST REFERENCE DESIGNATIONS
C_{23} L_7 R_{43}
NOT USED
R_{22}, R_{23}
C_{16}

come. However, it should be pointed out that two new standards may bring about some changes. These are ISO 1000, "S1 Units and Recommendation for the Use of their Multiples and Certain Other Units," and ANS Z 210.1 – 1976 (IEEE Std 268 – 1976), "Standard Metric Practice." The SI units are preferred by the authors and are becoming more commonly accepted.

Another helpful device is that shown in Table 7-4 which shows the highest number of each component, as well as omissions, in a diagram.

The reader may wonder why two resistors were not used. There could be several reasons for such an occurrence. One possibility would be that the original design was altered after testing or use and a particular component was taken out. If, later, another one is added, it is usually given a new number, not the number of the deleted one.

7-12 Waveforms

Some common waveform symbols are shown in Fig. 7-19. These are sometimes shown on an elementary diagram at possible test locations. Some common forms that are not shown are sawtooth, trapezoidal, rectangular, exponential, clipped, and triangular.

7-13 Mechanical Linkage and Other Mechanical Arrangements

In Fig. 7-15, on the left side, are two capacitors connected by means of dashed lines. The dashed (dotted) lines mean that the two devices are connected me-

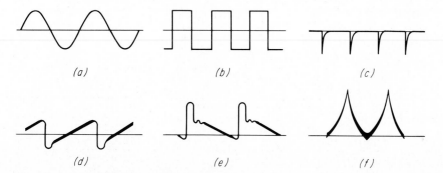

(a)   *(b)*   *(c)*

(d)   *(e)*   *(f)*

FIG. 7-19 Symbols of waveforms: *(a)* sine wave; *(b)* square wave; *(c)* trigger; *(d)* through *(f)* complex forms.

chanically, so that when one is turned, the other is turned to the same position simultaneously. Figure 7-20 shows two other situations in which mechanical connections are shown. Figure 7-20a is part of a diagram that is part pictorial, part schematic, and part connection. The mechanical path may be easily traced. The autopositioner works as follows:

1. The operator turns a remote-control switch to a certain position.

2. The relay (slow-operating in this example) is energized, lifts the pawl out of the sprocket wheel, and also turns on the motor.

3. The motor drives the autopositioner shaft until the position corresponding to the new position of the remote control is reached.

4. The relay circuit now opens, the pawl engages the stop wheel, and the energy to the motor is cut off.

Ganged tuning capacitors (Fig. 7-20b), like ganged switches, may be shown on a schematic diagram. Figure 7-20c shows mechanical connections between the attitude indicator and synchro of a DC-10.

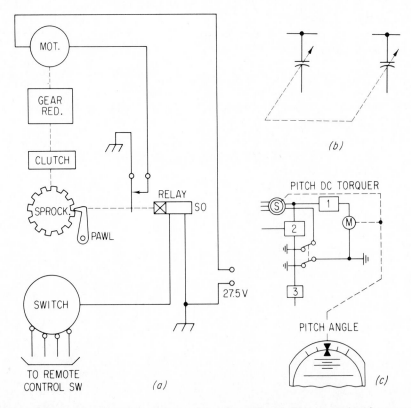

FIG. 7-20 Mechanical connections: *(a)* autopositioner; *(b)* ganged tuning capacitors; *(c)* attitude indicator.

7-14 Separation and Interruption

It is often desirable to show a separate package apart from the rest of the circuit or system. For instance, if the same package (or *pack* as it is sometimes called) is to be used several times, it may be drawn as blocks instead of as a schematic diagram. In such a scheme, the schematic diagram would have to be drawn once, but only once. In Fig. 7-21 three ways of drawing a pack schematically, using different terminal or interruption techniques, are shown. Figure 7-21*d* shows how the pack might be drawn diagrammatically as a block. Figure 7-21*a* shows the circuit as removed at terminals. Figure 7-21*b* simply encloses the pack with an optional enclosure line but does nothing about the terminals. (In all four figures, the lines, or terminals, have been identified by means of num-

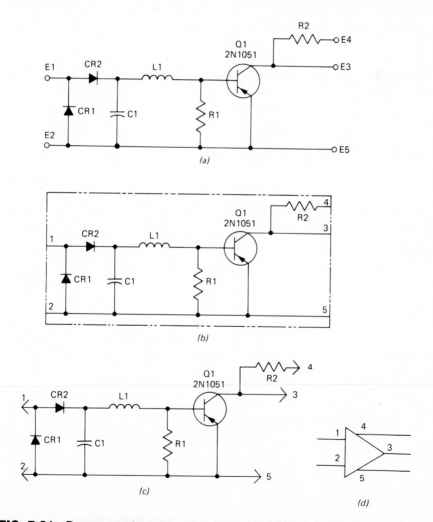

FIG. 7-21 **Representation of a circuit package separated from the rest of the system.** (From ANSI Y14.15, "Electrical and Electronic Diagrams.")

bers.) Figure 7-21*c* shows another method of separating a package. While this is a fairly widely used method, it is slightly confusing. The V-shaped lines at the end of each conductor do not represent any specific kind of connection in this scheme. Yet ANSI/IEEE Y32E uses this symbol to show a male contact. It is so used in other drawings in this book, especially in Chap. 10. The Vs, incidentally, are not supposed to be arrowheads. They are correctly drawn at a 90° angle.

Figure 7-22 illustrates other ways of interrupting groups of lines. Lines are generally grouped together in threes. If there are more than three lines to be shown, groups should be separated, as at the right in Fig. 7-22*a*. Spacing between lines should be a minimum of $\frac{1}{4}$ in., and between groups $\frac{1}{2}$ in. Brackets may be joined by means of dashed lines, as in Fig. 7-22*b*. Figure 7-22*c* shows hypothetical uses of circuit-return symbols. If the inverted triangle is used, letters and general notes (Fig. 7-22*d*) will probably be necessary. The chassis symbol and the first triangle (with letter F) serve the same purpose, and either one could be used. These symbols are usually oriented as shown here. If it is more convenient, they can be drawn at the ends of lines extending upward or to the right or left. In other words, the entire figure (Fig. 7-22*c*) could be turned at

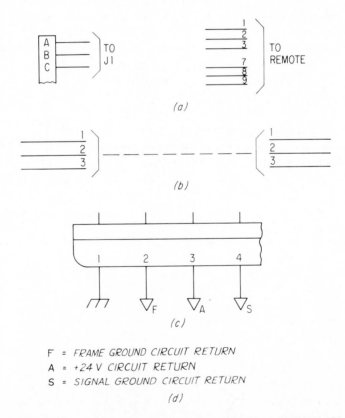

F = FRAME GROUND CIRCUIT RETURN
A = +24 V CIRCUIT RETURN
S = SIGNAL GROUND CIRCUIT RETURN

(d)

FIG. 7-22 **Interruption and circuit return.** (From ANSI Y14.15.)

90 or 180° from its present position. (Letters and numbers might be changed to read in a convenient manner, though.)

Figure 7-23 shows the schematic diagram for the preamplifier circuit of the Apollo moon camera. It is enclosed by the optional enclosure lines. The preamplifier is made up of discrete components, primarily to provide low-noise performance. The video signal from the camera is fed into the preamplifier. The no-dot system has been used.[1]

The camera is 17 in. long, including a zoom lens, and weighs 13 lb. It generates a field sequential color signal using a single-image tube and a rotating filter wheel. A ground-station color converter later changes the sequential color signal to a standard NTSC color signal. In addition to the camera there is a small (85-cm³) viewfinder monitor for the astronaut and a transmitter. Figure 6-2 shows the entire system in block form. The no-dot system has been used.[1]

7-15 Schematics Using ICs and Other Components

Since the age of microelectronics began, there has been an evolution from drawing schematics with only discrete symbols to drawing schematics with IC symbols and discrete component symbols. As microelectronics is able to put more and more devices on an IC, it becomes totally impractical to show the schematic of an IC on an overall schematic. These schematics combining ICs and discrete symbols presume that one understands the function of the ICs used, just as one would have to know the function of a tube, for example. The following figures, together with Fig. 7-17, will show a few schematics containing ICs and some basic ideas on how to draw them.

Figure 7-24 is a schematic of a two-phase clock for an 8080 microprocessor. This schematic uses discrete devices (resistor, capacitors, and a crystal) and ICs (NOR gates, buffers, and flip-flops). Actually, the NOR gates, the flip-flops and the buffers are each on one IC, respectively. Primarily, however, it is the timing diagram that we want to discuss when doing this type of schematic. Many digital circuit schematics utilize a timing diagram to show how the circuit works. The timing diagram can appear anywhere on the schematic, but the authors generally prefer it to be above or below its associated circuit, to make it easier to understand the circuit without having to look all over the diagram. In the circuit, the output of FF_1 output is half the frequency of the oscillator. O_1 is the OR-inverted output of the oscillator and FF_1. Gate 4 output O_1 is buffered to TTL levels (5 V). O_2 is the OR-inverted output of FF_1.

Figure 7-25 is an 8-bit analog-to-digital converter. This circuit utilizes two AND gates with inverted outputs, an inverter, four ICs, and several discrete components. The overall schematic uses the dot system for connections. The IC pins are shown within the IC symbol. If the pin functions are to be shown, they should be shown inside the IC symbol and the pin numbers outside. Manufacturers' model numbers are shown within the IC symbol, where it is preferred

---

[1] There are connections at points 9, 10, and 11.

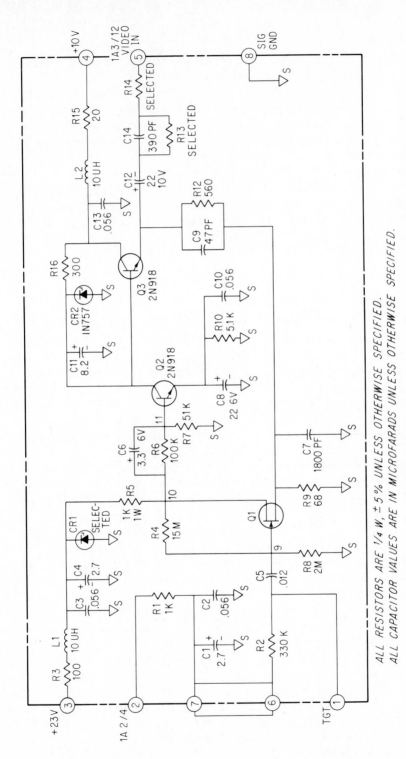

FIG. 7-23 Preamplifier circuit of Apollo television camera. (Westinghouse Electric Corp.)

ALL RESISTORS ARE ¼ W, ± 5% UNLESS OTHERWISE SPECIFIED.
ALL CAPACITOR VALUES ARE IN MICROFARADS UNLESS OTHERWISE SPECIFIED.

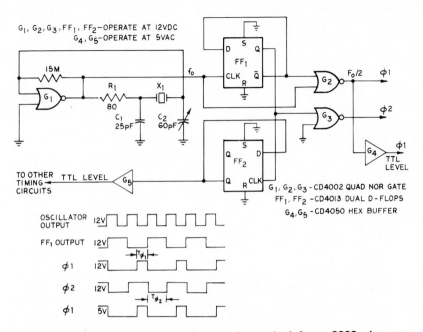

FIG. 7-24 **Schematic diagram of a two-phase clock for an 8080 microprocessor.**
(EDN Magazine, January 5, 1977, p. 50.)

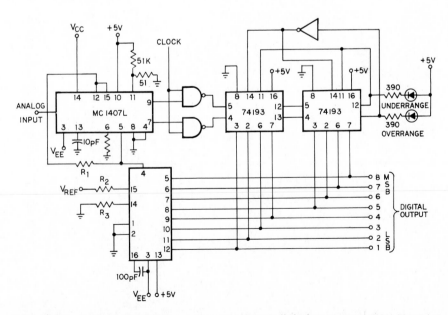

FIG. 7-25 **Schematic diagram of an analog-to-digital converter** (EDN Magazine,
June 20, 1975, pp. 118–120).

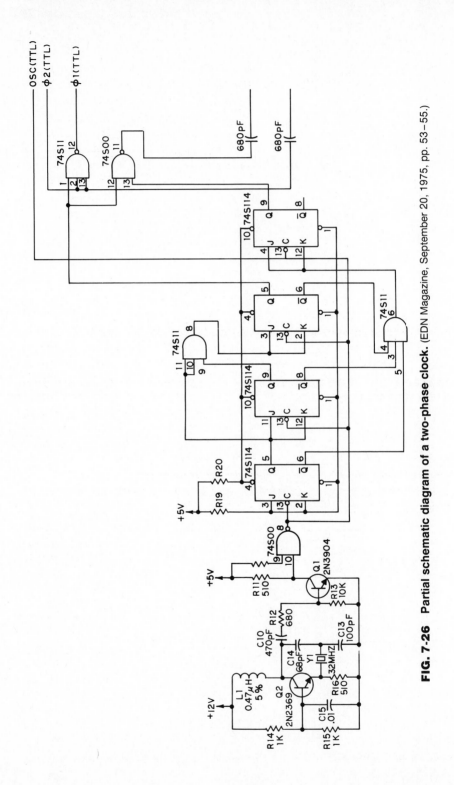

FIG. 7-26 Partial schematic diagram of a two-phase clock. (EDN Magazine, September 20, 1975, pp. 53–55.)

because there is less cluttering of the schematics. The voltage sources are not shown, which is generally typical for most schematics. The schematic flows from left to right with the analog input on the left and the digital output on the right. This circuit was used for teaching reasons and today it can be purchased on a single IC.

Figure 7-26 is a partial schematic of a two-phase clock for an 8080 microprocessor. This schematic is an example of how the functions are shown within the symbol of the ICs (J-K flip-flops) and the labeling of the pin numbers outside the IC symbol. The IC manufacturer's model number is shown outside the IC symbol. The schematic flows from left to right with oscillator on the left and the clock outputs on the right.

7-16 Hybrid Diagrams

Hybrid diagrams are just that; they can contain schematic diagrams, logic diagrams, wiring diagrams, pictorial drawings, mechanical connections, and pneumatic connections.

Figure 7-27 is a hybrid diagram combining most of the aspects of a hybrid drawing, including a schematic, wiring information, logic circuitry, and mechanical devices and their associated interconnections to electrical devices. Approximately the left third of this diagram contains a schematic. The upper center portion contains a pictorial view of a gyroscope and its mechanical connection to the various synchros. The diagram also contains logic gates (e.g., flag logic) and a schematic of the flag logic circuit. Along the right side of the diagram, there is wiring information showing the wire numbers connecting the jack pins. The reader will notice a looplike symbol of lines 49 through 51. This indicates that the three wires are twisted together. Somewhat below this is a similar symbol which indicates that wires 45 and 46 are twisted together and shielded.

SUMMARY

The schematic shows by means of graphic symbols the functions and connections of a specific circuit arrangement. Symbols for use in such a diagram are extensively covered in ANSI/IEEE Y32E and IEC Publication 117, and the preparation of the diagram itself is covered by ANS Y14.15. There are certain basic arrangements of transistors and tubes which are usually repeated in electronics circuit diagrams. With these and some types of interstage coupling, certain patterns are often discernible — usually early in the game, when one starts to plan an elementary diagram. Designation, or referencing of each component part of a circuit, is important. Certain standard abbreviations and prefixes are used for this purpose. Sufficient space must be made available near each component for referencing.

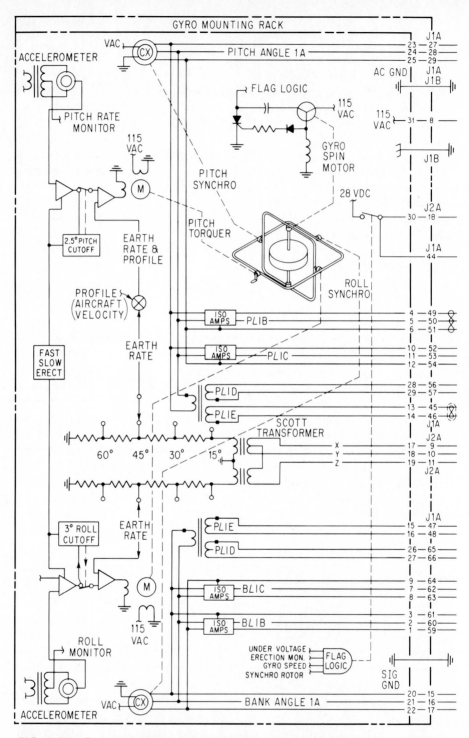

FIG. 7-27 Part of a system schematic of an aircraft altitude system. This is a hybrid drawing. (Douglas Aircraft Co.)

Conventional treatment is sometimes employed to eliminate the drawing of certain lines in order to avoid cluttering in a drawing. On the other hand, additional material, such as general notes and data on waveforms, is often added to a schematic diagram. A symbol is generally spaced midway along the span of circuit path on which it is drawn. Mechanical connections must be shown at times. Certain patterns are usually followed in laying out certain types of circuits in diagrammatic form. Radio circuits, for example, usually follow a *cascade* arrangement in which the signal path goes from left to right and the transistors are aligned horizontally. Auxiliary circuits, such as power circuits, are generally placed in the lower part of any schematic diagram. In circuits with ICs, the signal flow is generally left to right. However, due to multiple sections or functions on a single IC and several ICs within a circuit, there may be some "doubling back" to connect to a function of a previous IC. Overall balance and symmetry are other concepts that are observed in diagrammatic layout.

QUESTIONS

7-1. What set of standards governs the preparation of schematic diagrams?

7-2. In what way, or ways, is a schematic diagram different from a connection diagram?

7-3. How would you show the value 100,000 Ω on a circuit drawing? 150,000 pF?

7-4. To what sort of elements do the following terms refer:
  a. 12AV6
  b. Z_2
  c. 50 K
  d. 2N329
  e. CR_2

7-5. What is meant by the term *density* as it applies to schematic diagrams?

7-6. What is meant by the term *symmetry* as it applies to schematic diagrams?

7-7. Where should timing diagrams go on IC schematics?

7-8. For what purposes are schematic diagrams used?

7-9. Where are auxiliary circuits normally placed on the schematic diagram?

7-10. If a resistor is positioned on a vertical line (signal path), where would you place its identification?

7-11. Where would you place the identification for a vacuum tube in a schematic diagram of a circuit having cascade projection?

7-12. What letters are used to identify the three leads of a MOSFET?

7-13. What elements are often found in a hybrid diagram?

7-14. What do the parallel lines adjacent to the outline of a picture-tube symbol represent?

7-15. Define the prefixes micro-, milli-, kilo-, and mega-.

7-16. Is it possible to draw a schematic diagram using only one width of line and in so doing comply with the standards?

7-17. If you are assigned to make a finished drawing of a schematic diagram of a circuit, in what form is the information to which you will refer apt to be?

7-18. When an IC has both pin functions and numbers, where should each go?

7-19. What methods are used in coupling different stages together? (Name three.)

7-20. In what way may the number of connecting lines in a diagram be reduced?

7-21. What two shapes are generally used for IC modules?

7-22. State briefly, or use sketches to show, how you would separate a package from the rest of a circuit.

7-23. What is the difference between a picofarad and a micromicrofarad?

7-24. In which direction should the signal flow in a schematic?

7-25. What is the preferred method of where to show the IC manufacturer's model number in relation to the IC?

7-26. What types of drawing are found on a hybrid diagram?

PROBLEMS

7-1. Draw the schematic diagram of the all-channel CB monitor-receiver shown in Fig. 7-28. Use standard identification for each component. We have shown capacitor C_1 (10 pF) at the upper left. Each capacitor and resistor should have a similar reference designation.

Make any changes or improvements in the symbols or designations and values of components that you think would be appropriate. Add a suitable title and the following note: Unless otherwise specified all resistors are $\pm$ 5%, $\frac{1}{4}$ W. There is no connection at pin no. 6 of the amplifier LM 380. If well planned, this diagram, with a suitable title, will fit on an $8\frac{1}{2} \times 11$ sheet.

7-2. Draw the schematic diagram of a digital clock that is shown in Fig. 7-29. This will fit on an 11×17 sheet. Complete the drawing by showing resistors R_4 through R_9 and drawing the CRB1 full-wave bridge where indicated. Use ANSI practice in identifying the following components:

C_1 220 UF	S_1, S_2 .5 AMP
R_1 18K	Q_1-Q_8, 2N2222
R_2-R_{10} 10K	Q_9-Q_{16} 2N3866
	(Darlington)
CR_5 1N914	CRB1 WE460J

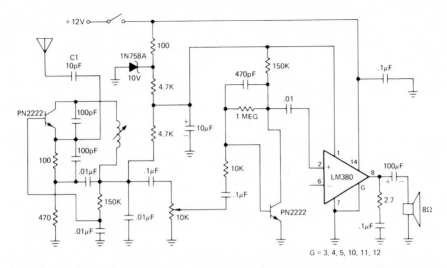

FIG. 7-28 **(Prob. 7-1.) Schematic diagram of CB monitor receiver.** (Kantronics, Inc.)

The corner pins are numbered on the Mostek clock chip. There are no connections to pins 13 through 17. The coil symbol at the left is part of transformer T_1, which has an iron core. The leads from the left winding of the transformer go to the 115-V ac supply.

7-3. Complete the schematic diagram in Fig. 7-30 by insertion of the correct symbols as indicated by standard notation. Add the identification of each component and the capacity if given below. Squares Z_1, Z_2, etc., are integrated circuits drawn as squares or rectangles with identification shown within the square. Video input is at 12. The letter U may be substituted for Z.

Z_1	SE501G	C_{10}	15 pF
Z_2	MD2905F	R_1	20 kΩ
Z_3	SE501G	R_2	510
Z_4	SE501G	R_5	51 kΩ
CR_1	IN914	R_6	10
CR_2	IN914	R_7	470
S	ground	C_{11}	6800 pF
C_1	3.3 μF	R_4	3.3 kΩ
C_2	3.3 μF	L_1	10 μH
C_5	22 μF	L_2	10 μH
C_6	0.056 μF	C_{14}	3.3 μF
C_7	22 μF		
C_8	056 μF		
C_9	15 pF		

This drawing will be slightly crowded on $8\frac{1}{2} \times 11$ paper and will have more than enough room on 11×17 paper.

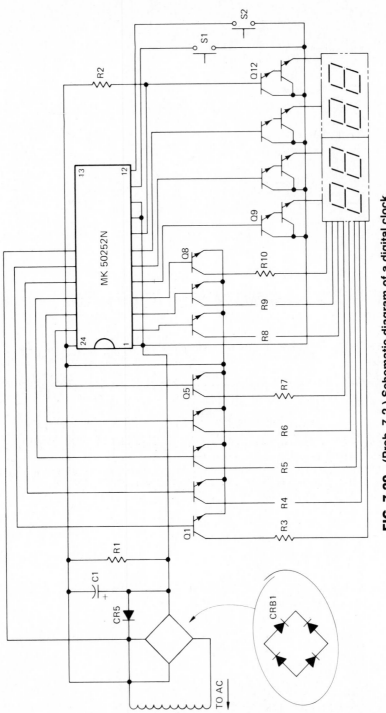

FIG. 7-29 (Prob. 7-2.) Schematic diagram of a digital clock.

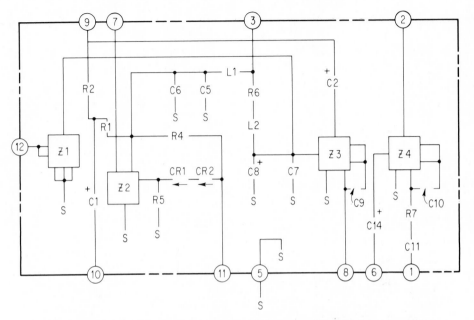

FIG. 7-30 (Prob. 7-3.) Postamplifier of a moon-walk television camera.

7-4. Complete the schematic diagram of the four-function calculator shown in Fig. 7-31. Major elements are the terminal (at top) that connects to the LED strip (not shown), the calculator chip TMC 972, and the keyboard module. The unfinished leads should be completed by drawing lines horizontally and/or vertically to the connector indicated. Use 11×17 paper. Use the standard battery symbol. Show stub lines at NCs (no connections).

7-5. Referring to Fig. 7-32, fill in the IC symbol with four 2-input, distinct-shape NAND gate symbols plus zero voltage and the positive voltage:

Pins 1 & 2	Inputs to NAND gate 1
Pin 3	Output of NAND gate 1
Pins 4 & 5	Inputs to NAND gate 2
Pin 6	Output of NAND gate 2
Pin 7	Zero voltage reference
Pin 8	Output of NAND gate 3
Pin 9 & 10	Inputs to NAND gate 3
Pin 11	Output of NAND gate 4
Pin 12 & 13	Inputs to NAND gate 4
Pin 14	Positive voltage reference ($+5$ V)

7-6. Redraw the high-fidelity amplifier shown in Fig. 7-18 to about $2\frac{1}{2}$ or 3 times the size that it appears in the book. Instead of the component identification system shown, use the ANSI system of identification shown in this chapter. Use 11×17 or 12×18 paper.

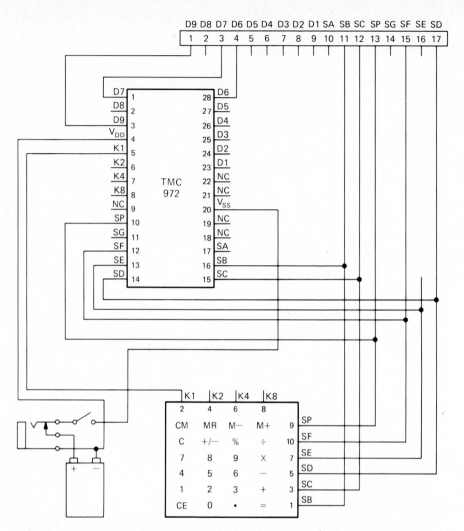

FIG. 7-31 (Prob. 7-4.) **Schematic diagram of a four-function calculator.** (Texas Instruments, Inc.)

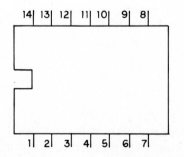

FIG. 7-32 (Prob. 7-5.) A four 2-input NAND-gate IC.

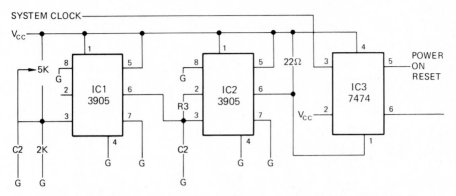

FIG. 7-33 (Prob. 7-7.) **Schematic diagram of a power on reset circuit.**

7-7. Redraw the schematic diagram shown in Fig. 7-33 to approximately twice the size it is shown in the book. Use standard identification for resistors, capacitors, etc. Add the following information on or near ICs 1 and 2: Pin 1, Trigger; 2, V_{REF}; 3, R/C; 4, GND; 5, V+; 6, C; 7, E; 8, LOGIC. On IC 3: Pin 1, CL; 2, D; 3, CLOCK; 4, GR; 5, Q; 6, $\overline{Q}$. Use $8\frac{1}{2} \times 11$ paper.

7-8. Figure 7-34 is a freehand sketch of an isolation circuit that is not complete. Make an instrument drawing of this circuit that is complete, uses correct

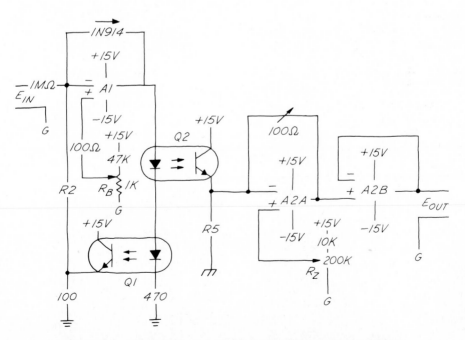

FIG. 7-34 (Prob. 7-8.) **An isolation circuit with optical couplers.**

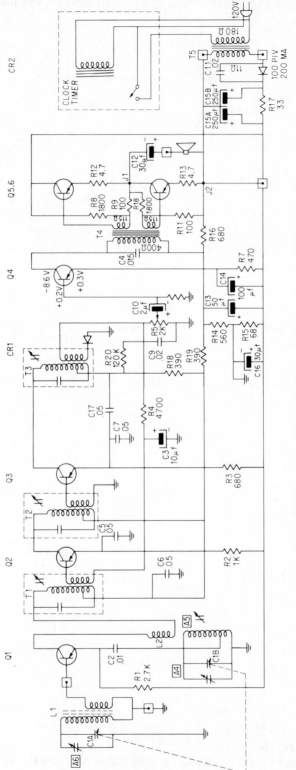

FIG. 7-35 (Prob. 7-9.) Schematic diagram of a transistor clock radio.

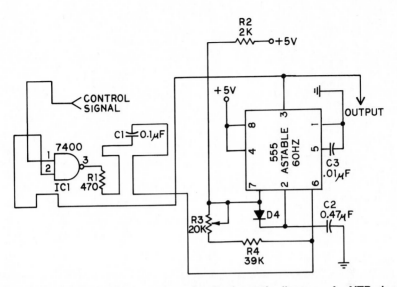

FIG. 7-36 (Prob. 7-10.) An unorganized schematic diagram of a VTR clock.

symbols in accordance with ANSI/IEEE Y32E, and has maximum identification of components. In the circuit, Q_1 and Q_2 are optical couplers, FCD 810 or equivalent, A_1 and A_2 are operational amplifiers (A_2 is separated into two parts, A_{2A} and A_{2B}), R_B is a bias resistor, and R_Z is a zero-adjust potentiometer. (R_2 and R_5 are 100 kΩ each.) Make any corrections which appear to be appropriate. Use a standard termination scheme for E_{IN} and E_{OUT}. This problem will fit on $8\frac{1}{2} \times 11$ paper; 11×17 paper would also be OK.

7-9.   Make a drawing of the clock-radio circuit shown in Fig. 7-35. Correct the symbols where obsolete or incorrect ones appear. Add identification for semiconductors as follows: Q_1, converter, 95101; Q_2, first IF, 95103; Q_3, second IF, 95102; CR_1, detector, IN295; Q_4, audio driver, 95201; Q_5 and Q_6, audio outputs, 95220; CR_2, rectifier, IN290. All resistors are $\frac{1}{2}$ W. The no-dot system has been used except at connections J_1 and J_2. Resistance values are in ohms unless otherwise noted, and all capacitance values are in microfarads. Use 11×17 or 12×18 paper.

7-10.   Rearrange the unorganized schematic of a video-tape-recording (VTR) clock (Fig. 7-36) so that it is in a logical systematic order with the input on the left and the output on the right.

7-11.   Figure 7-37 is a sketch of a six-transistor radio-receiver circuit. Many symbols are missing, but clues to their identity are given as reference designations or values. All transistors are PNP; their collector and emitter leads are identified. Capacitors at A_1 and A_2 are variable; no values are available. Capacitors in 455 kHz (KC is an older term) packages also have no values. Other capacitances are given as microfarads (μF) unless otherwise shown. Resistance

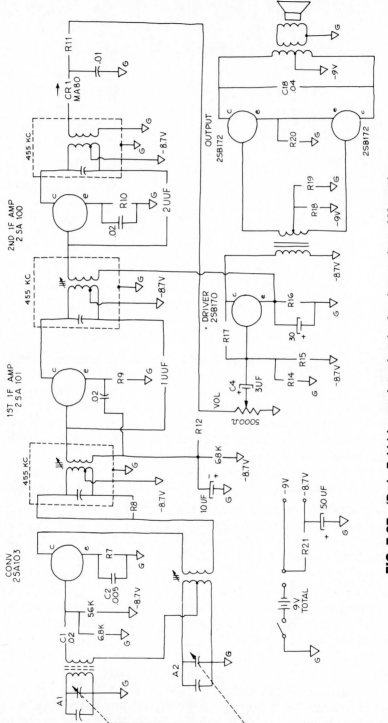

FIG. 7-37 (Prob. 7-11.) Incomplete sketch of a six-transistor AM receiver.

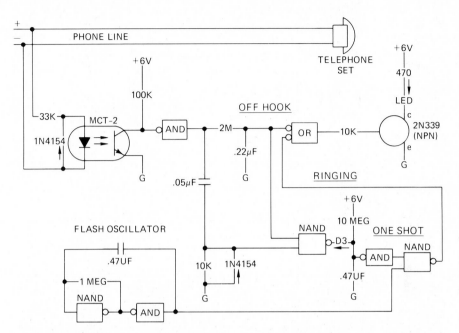

FIG. 7-38 **(Prob. 7-12.) A logic diagram and schematic diagram of an integrated semiconductor flip-flop circuit.** (Circuit from Texas Instruments, Inc.)

values not already shown are as follows:

R_7	2.2 kΩ	R_{12}	3900 Ω	R_{19}	680 Ω
R_8, R_{16}	100 kΩ	R_{14}	5600 Ω	R_{20}	5 Ω
R_9, R_{10}	1 kΩ	R_{15}, R_{18}	33 kΩ	R_{21}	100 Ω
R_{11}	470 Ω	R_{17}	220 kΩ		

Some symbols are incorrectly shown. Reference designations may not be in the best sequence. Correct and complete the schematic diagram of this circuit, using the standard symbols and reference-designation procedure. Use 12×18 paper or larger.

7-12. Figure 7-38 is the incomplete schematic diagram of a circuit that indicates at a remote location what the status of a telephone line is. LED T1L209 is dark when the line is not in use, on when the line is in use, and flashes when the phone is ringing. Complete the schematic by showing standard symbols for discrete elements and distinctive shapes for the logic elements. Elements should be properly identified. MCT-2 is an optical coupler. This is a tight fit on $8\frac{1}{2} \times 11$ paper.

7-13. Figure 7-39 is a sketch of the logic diagram and schematic diagram of a flip-flop circuit. Make a complete instrument drawing of this sketch, improving symbology or layout wherever you think it is advisable. Add the following

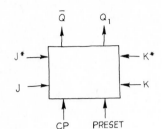

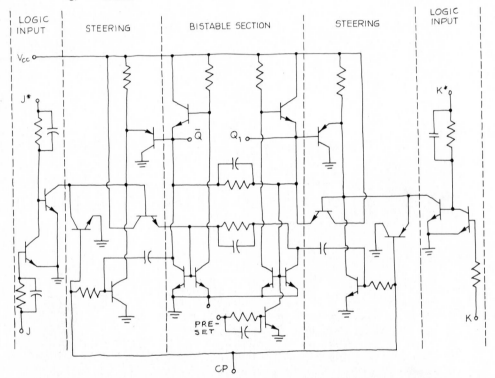

FIG. 7-39 (Prob. 7-13.) A block diagram and a schematic diagram of an IC flip-flop circuit. (Texas Instruments, Inc.)

information: total capacitance, 115 pF; total resistance, 70 kΩ; tunnels (crossover paths), 14; V_{cc}, pin no. 3; Gnd, pin no. 8. Use 11 × 17 or 12 × 18 paper.

7-14. Figure 7-40 shows an incomplete sketch of a satellite beacon oscillator. Many symbols are missing or incorrect, but clues are present as to their correct identities and values. All transistors are PNP; their emitter and collector leads are identified. Capacitors are identified as C_1, C_2, or with symbols, some obsolete or nonstandard. Complete the schematic diagram insofar as possible using standard symbols and reference designations.

Some values are R_1, 1780 Ω; R_2, 5210 Ω; R_3, 178 Ω; C_1, C_4, C_9, and C_{12}, 1 μF; C_3, 0.1 μF; C_5, C_8, and C_{11}, 0.01 μF; and C_6, 90 μF; and C_7, C_{10}, and C_{13},

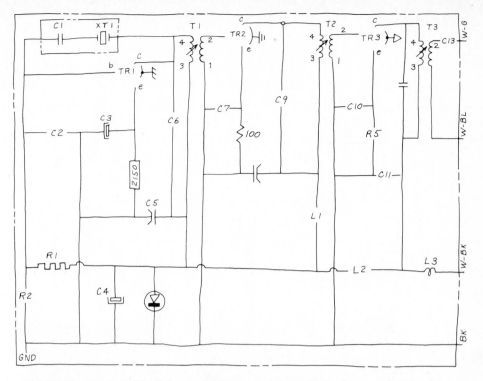

FIG. 7-40 (Prob. 7-14.) Schematic diagram of an oscillator for a communications satellite. (Rough sketch.)

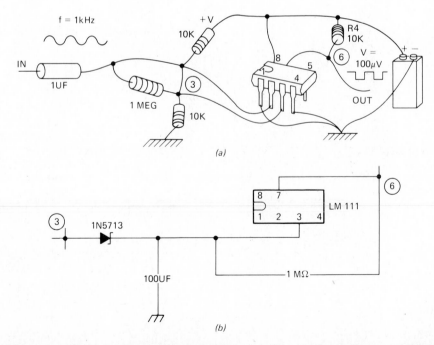

FIG. 7-41 (Prob. 7-15.) Integrated-circuit comparator used as a clipper: *(a)* pictorial diagram; *(b)* partial schematic diagram.

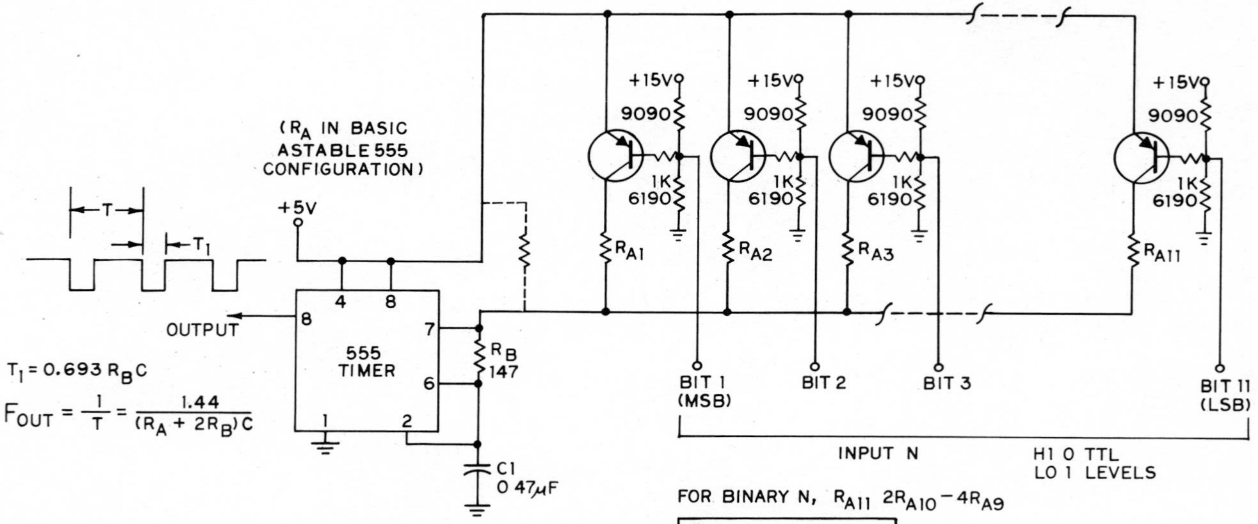

FOR BINARY N, R_{A11} $2R_{A10} - 4R_{A9}$

INPUT N

HI 0 TTL
LO 1 LEVELS

R_A	VALUES	
BIT	RES	BIT WT. HZ
1	7.21K	413
2	14.8K	207
3	29.8K	103
4	59.8K	52
5	120K	26
6	240K	13
7	481K	6.5
8	962K	3.2
9	1.9M	1.6
10	3.9M	0.1
11	7.7M	0.4

(R_A IN BASIC ASTABLE 555 CONFIGURATION)

555 TIMER

+5V

OUTPUT

$T_1 = 0.693 \, R_B C$

$F_{OUT} = \dfrac{1}{T} = \dfrac{1.44}{(R_A + 2R_B)C}$

R_B 147

C1 0.47μF

R_{A1} R_{A2} R_{A3} R_{A11}

+15V 9090 1K 6190

BIT 1 (MSB) BIT 2 BIT 3 BIT 11 (LSB)

FIG. 7-42 (Prob. 7-16.) Schematic diagram of a pulse generator. (EDN Magazine, August 20, 1974, p. 92.)

1000 μF. L_1, L_2, and L_3 are F54796; TR_1 is F54871; TR_2 is F54870; TR_3 is F54868; and CR_1 is F54857.

Use 11 × 17 or 12 × 18 paper.

7-15. Figure 7-41 has a pictorial view of an IC comparator circuit used as a clipper in *a*. An improvement to that circuit including a Schottky diode is shown in *b*. Make a complete schematic diagram of the improved circuit, including waveforms. Use 8½ × 11 paper.

7-16. Figure 7-42 is a schematic of a programmable pulse generator; however, bits 4 through 10 are missing. Complete the schematic on 11 × 17 paper, adding the missing bits and the values of all the timing resistors.

8

Microelectronics

One of the most fantastic and far-reaching developments of the twentieth century is that of microelectronics. It began in 1948 with the transistor and continues to grow and evolve at a phenomenal pace. It has made a new industry and many new manufacturing companies and has brought about a new technology in which graphics plays an important part. In this chapter the authors will show some of the drawing that is involved and some of the background that surrounds it.

8-1 Types of Integrated Circuit

An integrated circuit is a circuit in which devices of several to thousands of different types, such as capacitors, resistors, and transistors, are made on a single piece of material such as a small silicon chip and then connected to form a circuit.

Figure 8-2 lists the types of integrated circuit that have been developed to this point in time. Typical of many of these circuits is the wafer (Fig. 8-1), which includes many dice, each of which bears an individual integrated circuit (IC) chip. If we exclude the hybrid ICs for the moment, we notice that there are two fundamental processes, bipolar and metal-oxide semiconductor (MOS). The glossary that follows briefly describes the differences and variations among the processes shown in Fig. 8-2.

Bipolar: One of two fundamental processes for fabricating ICs and currently the more popular. A bipolar IC is made up of layers of silicon with differing electrical characteristics. Current flows between the layers when a voltage is applied to the "junction," or boundary between the layers.

TTL: Transistor-transistor logic, by far the most successful bipolar IC logic. It gets its name, as do other digital IC product families, from the way the components are combined to form the logic elements. Digital ICs solve problems by manipulating electrical signals that represent bits of information. Basic TTL is a mature product, but faster and lower-power versions are extending the life of TTL into the 1980s.

FIG. 8-1 **Partial view of a wafer on which many microchips have been fabricated.**
(Intel Corporation.)

ECL: Emitter-coupled logic, a bipolar IC family that uses a more complex design than TTL to speed up IC operations. It is costly, power-hungry, and difficult to use, but it could become important in the next generation of large computers because it is four times faster than TTL.

I²L: Integrated injection logic, a high-density bipolar logic design. It holds promise of producing ICs with a circuit density approaching MOS, the speed of TTL, and the low power requirements of CMOS.

MOS: Metal-oxide semiconductor is the second fundamental process for fabricating ICs and the fastest growing. The active area of a MOS chip is at the surface, where a gate electrode applies a voltage to a thin layer under it to create a temporary "channel" through which current can flow.

PMOS: The oldest MOS circuit technology uses a channel of P-type material, where the flow of current is made up of positive charges.

NMOS: The N-channel MOS, where the flow of current is made up of negative charges. It is two to three times faster than PMOS circuits.

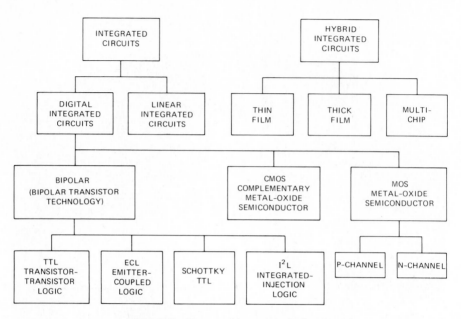

FIG. 8-2 Types of integrated circuit.

CMOS: Complementary MOS combines P- and N-channel transistors to create an IC that is as fast as a NMOS circuit but consumes very little power.

OTHER MOS: Variations of the MOS process are being developed. High-performance MOS (H-MOS) is one example. Transistors built on the slanting surface of a V groove (V-MOS) constitute another example.

8-2 Transistors, the Basic Building Block

Although this is not a book on electronic theory, Fig. 8-3 is a diagram of several transistors or transistorlike devices. The original bipolar PNP or NPN transistor has acquired some companions, but they are basically made of the same material, silicon being the most widely used. To enable the crystal to conduct electric current, small quantities of impurities are mixed with the pure silicon. If phosphorus is added, N type is the result. If boron is added, P type is produced.

By itself neither N- nor P-type material can accomplish a junction. But, when put together in various combinations and configurations, they provide a number of practical working semiconductor devices. It is the behavior of electrons and holes in the vicinity of the PN junction that gives bipolar junction transistors their unusual properties. In MOS transistors (MOSFETs), however, the active area is at the surface. The silicon controlled rectifier (SCR) is widely used in the electric-power and industrial fields and is discussed at more length in Chaps. 9 and 10.

Transistors are still widely available and used as discrete components. Along with diodes, capacitors, and resistors, they can be made to be part of an integrated circuit. A summary of semiconductor devices follows.

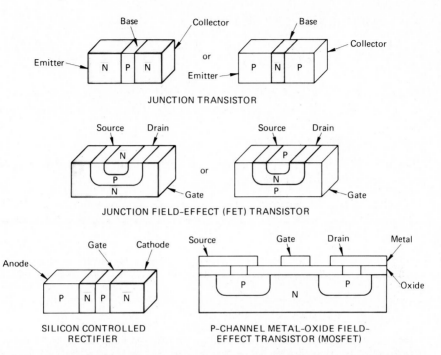

FIG. 8-3 Some basic semiconductors. From top to bottom: junction transistors, junction field-effect transistor (JFET), silicon-controlled rectifier (SCR), and MOS field-effect transistor.

1. Thyristors
Silicon controlled rectifier (Electrically controlled switch for dc loads.)
Triac (Electrically controlled switch for ac loads.)

2. Diodes (Exhibit very high resistance to current flow in one direction and little resistance in the other.)

3. Discrete transistors (Three-terminal devices used for amplifying or switching electrical signals.)
Bipolar
FET (field-effect transistor)
Unijunction

4. Charge-coupled devices (CCDs) (Storage devices in which packets of electric charge are moved across the surface of a semiconductor by electrical signals.)

5. Photosemiconductors (Emit or absorb electromagnetic radiation, usually in the visible or infrared band.)
Phototransistor
LED (light-emitting diode)
Photothyristor

6. Integrated circuits (To be defined and discussed later.)

7. Hybrid integrated circuits (To be defined and discussed later.)

8-3 An Integrated-Circuit Processing Glossary[1]

In describing the manufacture of ICs, we believe that presentation of the following terms will make it easier for the reader to understand the processes.

Alignment: A technique in the fabrication process by which a series of six to eight masks are successively registered to build up the various layers of a monolithic device. Each mask pattern must be accurately referenced to or aligned with all preceding mask patterns.

Die: A portion of a wafer bearing an individual IC, which is eventually cut or broken from the wafer. (The plural is *dice*.)

Diffusion: A high-temperature process involving the movement of controlled densities of N-type or P-type impurity atoms into a solid silicon slice in order to change its electrical properties.

Emulsion: A suspension of finely divided photosensitive chemicals in a viscous medium, used in semiconductor processing for coating glass masks.

Etching: A process using either acids or a gas plasma to remove unwanted material from the surface of a wafer.

Mask: A chrome or glass plate having the transparent circuit patterns of a single layer of a wafer. Masks are used in defining patterns on the surface of a resist-covered wafer.

Master mask: A chrome mask of a complete wafer's multiple images. It is used either in projection printing on a wafer or to contact-print additional masks.

Photoresist: A material that allows selective etching of a wafer when the wafer is photographically exposed. With a *negative resist*, the resist film beneath the clear area of a photomask undergoes physical and chemical changes that render it insoluble in a developing solution. In a *positive resist*, the same areas after exposure are soluble in the developing solution, so they disappear, permitting development of the exposed pattern underneath.

Reticle: A glass-emulsion or chrome plate having an enlarged image of a single IC pattern. The reticle is usually stepped and repeated across a chrome plate to form the master mask.

Step and repeat: A method of positioning multiples of the same pattern on a mask or wafer.

Stripping: A process using either acids or plasma to remove the resist coating of a wafer after the exposure, development, and etching steps.

Yield: The number of usable IC dice (chips) coming off a production line divided by the total number of dice going in. Yield tends to be reduced at every

[1] Reprinted from *Electronics*, July 21, 1977: copyright © McGraw-Hill, Inc., 1977.

step in the manufacturing process by wafer breakage, contamination, mask defects, and processing variations.

8-4 Integrated-Circuit Fabrication Process

Figure 8-4 shows the major milestones of the silicon IC fabrication process. Our description will begin at step 2, in which the crystal has been sawed into thin wafers which have been lapped and polished to a mirrorlike finish with no scratches or defects. Next, a layer of silicon dioxide (SiO_2) is grown on the surface to protect it from contamination and prevent the diffusion of impurities in some of the processes that follow.

The next step is to provide "windows" through which impurities can be deposited on the surface. Special photolithographic techniques are used to remove the oxide in certain selected regions. When the wafer is placed in a diffusion furnace at 1000°C and a gas containing the required impurity is passed over the surface, the impurities are deposited where there are windows and then are "diffused" into the silicon. This *buried layer* reduces the series collector resistance of transistors. Table 8-1 shows the masks that are used in the basic fabrication techniques. The windows mentioned above have been made using mask 1.

After the buried-layer diffusion is complete, the SiO_2 is removed from the wafer surface and a new layer of N- or P-type silicon is grown on the surface by means of epitaxial growth. In this step the wafer is exposed to an atmosphere containing a silicon compound at high temperature, and a new layer of silicon

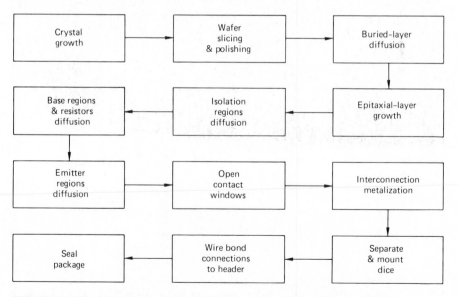

FIG. 8-4 **The steps in the manufacture of a silicon integrated circuit.** (From D. J. Hamilton and W. G. Howard, *Basic Integrated Circuit Engineering*, McGraw-Hill Book Co., New York, 1975.)

TABLE 8-1 The Basic Fabrication Sequence and Masks Required

	MASK 1	MASK 2	MASK 3	MASK 4	MASK 5	MASK 6	MASK 7
Process	Select buried-layer locations	Select isolation areas	Select base region	Select emitter region	Select capacitor locations	Select openings for ohmic contacts	Select areas where metal is removed
	Buried-layer diffusion	Isolation diffusion	Base diffusion	Emitter diffusion	Capacitor oxidation	Metallization	Gold backing if required
	Epitaxial layer and diffusion						

atoms is deposited in accordance with the crystal structure of the substrate. This *epitaxial layer* is also doped by including small amounts of the desired impurity in the gas stream during its growth.

As shown in Table 8-1 and Fig. 8-4, successive oxide growths, photolithography steps, and diffusions are undertaken to provide base and emitter regions, isolation walls, etc. The last steps include making windows for interconnections among the many devices now present on each circuit, followed by the metallization process. In this process a metal, often aluminum, is evaporated over the surface of the wafer; then unwanted portions are etched away and the interconnection contacts are left.

Finally the wafer is broken up into small dice, or chips, each having a single circuit or assembly, such as a microprocessor. Each chip is then mounted on a header, such as a dual-in-line mounting, and the aluminum contacts are connected to the header leads or pins by wire or one of the newer processes, such as film striping.

8-5 Photolithography

This process accomplishes two important functions: (1) reduction of the original mask drawing to the final mask size and (2) the photoresist process itself. The process of reducing the mask size also includes producing multiple images of the layout.[1]

The photoresist process involves making windows in the SiO_2 layer of the wafer. The photoresist is a light-sensitive coating that is placed over the oxide. The glass mask with the desired portion of the circuit pattern is placed over the photoresist. After exposure to ultraviolet light, the unexposed coating is dissolved; that which remains is chemically resistant to the solution or gas that is used to etch away the oxide. Later the remaining photoresist can be removed, so that additional masking and etching can be performed. The process is illustrated in Fig. 8-5.

Figure 8-6 includes, greatly enlarged, three steps of the diffusion process as they would appear if one were looking at the surface of an IC. (In other circuits, the shapes of these components may be different from those shown in Fig. 8-6.)

8-6 Drawing and Artwork for IC Design

Integrated circuits can be classified by the number of logic elements or "gates" they contain. Small-scale integration (SSI) refers to chips that have 100 or fewer gates. Medium-scale integration (MSI) refers to chips containing 100 to 1000 logic gates, while large-scale integration (LSI) refers to chips with more than 1000 elements. Very-large-scale integration (VLSI) includes any module that has more than 50,000 or 100,000 gates. Figure 8-7 shows the first mask of a series for an SSI. As shown in Fig. 8-8, this chip has 31 elements.

---

[1] Many identical circuit chips are made from one wafer (Fig. 8-1).

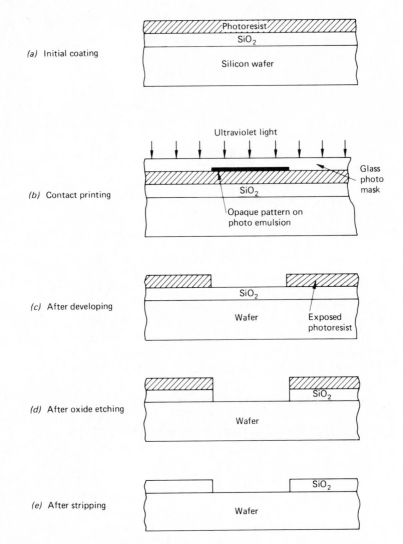

(a) Initial coating

(b) Contact printing

(c) After developing

(d) After oxide etching

(e) After stripping

FIG. 8-5 Steps in the photoresist process. (From D. J. Hamilton and W. G. Howard, *Basic Integrated Circuit Engineering,* McGraw-Hill Book Co., New York, 1975.)

The nature of IC design is such that devices may have different shapes in different circuits. Some other shapes are shown in Fig. 8-9. Figure 8-10 shows a series of masks for a latch circuit which is part of a microprocessor chip. The schematic diagram of this circuit is shown in Fig. 8-11, and the photograph of the LSI microprocessor chip that contains the latch is shown in Fig. 8-12.

The original drawing for the mask shown in Fig. 8-7 was drawn to a scale 150 times the actual size. Notice the grid system that is used to facilitate accuracy. Spacing between components is actually as close as 0.0005 in. at times; therefore, alignment errors cannot exceed a few *ten-thousandths* of an

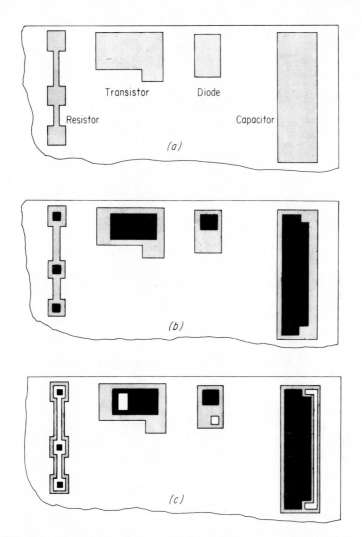

FIG. 8-6 Three steps in the manufacturing process of IC devices.

inch. Such accuracy is usually obtained only with computer-generated drawings. The actual size of the SSI chip (Figs. 8-7 and 8-8) is about 0.125 (3.2) × 0.250 in. (6.4 mm). It is in the form of a flatpack package with 14 external leads.

Original drawings for LSI chips such as the microprocessor are often drawn 500 times the actual size. The TMS 9900 microprocessor was produced using N-channel silicon-gate MOS technology. The MOS devices can be made with very small geometry, permitting high densities. For example, the memory space of the 9900 is 65,536 bytes (where one byte = eight binary units). Only a small part of the drawing for the entire unit is represented by the latch artwork (Fig.

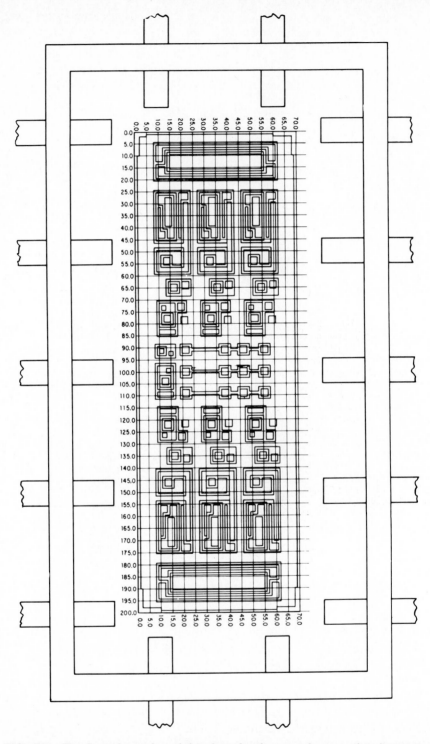

FIG. 8-7 The first of a series of drawings for the manufacture of an integrated circuit. Greatly enlarged. (Texas Instruments, Inc.)

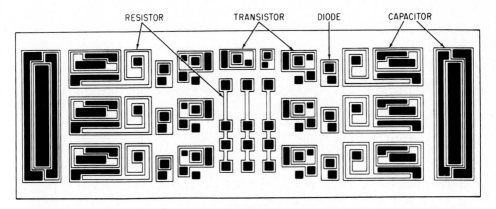

FIG. 8-8   The integrated circuit, greatly enlarged, (see Fig. 8-7) will appear about like this after several diffusion processes. (Texas Instruments, Inc.)

8-10). The microprocessor is housed in a ceramic dual-in-line package 0.90 (23) × 3.20 in. (81 mm) and having 64 pins.

We have enlarged the lower left portion of Fig. 8-10 in Fig. 8-13. The metal-oxide transistors are shown at 2 and E and at 5 and G, and the small section shows separation of the metal and diffused parts with gate oxide in between. Unfortunately the transistor symbols in Fig. 8-11, although used considerably, are not standard. In Fig. 8-14 we have shown two of these symbols at *a* and *c* and the standard symbols at their right. Not only are the symbols not standard, they do not have arrows indicating whether they are N or P type. This doesn't seem to bother the people who work with these diagrams day in and day out. But it is confusing to many who are not involved with the design and manufacture of these circuits.

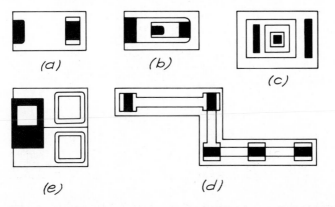

FIG. 8-9   Patterns of several elements after diffusion processes: *(a)* PNP transistor; *(b)* NPN transistor; *(c)* lateral NPN transistor; *(d)* resistor; *(e)* photonic IC with photodiode and transistors.

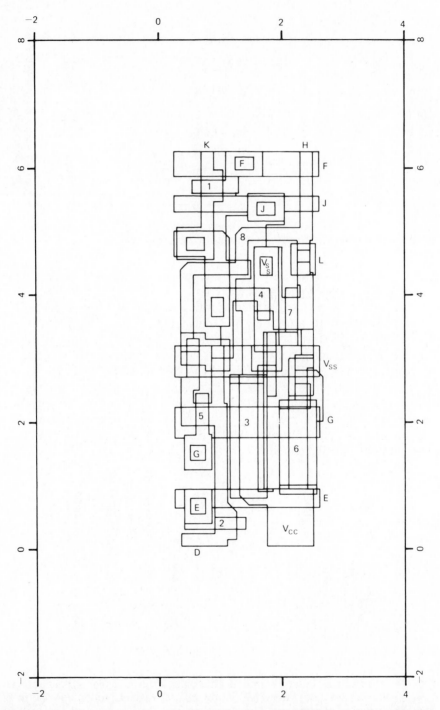

FIG. 8-10 The composite artwork for a latch circuit which is part of a microprocessor. (Texas Instruments, Inc.)

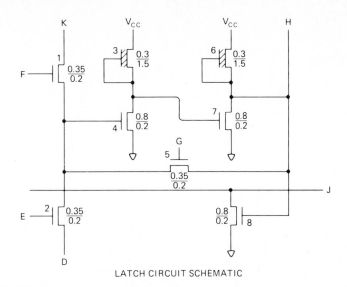

LATCH CIRCUIT SCHEMATIC

FIG. 8-11 The schematic diagram of the latch circuit shown in Fig. 8-10. (Texas Instruments, Inc.)

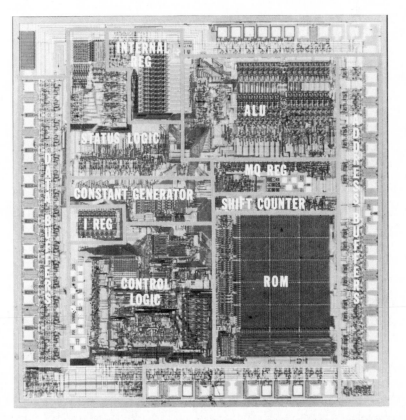

FIG. 8-12 Photograph of the TMS 9900 microprocessor greatly enlarged. Latch circuits, which are flip-flop units, are part of the register (REG) circuits. (Texas Instruments, Inc.)

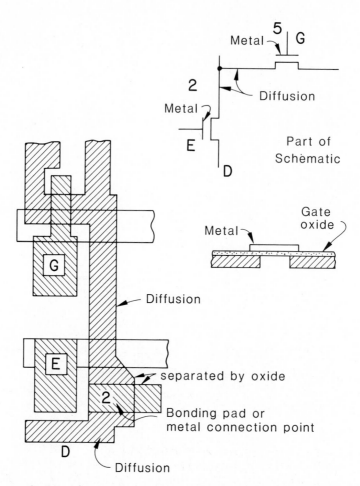

FIG. 8-13 Details of the MOS latch circuit shown in Figs. 8-10 and 8-11. (Texas Instruments, Inc.)

Artwork is made from each drawing on a plastic material called *Rubylith.* Cuts are made in this material—Mylar with a red plastic coating—and the coating is peeled off in the desired places. When the Rubylith artwork is photographed, the red areas will appear opaque to photographic film. The photograph looks pretty much like a negative.

In order to check the accuracy of medium- and large-scale IC drawings, debugging plots composed of all masks are made. Carefully following registration, the masks are superimposed on one another. Unwanted overlapping or less-than-minimum clearances can usually be seen.

Finally, to connect the elements of the circuit to their respective leads or other components (or both, in some cases) an aluminum pattern must be

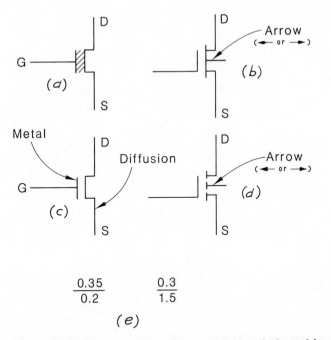

FIG. 8-14 Symbols used in preceding diagrams: *(a)* symbol used for depletion mode MOSFET; *(b)* standard symbol for the same MOSFET; *(c)* symbol used for enhancement-mode MOSFET; *(d)* standard symbol for this MOSFET; *(e)* numbers representing the ratio of the width to length of the source-to-drain conducting channel.

placed over the surface and in contact with the appropriate component. After openings for ohmic contacts have been made (mask 6 in Table 8-1 and step 8 in Fig. 8-4), the metal is evaporated over the entire surface of the wafer (step 9, mask 7), where it makes good contact with the silicon. The pattern could resemble that shown in Fig. 8-15. Unwanted metal is etched away, and the chip is ready to be mounted.

8-7 Packaging of Integrated Circuits

Until 1976 or so, two types of package were being used for most ICs: flatpacks and dual-in-line packages (DIPs), ceramic or plastic, although the transistor TO "can" provide a third alternative. Growing pin counts have been forcing some changes, particularly in LSIs and VLSIs. It has become necessary to find the right package for both the system and the chip.

One 3.20-in.- (81-mm-) long DIP has 64 pins. (The one in Fig. 8-16 is slightly smaller.) The DIP is reliable and easy to position on circuit boards. Since 3-in. bodies take up a lot of board space, however, some designers have turned to more economically sized packages. A redesigned square flatpack with

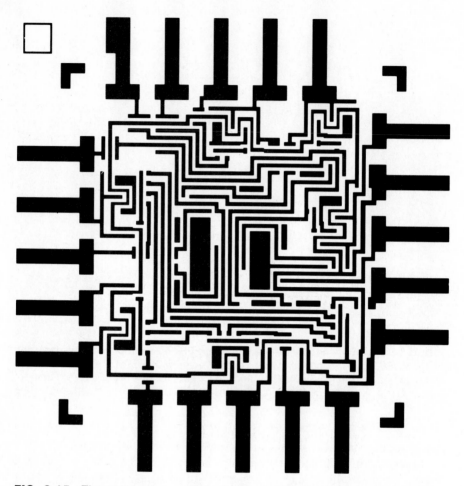

FIG. 8-15 The pattern for contact metalization (interconnections) of an integrated circuit. (Bell Telephone Laboratories, now part of AT&T Technologies.)

an equal number of leads on all four sides (up to 64 or 80 total) has been gaining acceptance. Multilayered DIPs have been in production, including one that has four rows of pins and is called a *QUIP* (quadri-chip).

A large number of ICs now are available in several packages. One example is the Motorola MC 68000, which can be obtained in a plastic DIP, ceramic DIP, or square chip carrier (see Figs. 1-28, 3-35, and 5-14).

Another approach is to use a film carrier, usually for several LSI units. The carrier, which is available in different shapes and sizes, is a multilayer package on the bottom of which is a pattern of "bumps" on 40- or 50-mil centers. There is a metal frame and base pad, making it possible to connect the internal IC pad to the external bumps through grooves in the sides. Another type of carrier uses film, including 35-mm film, which is quite compatible with the automation

FIG. 8-16 The 8748 microprocessor package. This 40-pin dual-in-line package is 3 in. (76 mm) long and 0.6 in. (15 mm) wide. (Intel Corporation.)

process. Figure 8-17 shows a chip carrier at *a* and the cross section of a solder bump at *b*. Small copper spheres are placed on the contact points of the circuit while it is still on the wafer. The wafer is then heated until solder melts over the spheres and forms a solder-coated bump on the contact pad as shown in Fig. 8-17*b*. When an individual IC is placed in a header and the assembly is heated, the IC becomes connected to the header through the bumps. Some variations of this appear in Fig. 8-18, where devices with 50-mil (1.27-mm) distances between leads can be placed on a board with the common 100-mil (2.54-mm) PC-board hole spacing. Thus LSI chips with leads only 50 mils apart can be through-connected to boards that have holes 100 mils apart (see Fig. 8-19).

8-8 Film (Hybrid) Circuits

A hybrid microcircuit is usually a mixture of etched or silk-screened circuit elements and individual (discrete) elements or standard IC chips. Manufacturing tolerances of resistors and capacitors are smaller in film techniques than in the diffusion process. Therefore, hybrid circuits combining silicon chips that

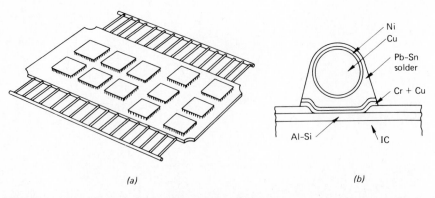

(a)   *(b)*

FIG. 8-17   A chip carrier and solder bump: *(a)* sketch of a board used for holding and assembling several chips; *(b)* greatly enlarged section of a solder bump.

have *active* devices (transistors and diodes) with film circuits composed of *passive* devices (resistors and capacitors), are sometimes the most appropriate design for a particular purpose. Hybridization may also minimize weight, space, and interconnections, as well as provide a larger range of values for some elements (see Table 8-2).

Figure 8-20 shows the same circuit manufactured as a thick-film device and as a thin-film device. *Thick-film* circuits are made by firing a paste onto an insulating substrate of a glossy or ceramic material. Formerly, expensive noble

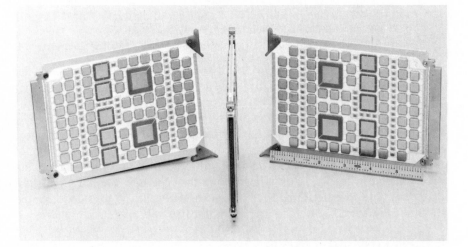

FIG. 8-18   A ceramic board populated by chip carriers and chip capacitors. "Front," edge, and "rear" views. Chip capacitors are the smaller rectangular shapes that are on the third vertical row "in" from each end, and also near the center of the board. The largest devices are ceramic multilayer chip carriers. (Texas Instruments, Inc.)

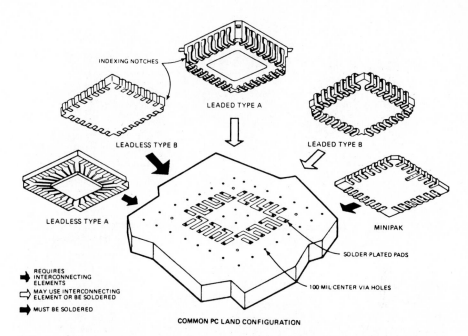

FIG. 8-19 Packaging for devices with leads or "bumps" at 50-mil (1.27-mm) centers. Such chip carriers can interconnect with standard circuit boards having 100-mil (2.54-mm) spacing. (Reprinted from *Electronics,* March 17, 1977; Copyright © 1977 by McGraw-Hill, Inc.)

metals were used, but alloys of lithium, nickel, and copper have recently been used successfully. The deposited material can be from $\frac{1}{2}$ (0.0127) to 2 mils (0.05 mm) in thickness. They are trimmed to the desired accuracy (0.5 to 2 percent) with lasers.

Two thin, parallel, light gray lines can barely be seen at the right of each dark resistor area in the upper photograph, where trimming has been performed. (The trimming has been done in the dark areas but does not show up in the photograph.)

Thin-film circuits are different in that the resistor and conductive films are sputtered or vacuum-deposited onto the substrate. The film can be deposited as

TABLE 8-2   Data on Film and Diffused Circuit Elements

	THICK FILM	THIN FILM	IC (DIFFUSED SILICON)
Manufacturing tolerance			
Resistors	0.50 to 10%	0.01 to 5%	5 to 20%
Capacitors	6%	3%	15 to 50%
Resistance values	$\geq 100\ M\Omega$	$10\ \Omega$ to $10\ M\Omega$	$100\ \Omega$ to $50\ k\Omega$
Thickness	0.013 to 0.05 mm	$\approx 0.00X$ mm	$\approx 0.00X$ mm
Resistive material	Lithium alloy	Tantalum alloy	Silicon, SiO_2

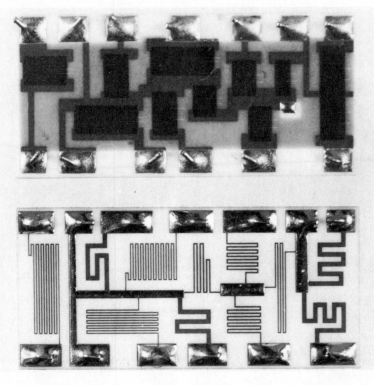

FIG. 8-20 Photographs of a thick-film circuit (top) and thin-film circuit (below). These devices, each about 1 in. (25 mm) in length, are the same as the circuit shown in Fig. 8.21*a*.

a pattern or as a film for subsequent photoprocessing and etching. Film thickness is a few micrometers, one micrometer being equal to 0.001 mm. Elements are trimmed to the desired size by such processes as abrasive trimming. As in thick-film technology, thin-film techniques are used to manufacture resistors, capacitors, and conductors. They have better resistor stability, resolution, and circuit-element density than do thick-film circuits. However, they are more expensive and cannot produce as large resistance values (see Table 8-2).

Figure 8-21 shows a schematic diagram of the resistor network that appears in the photograph (Fig. 8-20). The lower drawing (Fig. 8-21*b*, which is referred to in a later drawing) shows the outline of the substrate and its edge tolerances. Figure 8-22 includes the plan view of the thick-film circuit, a table, and some notes. The drawing shows the contacts along the top and bottom and the resistor and conductor areas as hidden lines. A coating to protect against unwanted contact with foreign objects is placed over the resistors and is shown by solid lines which outline its perimeter. A table is included that gives the desired resistance value, resistivity in ohms per square, and desired tolerances. The resistor sheet resistances range from 1 to 100 kΩ per square depending on the

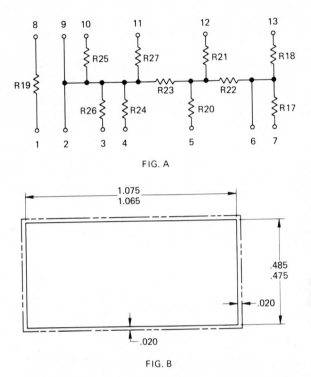

FIG. A

FIG. B

FIG. 8-21 *(a)* Schematic diagram of a circuit that has been produced in both thick- and thin-film form. *(b)* Outline drawing of the substrate.

film thickness. Tolerances are given in two categories, end of life (EOL) and initial trim. Thick-film resistances have a tendency to drift after they are put into use; hence there are two maximum allowable tolerances, the smaller of which is for the initial trim. Over the life of the circuit (perhaps 20 years), the values are expected to change no more than percentages listed. This drawing was drawn 10 times actual size.

Figure 8-23 includes the plan view of the thin-film circuit, drawn 10 times desired actual size, notes, and a table. If the resistors have a sheet resistance of, say, 75 Ω per square, the length of a resistor can be determined by an equation such as $L = (R/75)(W)$. Resistor R_{27}, with a width of 0.015 in. (0.38 mm), would have a length of $(2000/75)(0.015) = 0.40$ in. (10 mm). The designer has to select a route that will provide a 0.015-in.- (0.38-mm-) wide line that is 0.40 in. (10 mm) long. This has been done for each of the resistors in Fig. 8-21*a* and results in the patterns that are seen in the drawing and photograph. Note that the higher-value resistors have the narrower lines, 4 mils (0.1 mm). Line spacing has been kept equal to line width. In order to obtain accurate drawings and production models of the circuit, lines of this size are usually drawn by a computer or automatic drafting machine. Notes 3 and 4 of the drawing refer to Fig. 8-21.

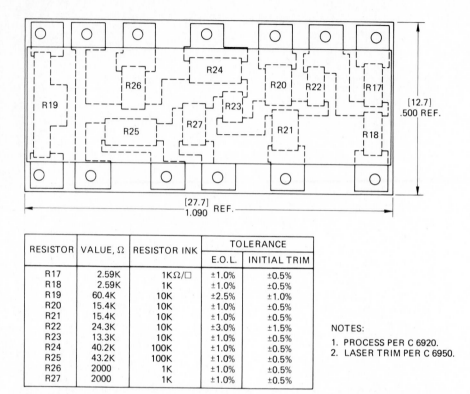

RESISTOR	VALUE, Ω	RESISTOR INK	TOLERANCE	
			E.O.L.	INITIAL TRIM
R17	2.59K	1KΩ/□	±1.0%	±0.5%
R18	2.59K	1K	±1.0%	±0.5%
R19	60.4K	10K	±2.5%	±1.0%
R20	15.4K	10K	±1.0%	±0.5%
R21	15.4K	10K	±1.0%	±0.5%
R22	24.3K	10K	±3.0%	±1.5%
R23	13.3K	10K	±1.0%	±0.5%
R24	40.2K	100K	±1.0%	±0.5%
R25	43.2K	100K	±1.0%	±0.5%
R26	2000	1K	±1.0%	±0.5%
R27	2000	1K	±1.0%	±0.5%

NOTES:
1. PROCESS PER C 6920.
2. LASER TRIM PER C 6950.

FIG. 8-22 Working drawing for a thick-film circuit. The schematic of this circuit is shown in Fig. 8-21.

Hybrid circuits having different combinations of manufacturing technologies are in extensive use. The hybrid laser preamplifier (Fig. 8-24) is a combination of thick-film resistors, thin-film capacitors, and IC chip and chip capacitors on a thick-film substrate. The small black rectangular objects are thick-film resistors, and the four (larger) gray rectangles are thin-film capacitors.

Figure 8-25 shows a power MOSFET with vertical integration through different manufacturing technologies. It is a power transistor with additional circuitry with which it will be more versatile and hence more useful. Another device, called SMART Power ™II, combines what the manufacturer calls CMOS logic and TMOS technology. It is an overvoltage and temperature-protection circuit that can replace three resistors, an SCR, and an integrated circuit (Fig. 8-26).

8-9 Custom-Built Chips and Gate Arrays

At the present writing about 25 percent of industry output is in custom-designed chips. Such a chip is usually incorporated in a single product or a family of closely related products. Custom-built chips are expected to reach 50 percent

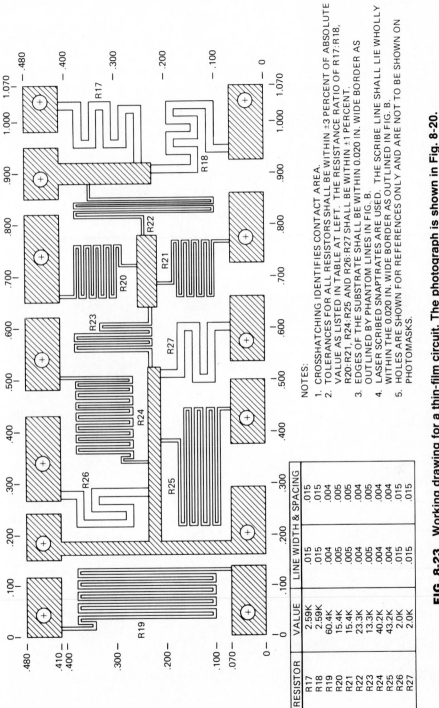

RESISTOR	VALUE		LINE WIDTH & SPACING	
R17	2.59K		.015	.015
R18	2.59K		.015	.015
R19	60.4K		.004	.004
R20	15.4K		.005	.005
R21	15.4K		.005	.005
R22	23.3K		.004	.004
R23	13.3K		.005	.005
R24	40.2K		.004	.004
R25	43.2K		.004	.004
R26	2.0K		.015	.015
R27	2.0K		.015	.015

NOTES:

1. CROSSHATCHING IDENTIFIES CONTACT AREA.
2. TOLERANCES FOR ALL RESISTORS SHALL BE WITHIN ±3 PERCENT OF ABSOLUTE VALUE AS LISTED IN TABLE AT LEFT. THE RESISTANCE RATIO OF R17:R18, R20:R21, R24:R25 AND R26:R27 SHALL BE WITHIN ±1 PERCENT.
3. EDGES OF THE SUBSTRATE SHALL BE WITHIN 0.020 IN. WIDE BORDER AS OUTLINED BY PHANTOM LINES IN FIG. B.
4. LASER SCRIBED SNAPRATES ARE USED. THE SCRIBE LINE SHALL LIE WHOLLY WITHIN THE 0.020 IN. WIDE BORDER AS OUTLINED IN FIG. B.
5. HOLES ARE SHOWN FOR REFERENCES ONLY AND ARE NOT TO BE SHOWN ON PHOTOMASKS.

FIG. 8-23 Working drawing for a thin-film circuit. The photograph is shown in Fig. 8-20.

FIG. 8-24 A hybrid laser preamplifier circuit. This is a combination of thick- and thin-film technologies and some chips. An IC chip is at the center, and a chip capacitor is at the right edge, halfway up. (Texas Instruments, Inc.)

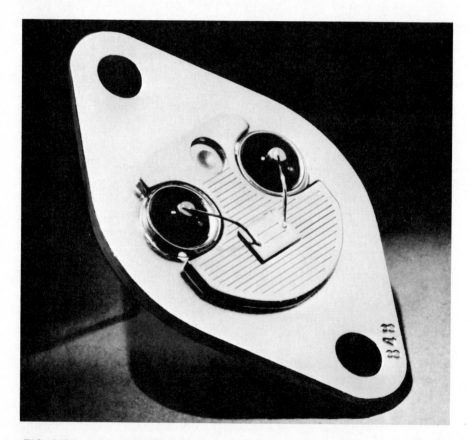

FIG. 8-25 A power MOSFET having vertical integration of different manufacturing technologies. (Courtesy of Motorola Semiconductor Products, Inc.)

FIG. 8-26 Another "smart" power transistor in which high-speed CMOS logic and high-current TMOS vertical power structuring are on a single chip. (Courtesy of Motorola Semiconductor Products, Inc.)

of IC volume by the 1990s. This means that a whole system can be put on a single chip. That has already occurred with a telephone exchange (PBX) where a chip replaced four circuit boards full of components.

Already available in large numbers and in many forms of circuits are *gate arrays*. These are standardized grids of transistors that are connected together in the last processing step. They are mass-produced in the manufacturing plant up to the final step or two. Then the gates or transistors are put together to suit the requirements of a customer. Three of many gate array circuits are shown in schematic form in Fig. 8-27.

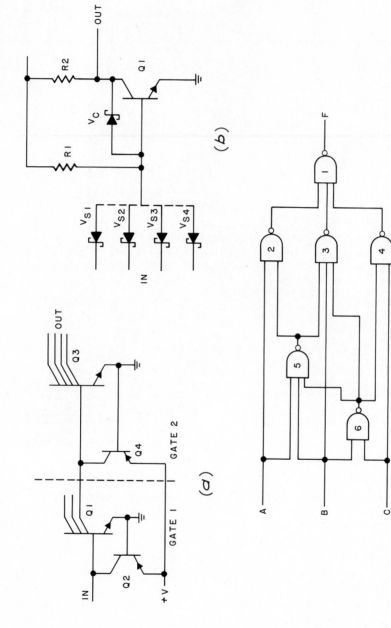

FIG. 8-27 Gate arrays: (a) multioutput single-input inverter; (b) circuit switching SDL network; (c) multi-input NAND network. (Texas Instruments, Inc.)

Another approach that is under heavy development at several large companies is called *wafer-scale integration.* Instead of cutting a large wafer into many chips, why not use the whole area for a single superchip? That is what one organization is doing: developing a $2\frac{1}{2}$-sq-in. chip that will replace approximately 100 conventional chips. The result is tremendous speed, but also terrific heat; the latter requires a solution not yet forthcoming.

8-10 Microprocessors and Microcomputers

The development and maturity of LSI technology made possible the design and quantity production of microprocessors. A *microprocessor* provides on one chip the CPU, or central processing function, of a digital computer. (See the glossary in this chapter for microprocessor terms.) One microprocessor chip might do the work of 80 standard integrated circuits. A refinement of the microprocessor has resulted in the development of a chip called a *microcomputer*, which has most of the functions of a digital computer. Actually the two terms are rather loosely applied, and there will probably be some confusion about these titles for years to come. The term *microcomputer* is sometimes applied to *personal computers*, partly to distinguish them from *minicomputers*, which are larger.

The famous 8080 microprocessor has an arithmetic logic unit (ALU), accumulator, registers (three types), and control logic and can address up to 512 input-output ports. Memory is minimal. The TMS 9900 microprocessor has an ALU, three registers, control logic, ROM, and input-output (I/O) control lines. It is usually augmented with other chips, including RAM. The 8748 microcomputer has a CPU, a clock circuit that times and coordinates data flow, ROM and RAM, I/O circuits, and expansion circuitry which enables the user to extend the capabilities of the chip. As powerful as this chip is, it is usually mounted on a board with still other components (Fig. 8-30).

In Fig. 8-28 we have two photographs of the Motorola MC 68000 microprocessor, which is 246 × 281 mils in size. It is the brain of the Apple Macintosh personal computer. It is a 16-bit microprocessor with 32-bit registers, a 32-bit program counter, and a 16-bit status register. It is also one of a family of processors, some of which are:

MC 68010   Virtual memory processor

MC 68450   Direct memory access controller

MC 68652   Communications controller

MC 68881   Floating-point coprocessor

MC 68020   Enhanced 6800 microprocessor

These microprocessors greatly extend the performance and capabilities of the 68000 family. The microprocessor is also used in the IBM 370 XP computer and in the Apollo engineering workstations, as well as some other workstation applications.

(a)

(c)

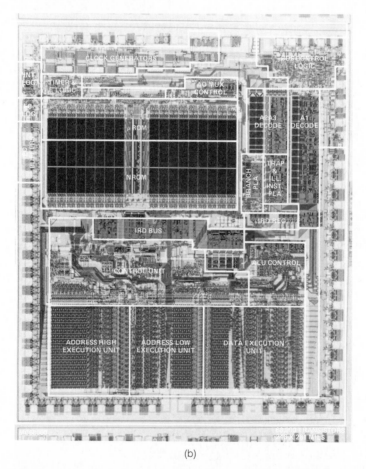

(b)

FIG. 8-28 The Motorola MC 68000 microprocessor: *(a)* the .24 × .28 in chip (lying on a dime); *(b)* an enlarged photograph of the chip; *(c)* the Apple Macintosh utilizing the 68000 microprocessor. **Shown are the main unit, detachable keyboard, and mouse pointing device.** (Courtesy of Motorola Semiconductor Products, Inc. and Apple Computer, Inc.)

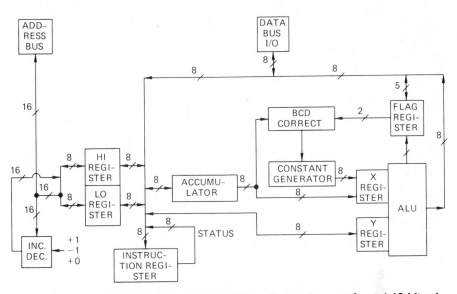

FIG. 8-29 Functional diagram of the 8080A microprocessor: 8- and 16-bit address busses allow direct addressing of 64 kilobytes of memory. Numbers indicate the number of data bits of the various busses. The block at the lower left represents the incrementer-decrementer unit. (Intel Corporation.)

A functional diagram of an older microprocessor is shown in Fig. 8-29.

It is very common to find combinations of microprocessors or microcomputer chips and other devices mounted on a PC board. These single-board computers can perform many simultaneous or sequential operations for computation or for process control. Figure 8-30 is the photograph of such a board. Such arrangements provide essentially the same functions as a minicomputer. This is a low-cost (hundreds of dollars) way to control many parts of a system or process, and to provide convenient monitoring (video tube, LED), input (tele-

FIG. 8-30 The SBC 80/10 single-board microcomputer. It contains an 8080 CPU, 1 kilobyte of RAM, 4 kilobytes of EPROM, and 48 programmable I/O lines. (Intel Corporation.)

type, sensors), and ease of programming (FORTRAN and BASIC languages and floating decimal point). A functional flow diagram of the SBC 80/10 is shown in Fig. 6-6. This diagram is explained in Sec. 6-3, which also gives an overview of how a microprocessor works by means of a general flow diagram.

Microprocessor Glossary

CPU: Central processing unit, the heart of a microcomputer. It contains the arithmetic logic unit, control circuitry, and registers for the temporary storage of data.

ALU: Arithmetic logic unit, that part of the chip that performs all the arithmetic using binary techniques. Most ALUs perform boolean logic operations and have shift capabilities.

Input-Output (I/O) Circuits: These are "ports" to keyboards, printers, cathode-ray-tube, sensors in industrial and automotive control systems, and other possible devices.

Register: A temporary storage unit. Program counters and instruction registers have dedicated uses. Accumulators have more general purposes.

RAM: Random-access memory, a volatile memory requiring electric power to retain data. Data can be stored for manipulation by other parts of the processor or computer.

ROM: Read-only memory, a nonvolatile memory in which data are stored permanently; no electric power is required to retain the data.

PROM: Programmable read-only memory.

EPROM or UV EPROM: Erasable programmable read-only memory, in which data can be erased by ultraviolet (UV) light.

EEPROM: Electrically erasable programmable read-only memory.

Bits and Bytes: A bit is a binary digit, a 0 or a 1. A byte is an 8-bit "word" or unit. A kilobyte represents 1000 bytes or 8000 bits. Microprocessors that handle 8-bit words (bytes) are suitable for tasks such as educational, automotive, or home control systems. Devices that handle 16-bit words are designed for even more sophisticated systems, such as navigation or data processing. Microprocessors that handle 32-bit words are in newer, faster computers and systems.

8-11 Applications of Microprocessors

It is impossible to give the reader a comprehensive idea of the many present and future uses of microprocessors. Rather, we can present only a few examples in this and the next two chapters. From a control standpoint these low-cost microcircuits are making switches, gears, levers, springs, and relays obsolete in many applications.

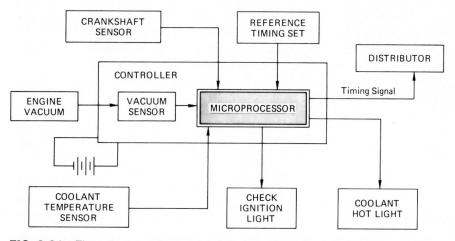

FIG. 8-31 **Flow diagram of a spark-advance system in a car.** (Reprinted from *Electronics,* September 29, 1977; Copyright © 1977 by McGraw-Hill, Inc.)

Automotive control is one area that readers will appreciate. The Delco-Remy Misar system includes four inputs: crankshaft position, coolant temperature, reference timing, and manifold vacuum. Basically, this system "looks up" the optimum spark advance in a memory map that corresponds to the driving conditions and passes this information to the distributor. Figure 8-31 shows this system in diagrammatic form. Another approach puts both spark advance and exhaust gas recirculation under the control of a microprocessor. Table 8-3 shows the computer command control system for a six-cylinder engine of the 1980s.

Other applications are in the area of industrial controls. One such arrangement checks the dimensions of sheet-metal parts, reads a binary digital signal determining whether each part is within the allowable tolerance or not, and provides whatever statistical analysis is required. Another, incorporated in a cash register, drives a seven-segment digital readout, prints a receipt with the amount of the purchase on it, and makes change — all at practically the same instant.

TABLE 8-3 Emissions Control System [Computer Command Control (CCC)]

INPUTS		OUTPUTS
Coolant temperature sensor		Air diverter valve
Oxygen sensor		Air select valve
Vacuum sensor	CCC	Fuel metering solenoid
Throttle position sensor	with	Check engine indicator
Idle speed control throttle switch	programmable	Solenoid valve assembly
Assembly-line diagnostic link connector	PROM	Idle speed control indicator
Barometer pressure sensor		
Vehicle speed sensor		
Park-neutral switch (In) Electronic spark timing distributor (Out)		

Programming a microprocessor system may require quite a bit of effort. In fact, because of the low cost of these circuits, the programming may cost more than the initial cost of the hardware. Program writing is often done in machine language or mnemonic (assembly) language. To use either, the user must develop an understanding of the language and how the processor works and skill in using the language and processor. Typical instructions might be:

INSTRUCTION	DESCRIPTION
INXB	Add 1 to BC register pair
JNC	Jump if C flag is false
DCXB	Subtract 1 from BC register pair
OUT	Output data from accumulator
STA	Store A value at direct address

Here is how the HERO I robot by Heathkit is programmed to say "HELLO":

INSTRUCTION	DESCRIPTION
1B	H
3B	E (as in get)
18	L
35	O (as in board)
37	OU (as in you)

The robot, which has a Motorola 6808 microprocessor, utilizes a two-digit, or digit-letter, code for each of 64 different sounds (phonemes). These can be put together to make words and sentences.

These are a few of more than a hundred such instructions for a particular microprocessor system. By spending more money for hardware or software, it is often possible for a user to have a system that utilizes a language that is easier to learn and work with, such as FORTRAN and BASIC. In BASIC, for example, 2 times 3 is simply written or typed 2*3. PRINT 2*3 produces the answer, 6, on a typewritten sheet or video screen.

Personal computers represent a logical development in the use of microprocessors. Because of their low cost, compactness, and versatility, they are well suited for use around the home. They have been used to control lighting and sound throughout a house, for video-type games, to plan menus for those on a diet, to lock the front door at night, to keep track of investments, to keep telephone numbers and Christmas card lists up to date.

Personal and professional (desk-top-size) computers are also used for word and data processing. With special commercially written programs, usually on floppy diskettes (disks), they permit the user to type material, then see it on a monitor, then correct mistakes, rearrange material, etc., and finally have it printed exactly as desired. Also, using the coordinates of the computer screen, the user can make drawings appear on the monitor. If desired, these can be printed on a plotter or dot-matrix computer. The following program in BASIC language will draw a ground symbol on an IBM PC screen:

INSTRUCTION	DESCRIPTION OF LINE
90 LINE (480,115) to (480,119)	Short vertical line at top of symbol
100 LINE (467,119) to (493,119)	Topmost and longest horizontal line
110 LINE (471,122) to (489,122)	Middle horizontal line
120 LINE (475,125) to (485,125)	Lowest and shortest horizontal line

The screen on which this symbol was drawn has 640 elements from left to right and 200 from top to bottom. Unfortunately, these must be divided by the dimensions of the "tube," viz.: $(640/12) = 53$ points to the inch and $(200/10) = 20$ points to the inch. This must be kept in mind when writing the program. The ground symbol was drawn in the right part of the screen as part of a series (menu) of symbols. The top of the symbol is at horizontal coordinate 480 and vertical coordinate 119. (Zero is at the top.) Other equipment makes drawing less complicated. The mouse and the joystick (toggle) are examples. The Cascade system developed for McGraw-Hill is another example. It includes extensive software.

8-12 Chip Wars and the Learning Curve

As the reader has learned from reading this chapter and elsewhere in the book, a battle is continuing between U.S. chipmaking companies and their Japanese counterparts. Although the IC industry was begun in the United States, the Japanese are a strong No. 2 and are winning many customers in the United States. By the end of 1979, the Japanese had won 42 percent of the U.S. market in what was then the latest generation of memory, the 16K RAM. They did better in 1980 with the 64K RAM, and they may be doing it with the 256K, which is the generation that follows the 64K RAM. The same sort of economic warfare prevails with consumer products that use these and other semiconductor devices.

The good news is that the user benefits, because as the *density* (the number of functions) on the chip increases, the cost per function decreases. This has been possible because production costs fall as volume increases, workers gain experience, and the proper investment in new plant and automated equipment is made. The result is known as "learning curve pricing." The basic concept is that whenever the total number of products doubles, the unit price reduces by a constant percentage. The density of ICs was 50 transistors in the 1960s and will be as many as 20 million transistor functions in the 1990s.

Figure 8-32 is a learning curve graph with four curves that apply at various times and places in the electronics industry. The 70 percent ratio curve has been drawn through two points, one at 10M and $1 and one at 20M and $0.70, and then extended in a straight line across the logarithmic paper. It is the steepest of the curves and will provide the best improvement. In fact, it has an *improvement rate* or *ratio* of

$$m = \frac{\log 0.70}{\log 2} = \frac{-0.155}{0.301} = -0.515 \qquad (8.1)$$

which is negative but is generally used in a positive sense since it represents an improvement in costs. Tables have been constructed to provide unit costs or times and cumulative costs or times. Table 8-4 shows excerpts from two tables. It is used as follows: If 10 hours were required to produce the first unit with an 80 percent learning curve, production of the 10th unit would require 0.4765×10 or 4.765 hours (using the upper part of the table). According to the values given in the lower part of Table 8-4, if it cost $10 to produce the first unit at 75 percent, production of 10 units would require 5.589×10 or $55.89. (The numbers in the table may be considered to be hours or dollars.)

Using the graph in Fig. 8-32, we see that with the 70 percent learning curve, increase in production from 100 million units to 500 million would reduce the cost from $0.31 to a little less than $0.14. Since the graph is not very accurate, we can go to Table 8-4 and find comparable numbers, 0.3058 and 0.1336 in the upper part.[1] It is highly probable that any particular unit will not cost $1 each for 10M production as the graph is drawn. This means that one would have to apply a factor, or ratio, after obtaining the numbers from the graph or table.

It should be stressed that the learning curve applies only when large volume is assured. Within a large company there may be several different learning rates

[1] $\dfrac{1336}{3058} = 0.4368 \qquad 0.4368 \times 0.31 = \$13.54$

TABLE 8-4   Learning Curve Tables

		70%	75%	80%	85%
	1	1.0000	1.0000	1.0000	1.0000
	2	0.7000	0.7500	0.8000	0.8500
	3	0.5682	0.6338	0.7021	0.7729
	4	0.4900	0.5625	0.6400	0.7225
	*				
UNIT	10	0.3058	0.3846	0.4765	0.5828
VALUES	*				
	50	0.1336	0.1972	0.2838	0.3996
	*				
	100	0.0935	0.1479	0.2271	0.3397
	*				
	200	0.0655	0.1109	0.1816	0.2887
		70%	**75%**	**80%**	**85%**
	1	1.000	1.000	1.000	1.000
	2	1.700	1.750	1.800	1.850
	3	2.268	2.384	2.502	2.623
	4	2.758	2.946	3.142	3.345
	*				
CUMULATIVE	10	4.931	5.589	6.315	7.116
VALUES	*				
	50	12.31	15.78	20.12	25.51
	*				
	100	17.79	24.18	32.65	43.75
	*				
	200	25.48	36.80	52.72	74.79

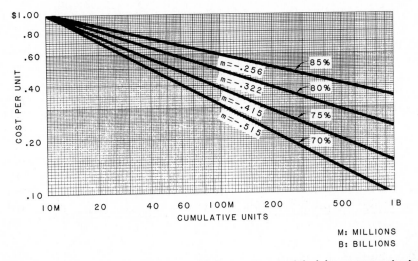

FIG. 8-32 The learning curve graph with four curves and their improvement rates (*m* values).

for different products. In general, the same learning curve applies to a "family" of units. This might mean a microprocessor and its related chips or a family of surface-mounted resistors.

SUMMARY

The development of integrated-circuit manufacturing has changed the face and substance of the electrical and electronics industry as well as the lives of millions of users of these products in the United States and throughout the world. The diffused silicon process—and later metal-oxide technology—made possible medium-scale integration (100 to 1000 transistor functions), large-scale integration (more than 1000 functions), and VLSI (100,000 or more).

Manufacture of some circuits, namely, thin-film and thick-film, is more analogous to the printing process than to the diffusion process. They also provide more accurate values of such passive devices as resistors. Accuracy of manufacture requires extremely high precision in drawings for integrated circuits, with the result that most are generated by the computer. These drawings are usually executed many times the size of the circuit they help produce.

Such consumer products as digital watches, pocket calculators, microprocessors, and microcomputers have been made possible because of LSI technology. Microprocessors are usually employed in dedicated situations, whereas microcomputers can be programmed for different uses. Industrial applications will be presented in the next chapter.

A major trend is toward the combination of technologies to produce hybrid circuits. Thin film and thick film have been combined with silicon-chip technology to great advantage. Custom and semicustom silicon chip technology is gaining in market penetration. The learning curve applies to much of microelectronics manufacturing.

QUESTIONS

8-1. What are the advantages of thin-film circuits over integrated silicon circuits?

8-2. How much thicker is a typical thick-film resistor than a thin-film resistor?

8-3. What drawings would be required for the manufacture of a thin-film circuit? What else would probably be included in the drawings?

8-4. What are three specific problems brought on by the manufacture of great numbers of components?

8-5. Why is CMOS technology so popular in the production of silicon chips?

8-6. Why is computer-aided drafting used for drawings of ICs?

8-7. What are two functions of the photolithographic process?

8-8. What are some typical widths of thin-film resistors?

8-9. How is it possible to obtain a series of several circuits from one integrated semiconductor pattern that has capacitors, diodes, resistors, and transistors?

8-10. What are five drawings, of a series, required for production of a completed integrated- (semiconductor-) circuit package?

8-11. What is your concept of hybridization insofar as this chapter is concerned?

8-12. What trends do you identify in electrical drawing for miniature circuits and devices in the next decade?

8-13. What is the difference between a RAM and a ROM? Between a PROM and an EPROM?

8-14. List five functions in a large office that might be controlled by a microcomputer or microprocessor.

8-15. When does the learning curve concept *not* apply in electronics manufacturing?

PROBLEMS

8-1. Layouts of three masks for a multiemitter transistor are shown in Fig. 8-33. The upper right corner of the collector diffusion mask is shown. The

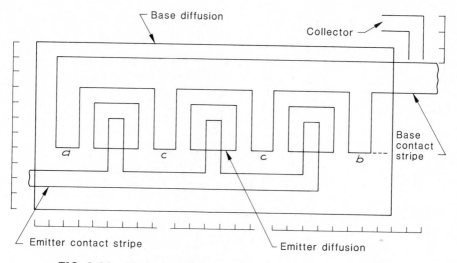

FIG. 8-33 (Prob. 8-1.) Partial mask layout for an IC transistor.

collector is a U-shaped area that comes down each side and stops even with lines a and b. The scales around the edges have marks that are $\frac{1}{2}$ in. apart (on the final drawing). Horizontal distances at c are $\frac{3}{4}$ in. Make a drawing of all the masks, including the collector of this IC transistor on 11×17 or 12×18 paper.

8-2. Metal-oxide semiconductor technology is somewhat different than the bipolar method for IC manufacture. The following steps are employed: (1) the initial slice is N-type oxidized silicon; (2) the first photoresist process cuts a window in the oxide; (3) the surface in the window is reoxidized (thin layer); (4) windows for the source and drain are cut by the second photoresist process; (5) boron is diffused "in" to form the source and drain; (6) the oxide in the main window is stripped off; (7) pure oxide is formed for the gate region; (8) windows for the source and drain are cut by the third photoresist process; and (9) aluminum contacts are deposited and then defined by the fourth photoresist. Make a flow diagram of a MOSFET fabrication process similar to that shown in Fig. 8-4.

8-3. Figure 8-34 shows the schematic diagram of a thin-film resistor and a scale drawing that is complete except for the resistors. It is desired to fabricate resistors 0.02 in. (0.51 mm) wide. Compute the length of the resistors using the equation $L = [(R - 120)/40]W$.

Make a table for the resistors, including value in ohms, width, and length for each. Make a scale drawing of the complete thin-film layout using these widths and lengths. A suggested scale is 1 in. = 0.20 in. Including the 0.2 grid system as shown should enable you to make a reasonably accurate scale drawing. Use $8\frac{1}{2} \times 11$ paper for the layout only, 11×17 paper for layout, table, and schematic.

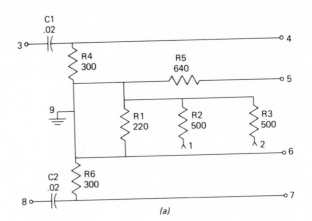

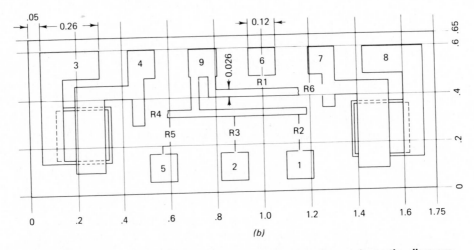

FIG. 8-34 (Prob. 8-3.) Thin-film graphics: *(a)* complete schematic diagram; *(b)* scale drawing, incomplete. (Original scale was 1 in. = 0.20 units or 5:1.)

8-4. The integrated-circuit bar diagram shown in Fig. 8-35 can be connected to form an incline dc amplifier, circuit shown in Fig. 7-4*b*. Draw the mask that would be required for metallization that would produce the amplifier circuit. Your instructor may require you to draw the complete IC also. Suggested scale: 1 in. = 10 units (mils), shown around the edges. Let the conductor paths be 1.5 units wide. The capacitors have the same value. Use any shape for resistors, except that R_2 and R_5 should have same shape ($8\frac{1}{2} \times 11$).

8-5. Figure 8-36 is a perspective drawing of the architecture of a microprocessor chip. (*a* is the CPU; *b*, the clock circuit; *c*, memory circuitry; *d*, expansion circuitry; and *e*, I/O ports.) Rough outside dimensions are 4.5 × 5.6 units. Some other rough dimensions are (a) 1.4 × 2, (c) 1.4 × 1.3, and (d) 3 × 0.5. Estimate any other dimensions you need and make an isometric drawing of this

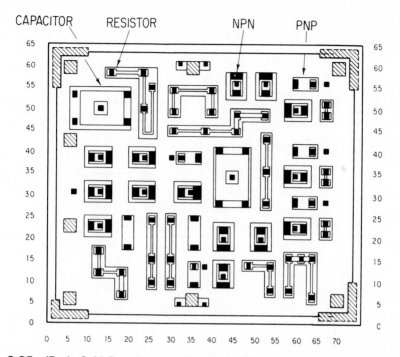

FIG. 8-35 (Prob. 8-4.) Bar diagram of a bipolar SSI. Typical devices have been labeled. Units around the edge are mils.

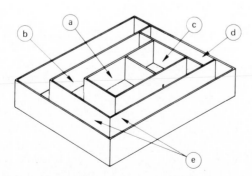

FIG. 8-36 (Prob. 8-5.) A perspective drawing of the "architecture" of a microprocessor chip: *(a)* CPU; *(b)* clock circuit and other auxiliary circuits; *(c)* memory; *(d)* expansion circuitry; *(e)* I/O ports.

chip. Add the letters and leaders much as they appear in the figure, and then write a legend in the lower part of your drawing describing the five parts (a through e).

8-6. Figure 8-37 includes the incomplete flow diagram of a cash register system. Make a flow diagram of the completed diagram in which item 1 is the 8085 CPU, 2 is the 8155 RAM-I/O-timer, item 3 a receipt ticket printer, item 4 an 8355 RAM, and item 5 a change dispenser. (Use $8\frac{1}{2} \times 11$ paper.)

8-7. The flow diagram of a typical microcomputer is presented in Fig. 8-38. Draw the flow diagram with the following additions: At (1) add the Keyboard Display block with flows to Keyboard Switches & Sensors and the Alpha-Numeric Display blocks as shown in the upper part of the figure. At (2) add the Standard Interface block with flows to PROMS, etc., as shown in the upper part of the drawing. Add light shading to the CPU and data bus. (Size: 11×17.)

8-8. Figure 8-39 shows a rudimentary drawing of the dual-displacement engine. Sensors measure temperature, transmission gear, throttle position, and engine load and speed. At a certain condition the control module shuts down three of the cylinders. Draw the outline-schematic drawing of this system, showing the engine, sensors, microcomputer, control signals, and connecting lines. (Use $8\frac{1}{2} \times 11$ paper.)

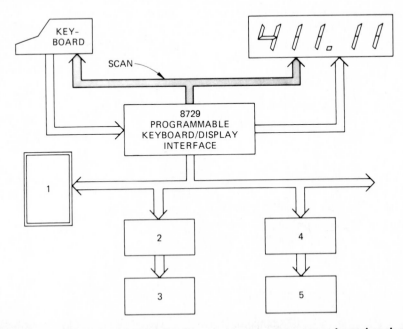

FIG. 8-37 (Prob. 8-6.) Incomplete diagram of a microprocessor-based cash register. Shaded busline is for internal scanning. The other busses are for control and data transmission.

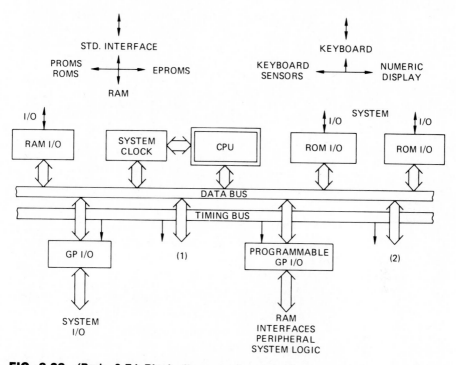

FIG. 8-38 (Prob. 8-7.) Block diagram of a typical microcomputer. Additional blocks are to be added at 1 and 2.

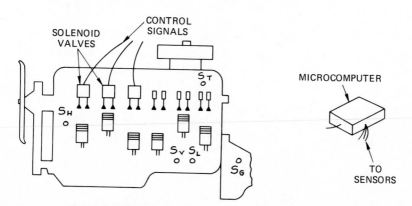

FIG. 8-39 (Prob. 8-8.) Rudimentary outline of a six-cylinder dual-displacement motor and pictorial view of a microcomputer. Approximate sensor locations are shown as follows: S_H, temperature; S_V, engine speed; S_L, load; S_G, gear; and S_T, throttle.

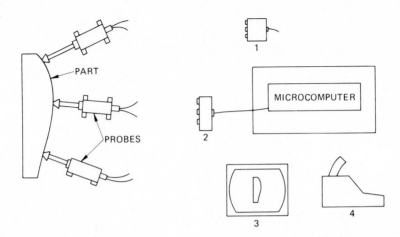

FIG. 8-40 (Prob. 8-9.) A gaging system for manufactured parts. (A single-board microcomputer is used.)

8-9. Figure 8-40 shows a gaging system which is used to determine whether or not a part is within acceptable tolerance. Item 1 is a junction box for a line leading from each probe. Item 2 is a manifold which also has a line from each probe. The manifold and junction box each feed into an 80/20 microcomputer board. The board has outputs to video display (3) and printer (4). Show a schematic flow diagram of this complete system. (Use $8\frac{1}{2} \times 11$ paper.)

8-10. How many minutes would be required to provide just the 50th transponder if 2.3 h were required to produce the first one at (a) 75 percent learning rate and (b) at an 85 percent learning rate?

8-11. If it cost $6.90 to assemble the first transceiver unit, what would it cost to assemble the 100th unit at (a) an 80 percent learning curve rate and (b) an 85 percent rate?

8-12. If it takes 69 minutes to manufacture the first smart power transistor, how many more minutes would be required to make the first 10 units at an 85 percent rate than it would at the 80 percent rate?

8-13. In order to arrive at a competitive price for a new surface-mounted device, the manufacturer estimates it would cost $1.75 per unit if 10 million were to be made. If a 75 percent learning curve could be followed, how much would it cost per unit if sales (and production) could be increased to 60 million?

9

Industrial Controls

9-1 Introduction

In recent years, there has been a steady evolution toward the use of solid-state components in industrial control. This does not mean that all electromechanical devices will be replaced, but they will share a smaller part of the industrial control spectrum in the future.

This chapter covers the drawings encountered in many different industrial control situations, from basic motor control to computer control. Almost all industrial control drawings follow some drawing standard or recommended practice, and these standards will also be described.

In general, as mentioned previously in the text, there are several types of drawing that will be associated with all types of control. They are: (1) the elementary diagram or schematic, (2) the wiring diagram, (3) the layout, (4) the block plan, (5) the parts list, and (6) the assembly drawings. However, some types of industrial control have associated specialized drawings, which will also be covered. An interesting phenomenon takes place in industrial control drawings; namely, as the control circuits become more sophisticated and complex, the drawings tend to become simpler. This will become obvious as the chapter proceeds.

A word of caution should be inserted here. Although this chapter presents industrial control drawings in discrete categories (e.g., electromechanical, solid-state logic, etc.), in reality, these types of control are sometimes used together. This can, in some instances, present nightmares to the drafter putting them down on paper. Basic concepts will show how to ease this interfacing chore.

9-2 Basic Motor Control

Most designers and drafters doing industrial control work start by "cutting their teeth" on some type of motor control circuit. Before delving into the types of drawing used in motor control circuit design, it is a good idea to first understand the functions of motor control and some of the components associated with motor controllers. There are certain definite functions that are performed or governed by motor control. These are: (1) starting, (2) stopping, (3) running, (4)

321

speed regulation, and (5) protection. Starting technically refers to rotating the motor from zero speed to maximum (breakdown) torque speed. Running refers to propelling the motor from maximum torque speed to load speed (usually faster). These two functions are accomplished by one device in most motor control circuits. Speed regulation is done by some device such as a rheostat in a motor control circuit. Motor control circuits provide protection for both the power source and the motor. An example of one of the motor control functions is "reduced voltage starting," which reduces the current inrush during the starting period of a motor or machine. This eliminates or minimizes the shock of a quick start on the driven machine, the reduction of voltage in the line or system to which the machine is connected, and the dropping out of synchronism of synchronous motors on the line. Reduced-voltage starters, which provide smooth accelerated starting without a serious drop in line voltage, are available.

A large motor in a New England plant requires 56 s starting time under normal load. An oil-well pump in Texas will suffer serious damage if its rotor locks and the motor is not tripped from the line in 20 s. A conveyor drive motor in a Florida potash plant can withstand 25 percent overload for 30 min, but a compressor motor in Missouri may burn up in 3 min at 25 percent overload. Each of these motors must be protected from abnormal overload currents, which creep in from time to time.

An electric motor controller is a device, or group of devices, which controls the electric power delivered to the motor. Motor controls range in complexity from a simple manual motor starter to an elaborate motor control center. Some basic control devices which may be built into motor controllers, as well as other industrial controls, are described in the following paragraphs.

A circuit breaker has the function of interrupting the flow of power in a circuit under normal and abnormal conditions. Although it might be used as a power (disconnect) switch, its primary function is to protect the line against abnormally high currents. Hence, it is designed for infrequent operation only. Just as frequently, a line switch and a set of fuses are used in place of a circuit breaker. The switch is used as the disconnect, while the fuses protect the line against faults.

A contactor is a device for repeatedly establishing and interrupting an electric power circuit. In essence, it is a specialized type of relay. Usually operated magnetically, it can be operated in line or can be remotely governed by pilot devices or relays. The overload relay is used to interrupt maximum overload current or to remove the power supply from a starter and motor under a normal overload. Pushbutton switches are used to energize or deenergize the motor controller, and thus start and stop the motor, respectively. Transformers are used to reduce the line (power-source) voltage to the control-circuit voltage. Sometimes, transformers are not used, but most designers prefer to have their control voltages at 120 V ac or less for safety reasons. Many motor controllers have indicating, or pilot, lights on them to show whether the motor is running or stopped.

The types of drawing encountered when doing motor control work are the elementary diagram, the wiring diagram, the layout, the motor-control-center layout, and the motor-control-center schedule. Figure 9-1 shows two types of motor controller. Figure 9-1 shows the elementary diagram of a motor controller. This diagram may appear a little strange to the reader because of the unusual symbols. Unusual symbols can appear because of two factors: (1) different devices are often used in industrial controls than in computer and communication equipment, and (2) different, or alternative, symbols are often used for components that are commonly used. A number of alternative symbols are used in diagrams of this type. These include the contacts and heaters. The unit shown in this diagram is a combination motor starter. Combination means that it contains both the motor starter and a power-source disconnect (switch). Usually in industrial control, combination motor starters are used.

The motor control circuit in the elementary diagram is divided into two distinct circuits, the load circuit and the control circuit. The load circuit, which provides the utilization power to the motor, is drawn horizontally from left to right. The control circuit, which contains the operating coil of the contactor and provides the motor control, is drawn so that devices are arranged horizontally

(a) (b)

FIG. 9-1 Two types of motor controller: *(a)* switch-type combination starter; *(b)* synchronous motor starter. (Allis-Chalmers Manufacturing Company.)

in vertical rows like a ladder. This drawing shows only the first "rung" of the ladder. This motor control circuit operates in the following manner. The disconnect switch (DISC) is closed and usually remains closed all the time unless it is necessary to disconnect the motor from the line circuit. With the disconnect switch closed, voltage is applied via the control power transformer to the control circuit. The start pushbutton (2PB) is depressed and energizes the contractor coil, and its power contacts apply lone voltage to the motor, which it starts and runs. In parallel with the start pushbutton is a contact, which is also on the motor starter, that keeps the contactor coil energized when the pushbutton is released—in other words, it "holds on" the contactor. The three overload relays (OLs, also called heaters) in the load circuit sense the amount of current going to the motor. The overload relays are temperature-sensitive devices. If the load current is too large or the circuit is overloaded, the temperature of sensitive coils becomes too hot, causing the temperature contacts in the control circuit to open up, and shutting down the motor. Because overload relays are temperature-actuated devices, care must be exercised in their selection, with attention to the temperature characteristics of both the relay and the motor.

There are many ways to control motor-starter contactors or relays. Figure 9-3 shows seven typical methods of industrial control. If these methods were used to control the motor in Figure 9-2, the devices—knife switch and pressure switch, for example—would appear in the control circuit. Circuits (*a*) through (*e*) are two-wire control. Circuits (*f*) and (*g*) are three-wire control, using a contact from some other device such as a relay for automatic control.

The wiring diagram in Fig. 9-4 shows exactly how the motor starter in Fig. 9-2 is wired. This type of diagram helps the electrician install and maintain the

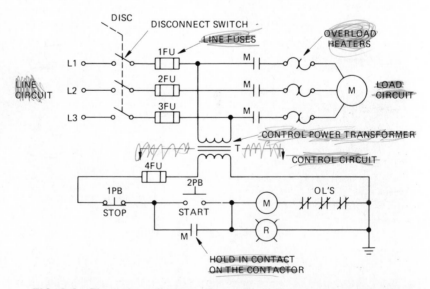

FIG. 9-2 Elementary diagram of a pushbutton motor control circuit.

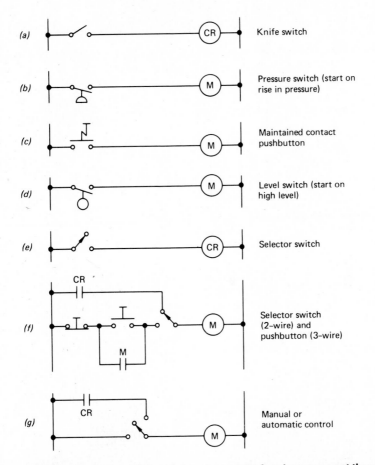

(a) — Knife switch

(b) — Pressure switch (start on rise in pressure)

(c) — Maintained contact pushbutton

(d) — Level switch (start on high level)

(e) — Selector switch

(f) — Selector switch (2–wire) and pushbutton (3–wire)

(g) — Manual or automatic control

FIG. 9-3 Typical methods of industrial control: parts *f* and *g* represent three-wire control; the others, two-wire control.

motor control circuitry. Most electrical-control manufacturers have booklets available containing many of the motor control wiring diagrams. These booklets are of invaluable aid to the designer-drafter in designing and laying out the motor control circuits.

Many times an engineering job will require that many motors be controlled from a centralized location. In order to handle this requirement, a motor control center is a convenient, economical, versatile, and designable package that allows an engineer to put a multitude of different types of motor controller in one enclosure. It is a compact, floor-mounted assembly, comprised principally of combination starters. It consists of one or more vertical sections, with each section having a number of spaces for motor starters. The number of spaces is determined by the horsepower ratings of the individual starters. For example, a starter that will control a 10-hp motor will take up less room than a starter that will control a 100-hp motor. The motor control centers are built so

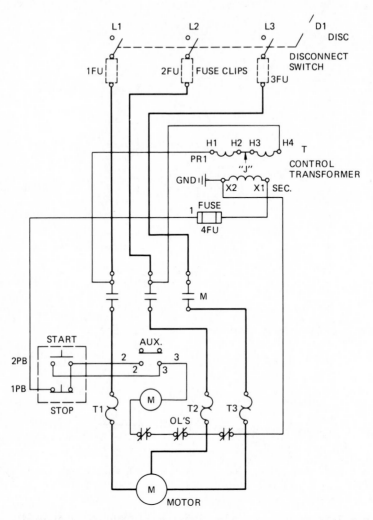

FIG. 9-4 Wiring diagram of a pushbutton motor control circuit.

that they will have starters on one side (front-of-board construction) or both sides (back-to-back construction). Basically, motor control centers have the main power busbar running horizontally and continuously through all sections and individual section busses running vertically. The motor starters plug into the section vertical busses. The drawings encountered in designing and specifying motor control centers, besides the elementary and wiring diagrams just discussed, are the motor control center layout, schedule, and single-line diagram.

Figure 9-5 shows a typical motor control center (front-of-board construction) layout. This drawing is done to scale. Each section is numbered (1, 2, etc.),

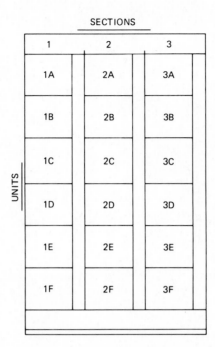

FIG. 9-5 Motor control center layout having three vertical sections.

and each space is lettered (A, B, etc.). If the motor control center were of back-to-back construction, two layouts would be required and an additional letter would appear in each section to designate front (F) or rear (R). The height of the space is determined by the size of the motor starter; the vertical section widths, heights, and depths are determined by the manufacturer, but are in accordance with National Electrical Manufacturers Association (NEMA) Standards. Figure 9-6 shows the motor control center depicted in the layout of Fig. 9-5. Associated with the motor control layout is the motor control center schedule shown in Fig. 9-7. This schedule is developed by the designer to show the "nuts and bolts" of what types of motor controller go into the motor control center. The motor control center schedule usually appears on the same drawing as the layout. Along the left margin is listed the unit or space number, and adjacent to this number are all the particulars associated with the motor controller going in that space. For example, the motor controller going in space 1B is a combination motor starter (CMS), full-voltage (FV), nonreversing (NR), with 3 poles, a 30-A switch, 1.4-A fuses, 1 normally open contact-on contactor, a red "ON" pilot light, and a three-position selector switch for three-wire control (probably hand and thermostat). The controller is for a motorized steam unit heater.

The single-line diagram is a drawing which represents three-phase power and controls as a single line. It will be discussed at length in Chap. 10.

FIG. 9-6 Photograph of a motor control center.

MOTOR CONTROL CENTER No. 40–A

BUS	UNIT NUMBER	DESCRIPTION	SWITCH RATING (AMPS)	FUSE	NEMA SIZE	MOTOR H.P.	AUX. CONTACTS N.O.	N.C.	START-STOP P.B.	PILOT LT. (COLOR)	SEL. SWITCH 2 POS.	3 POS.	NAMEPLATE	REMARKS	WIRING DIAGRAM NUMBER	DRAWING NUMBER
	1A	INCOMING LINE	–	–	–	–	–	–	–	–	–	–	INCOMING LINE		–	
	1B	CMS FV NR	3P-30A	1.4	1	3/4	1	–	–	R	*	–	UNIT HEATER NO. 1		WD-2	
	1C	CMS FV NR	3P-30A	1.4	1	3/4	1	–	–	R	*	–	UNIT HEATER NO. 2		WD-2	
	1D	CMS FV NR	3P-30A	–	1	–	2	–	–	R	*	–	SPARE		–	
	1E	FUSED SWITCH	3P-60A	60	–	–	–	–	–	–	–	–	WELDING RECEPT.		WD-4	
	1F	BLANK	–	–	–	–	–	–	–	–	–	–	–		–	
3φ 3W 600A MAINS	2A	CMS FV NR	3P-30A	8	1	5	2	–	–	R	*	–	ROOF EXH. FAN NO. 1		WD-8	
	2B	CMS FV NR	3P-30A	8	1	5	2	–	–	R	*	–	ROOF EXH. FAN NO. 2		WD-8	
	2C	CMS FV NR	3P-30A	8	1	5	2	–	–	R	*	–	ROOF EXH. FAN NO. 3		WD-8	
	2D	CMS FV NR	3P-30A	5	1	3	2	–	–	R	*	–	MACH. NO. 5 OIL PUMP		WD-6,-7	
	2E	CMS FV NR	3P-30A	5	1	3	2	–	–	R	*	–	MACH. NO. 6 OIL PUMP		WD-7	
	2F	BLANK	–	–	–	–	–	–	–	–	–	–	–		WD-7	
	3A	CMS FV NR	3P-60A	15	1	10	2	–	–	R	–	–	REFRIG. PUMP NO. 1		WD-9	
	3B	CMS FV NR	3P-60A	15	1	10	2	–	–	R	–	–	REFRIG. PUMP NO. 2		WD-9	
	3C	CMS FV NR	3P-30A	8	1	5	2	–	–	R	*	–	PUMP ON. UNIT		WD-5	
	3D	CMS FV NR	3P-60A	15	1	10	2	–	–	R	–	–	REFRIG. PUMP NO. 3		WD-9	
	3E	CMS FV NR	3P-60A	15	1	10	2	–	–	R	–	–	REFRIG. PUMP NO. 4		WD-9	
	3F	BLANK	–	–	–	–	–	–	–	–	–	–	–		–	

FIG. 9-7 Motor control center schedule.

9-3 Electromechanical Controls

Basically, an electromechanical control is a device that is electrically operated and has mechanical motion, such as a relay, servo, solenoid, etc. The motor starter contactors discussed in the previous section are electromechanical devices. Today, the majority of the devices used in industrial control are electromechanical, although, as mentioned, a larger percentage of the devices are becoming solid-state. Builders and users of industrial equipment have long realized that the safety of the operator and the continuance of production are of primary importance in the design and purchase of such equipment. Good drawings help promulgate such equipment.

In electromechanical design, the standards for the drawings that are most widely accepted and most frequently encountered are the Joint Industrial Council (JIC) "Electrical Standards for Mass Production Equipment" (EMP-1-67) and "General Purpose Machine Tools" (EGP-1-67). Although these standards are strictly advisory and their use is entirely voluntary, they do provide an excellent and logical basis for industrial control drawings. These standards will be the basis for doing electromechanical industrial control drawings in this text. There are some slight differences in these standards, and they should be consulted before using them.

To begin with, the JIC standards describe the drawing's size, the device designations and symbols to be used, the drawing arrangements, and the drawings required. The drawing's size should be a multiple of $8\frac{1}{2} \times 11$ or 9×12 in. with a maximum size of 24×36 in. Multiple drawings must be cross-referenced.

Device designations are shown in Table 9-1. Device designations cannot be used for other purposes. If special observations must be used, they should be listed on the diagrams. The device symbols are found in Appendix C. If no symbol is listed, then the prevailing symbol used in ANSI 32.E should be used. Logic symbols should be in accordance with NEMA Standard IC-1.

The drawings for a complete industrial control job should include an elementary diagram, a block diagram, a logic diagram, a sequence of operations, a panel layout, an interconnection diagram, a stock list (bill of materials), an electrical layout, and a foundation drawing. Each one of these will be shown for the same piece of equipment in order to provide the reader with a sense of continuity and deeper understanding.

The elementary diagram, shown in Fig. 9-8, is the key to a set of industrial control drawings. The elementary diagram symbolically represents the control circuit and graphically shows the observer how the control system works. The elementary diagram is similar to the elementary diagram of the motor starter, with the source of power starting on the top left, is drawn horizontally, and connects the source circuit device [a disconnect switch, the contactor contact, and the overload coils (heaters)] and the loads (motors) at the top right. The control part of the diagram is drawn horizontally between two vertical lines, which represent the control power, from the source of the control power to the

TABLE 9-1   Device Designations

A	Accelerating contactor or relay	MTR	Motor
AM	Ammeter	MN	Manual
AU	Automatic	OL	Overload relay
BR	Brake relay	PB	Pushbutton
CAP	Capacitor	PL	Plug
CB	Circuit breaker	PLS	Plugging switch
CR	Control relay	PS	Pressure switch
CRA	Control relay automatic	R	Reverse
CRE	Control relay electronic energized	REC	Rectifier
CRH	Control relay manual (hand)	RECP	Receptacle
CRM	Control relay master	RES or R	Resistor
CT	Current transformer	RH	Rheostat
CV	Counter voltage	S	Switch
D	Down	SOC	Socket
DB	Dynamic braking contactor or relay	SOL	Solenoid
DISC	Disconnect switch	SCR	Series control relay
DS	Drum switch	SS	Selector switch
ET	Electron tube	SSW	Safety switch
F	Forward	T	Transformer
FLS	Flow switch	TB	Terminal board
FS	Float switch	THS	Thermostat switch
FTS	Foot switch	TR	Time delay relay
FU	Fuse	TVM	Tachometer voltmeter
GRD	Ground	Q	Transistor
IOL	Instantaneous overload	U	Up
LS	Limit switch	UCL	Unclamp
LT	Lamp	UV	Undervoltage
M	Motor starter	VM	Voltmeter
MB	Magnetic brake	VS	Vacuum switch
MC	Magnetic clutch	WM	Wattmeter
MCS	Motor circuit switch	X	Reactor

Relays:		Examples:	
	General use		CR 1CR 2CR
	Master		CRM
	Automatic		CRA
	Electronically energized		CRE 1CRE 2CRE
	Manual (hand)		CRH
	Latch		CRL 1CRL 2CRL
	Unlatch		CRU 1CRU 2CRU
	Timers		TR 1TR 2TR
	Overload relay		OL 1OL 2OL
	Motor starters		1M 2M etc.

Source: JIC "Electrical Standards for Industrial Equipment."

last control device required in the circuit. As can be seen, this method of drawing control circuits between two vertical control-power lines produces a diagram similar to a ladder; thus this is often called a "ladder" diagram. The control sequence starts on the top rung of the ladder and ends on the bottom rung, always going left to right.

Under selected control devices and on the right-hand side of the ladder diagram, there are terms describing the command and status functions, e.g.,

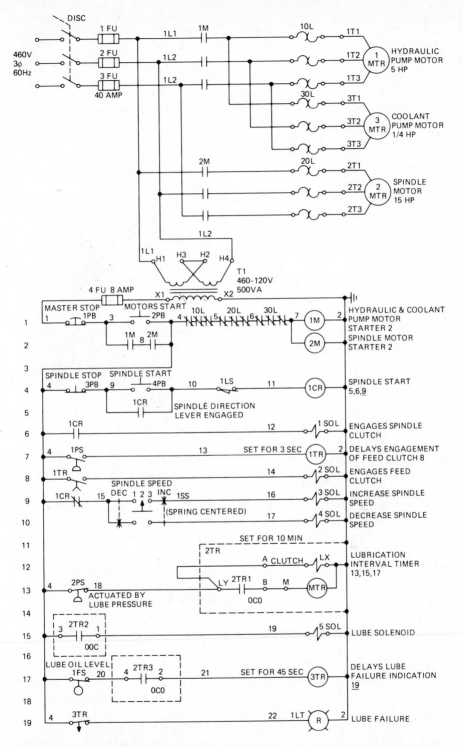

FIG. 9-8 **Elementary or ladder diagram.** [Courtesy of Joint Industrial Council (JIC).]

master stop. Associated with the right-hand-side control device and found under the descriptive terms and on the left-hand side of the ladder diagram are numbers called *line numbers.* They increase consecutively from top to bottom. Line numbers are used for reference: control devices are referenced to the circuits they control by these line numbers, in order to make it easier to understand and follow the circuitry. The numbers between the devices are conductor (wire) numbers, and these numbers should *not* be repeated. Experience has shown the authors that minimum spacing should be $\frac{3}{8}$ in. between lines and $\frac{1}{2}$ in. between components and that minimum component size (circle diameter) should be $\frac{1}{4}$ in., for clarity and ease of reading.

In order to draw an elementary diagram, a number of rules must be observed. Some of these rules are as follows. Actuating coils of control devices, i.e., relays, solenoids, etc., are to be shown at the right-hand side. All contacts should be shown between these coils — not necessarily on the same line — and the left vertical line. If subassemblies are used and their internal wiring diagrams are shown on separate drawings, the subassemblies should be represented on the elementary diagram as rectangles, with actuating coils or contacts shown. The control-device symbols are shown and numbered in the order in which the control sequence takes place. Services *must* be shown in the deenergized position, with the utilities (e.g., water to a flow switch) turned off and the equipment at its normal starting position. Only device contacts that are to be used should be shown. Contacts of multiple-contact devices should be on the line of the elementary diagram, where they are connected in a circuit, and all except contactors and control relays should be connected with dotted lines. Continuity of device descriptions should be maintained through all drawings required. Device functions should be shown adjacent to the respective device symbol.

Descriptive terms for command and status terms should be written in present or past tense. The values of electronic components, i.e., resistors, capacitors, etc., should be listed. If electronic diagrams are used, pertinent information for maintenance troubleshooting should be listed. Although quite a number of rules for elementary diagrams are presented here, do not become anxious or overconcerned. These rules are really just good common sense and will become "old hat" after you have done several elementary-diagram drawings.

Sometimes block diagrams are used to supplement elementary diagrams, when the complexity of the control systems warrants. On a block diagram, each block should be identified and cross-referenced in such a manner that circuitry can be easily located on the elementary diagram.

Logic diagrams are used in conjunction with the elementary diagram when static control or logic modules are provided. These diagrams will be discussed in detail in the next section of this chapter.

Associated with, and on the same drawing as, the elementary diagram is a verbal description, explaining how the circuit works. This description, called the *sequence of operations,* is shown in Fig. 9-9. Typing of the sequence of operations is preferred. Basically, the sequence of operations indicates the

SEQUENCE OF OPERATION

A. MACHINE OPERATION: PRESS "MOTORS START" PUSHBUTTON "2PB". MOTORS START.

B. SELECT SPINDLE SPEED BY TURNING SELECTOR SWITCH "1SS" TO "INC", ENERGIZING "3 SOL", TO INCREASE OR TO "DEC", ENERGIZING "4 SOL", TO DECREASE SETTING.

C. WITH CORRECT SPINDLE DIRECTION SELECTED LIMIT SWITCH "1LS" IS ACTUATED. PRESS "SPINDLE START" PUSHBUTTON "4PB" ENERGIZING RELAY "1CR" WHICH ENERGIZES "1 SOL". SPINDLE STARTS AND PRESSURE SWITCH "1PS" IS ACTUATED. "1PS" ENERGIZES "1TR" AND AFTER A TIME DELAY "2 SOL" IS ENERGIZED PERMITTING MOVEMENT OF MACHINE ELEMENTS AT SELECTED FEED RATES.

D. PRESSING "SPINDLE STOP" PUSHBUTTON "3PB" STOPS SPINDLE AND FEEDS MOVEMENTS SIMULTANEOUSLY.

E. LUBRICATION OPERATION:

F. PRESSURE SWITCH "2PS" IS CLOSED.

   1. TIMER "2TR" CLUTCH IS ENERGIZED WHEN MOTORS START.

   2. CONTACT "2TR-1" CLOSES AND ENERGIZES TIMER MOTOR "MTR" STARTING LUBE TIMING PERIOD.

   3. CONTACT "2TR-3" CLOSES AND ENERGIZES TIMER "3TR".

G. TIMER "2TR" TIMES OUT.

   1. CONTACT "2TR-1" OPENS, DE-ENERGIZING TIMER MOTOR "MTR".

   2. CONTACT "2TR-2" CLOSES, ENERGIZING "5 SOL".

   3. CONTACT "2TR-3" DE-ENERGIZING TIMER "3TR".

   4. LUBRICATION PRESSURE ACTUATES PRESSURE SWITCH "2PS", DE-ENERGIZING AND RESETTING TIMER "2TR". CONTACTS "2TR-1", "2TR-2" AND "2TR-3" OPEN.

   5. CONTACT "2TR-2" OPENING, DE-ENERGIZES "5 SOL".

H. REDUCED LUBRICATION PRESSURE DE-ACTUATES PRESSURE SWITCH "2PS" AND SEQUENCE REPEATS.

SWITCH OPERATION

1LS (4) ACTUATED BY SPINDLE DIRECTION LEVER ENGAGED
1PS (11) OPERATED WHEN SPINDLE CLUTCH ENGAGED
2PS (13) OPERATED BY NORMAL LUBE PRESSURE
1FS (16) OPERATED BY ADEQUATE LUBE SUPPLY

FIG. 9-9 Sequence of operations. [Courtesy of Joint Industrial Council (JIC).]

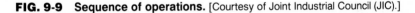

progression of operation of the devices shown on the elementary diagram and, thus, tells how the control circuit works. Graphical representations, such as bar charts, may be used to supplement the written description.

The panel layout and interconnection diagram are shown in Fig. 9-10. These diagrams are of the same control system shown in the elementary diagram in Fig. 9-8. The panel layout shows the general physical arrangement of all devices inside the control-panel enclosure. The panel layout is used to tell the tradesperson how to lay out and construct the control system. Devices are represented by squares, rectangles, and, sometimes, circles, and bear the same identification they had in the elementary diagram. Spare panel space is dimensioned. The panel layout also shows the layouts of the remote console or operator stations. These stations contain the switches that the operator uses to work the control system and are often called pushbutton stations.

The interconnection diagram, also shown in Fig. 9-11, shows all terminal boards and their associated identification within the electrical control system.

PANELS AND CONTROL STATION LAYOUT

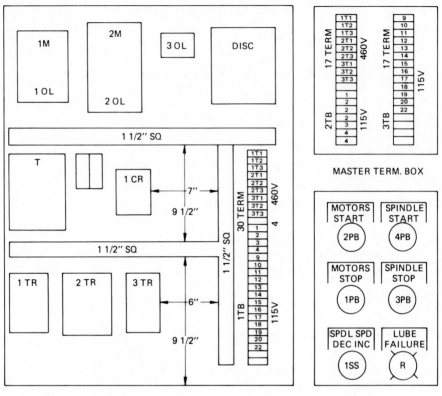

FIG. 9-10 Panel layout and interconnection diagram. [Courtesy of Joint Industrial Council (JIC).]

Terminal-board numbers must correspond exactly to the number of the wire which attaches to the board.

The authors have seen several variations of the panel layouts that have been used successfully. One is to lay out each component on the panel, showing the coils, contacts, etc., of each device and the wire number associated with it. Another method is to show the interconnect wiring between the main control panel and the various operator stations. This method uses parts of a highway diagram and labels the wires' numbers at the individual stations.

The stock list, or bill of materials, shown in Fig. 9-11, tells what devices to use for the construction and maintenance of the control system. The stock list shows each device by designation, quantity, manufacturer's name and model (and, sometimes, serial) number, and any other information necessary for construction and maintenance replacement. The stock list should appear on the same drawing as the elementary diagram or panel layout, but it may appear on a separate drawing because of its size. If it appears on the elementary diagram

SYM.	QUAN.	STOCK NO.	DESCRIPTION
2TR	1		TIMER, 15 MIN, EAGLE NO. HP518
1FS	1		FLOAT SWITCH, ALLEN BRADLEY NO. 840-1A2
1TR	1		TIMER, 60 SEC, TDOE, GE NO. CR122 A04022AA
1CR	1		RELAY, 4PST, 2NO-2NC, A-B NO. 700-N400
1-5 SOL.	5		SOLENOID VALVE, 125 PSI, ASCO NO. 8210D1HW
1LS	1		LIMIT SWITCH, SPDT, MICROSWITCH NO. LSA1A-1A
1PS	1		PRESSURE SWITCH, 0-30 PSI, A-B NO. 836T-T251J
1PB	1		OPERATOR SQ-D NO. KR-5R
3PB	1		OPERATOR SQ-D NO. KR-1R
1,3PB	2		CONTACT BLOCK SQ-D NO. KA-3
4FU	1		FUSE, 5A, 130V, BUSS NO. KAA-5
2MTR	1		MOTOR, 15HP, 460V, 3600 RPM, GE NO. 5K284BN322
1MTR	1		MOTOR, 5HP, 460V, 1200 RPM, GE NO. 5K215BL305
1M,2M	2		STARTER, ALLEN-BRADLEY, SIZE 1, NO. 709-B00103
T	1		TRANSFORMER, 750VA, SORGEL NO. 750SV1B
1-3FU	3		FUSE, 40A, 600V, BUSS NO. FRS-40
DISC	1		SAFETY SWITCH, 3P, 60A, 600V, SQ-D NO. ARD-13
	1		PANEL, HOFFMAN NO. A30P24
	1		ENCLOSURE, HOFFMAN NO. A30S2508LP

FIG. 9-11 Stock list or bill of materials.

or panel layout, entries should be made from bottom to top to avoid running out of room. If it appears on a separate sheet, entries should be made from top to bottom, left to right.

The electrical layout, in Fig. 9-12, shows the outline of the equipment that the control system is to control. It shows the location of the main control panel, the operator's console, and other remote-control devices, such as limit switches, whose location cannot be readily determined from the elementary diagram. Also shown are accessory units, such as hydraulic power units, which are not attached to the equipment, and their relative locations. All devices shown on the electrical layout have the same identification as they have on the elementary diagram.

9-4 Motors

Having discussed motor control and electromechanical controls, we should now discuss motors and how to draw them. Motors are available in all sizes and types from the tiny fractional horsepower motor found in the household refrigerator to extremely large motors (6000 hp and above) found on large pumps in electric power plants. Motors consume 35 percent of all the electric power generated in the United States. Because motors use this large percentage of energy consumed, many organizations from all facets of public life [Department of Energy, GSA, trade associations (NEMA), technical societies (IEEE,

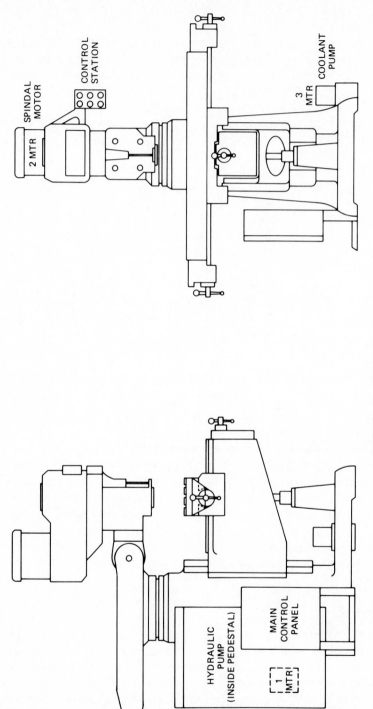

FIG. 9-12 Electrical layout.

SPINDAL MOTOR

CONTROL STATION

2 MTR

3 MTR

COOLANT PUMP

HYDRAULIC PUMP (INSIDE PEDESTAL)

MAIN CONTROL PANEL

1 MTR

ASHRAE), and manufacturers] are considering ways to improve motor efficiency. There are over 340 companies producing motors and generators at approximately 450 different plants. Over 60 percent of the motors manufactured are used in equipment manufactured by industry while the remaining 40 percent are used by industry for "in-house" use.

Figure 9-13 shows several types of motor, some of which are discussed in this chapter. Depending on which text is read, motor classification may differ slightly; however, in this text the following classifications are used: ac motors, dc motors, definite, and special-purpose motors. There are divisions or types in each of these classifications.

Alternating-current motors may be divided into three categories: synchronous, asynchronous, and single-phase motors. Synchronous motors are inherently constant speed because they operate in an exact synchronism with the line voltage frequency (in the United States, 60 Hz; in other parts of the world, 60 or 50 Hz). Therefore, a motor operating within line voltage frequency of 60 Hz could operate at a multiple of only 60, such as 540, 720, 960, 1200, 1800, and 3600 rpm. Synchronous motors are made in all sizes from subfractional, self-excited units to large horsepower (6000-hp +) dc-excited motors. Synchronous motors are used where a constant speed is mandatory or efficiency is required,

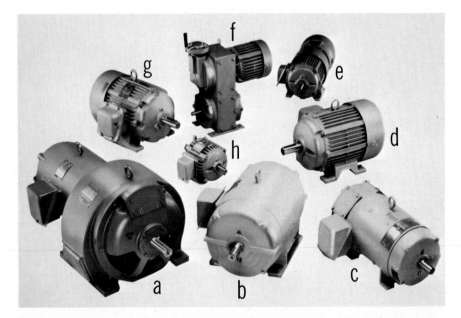

FIG. 9-13 Photograph of various types of motors: *(a)* eddy current clutch motor; *(b)* squirrel-cage induction motor (dripproof enclosed); *(c)* dc motor; *(d)* squirrel-cage induction motor (totally enclosed — fan-cooled); *(e)* brake motor; *(f)* mechanical variable-speed drive motor; *(g)* squirrel-cage induction motor (explosion-proof); *(h)* squirrel-cage induction motor (totally enclosed, nonventilated). (Louis-Allen.)

power factor correction is required, or a combination of all these reasons. Figure 9-14 shows the symbol for dc-excited synchronous motor and two types of asynchronous motors (squirrel cage rotor and wire wound rotor).

The asynchronous motor does *not* operate in exact synchronism with the line voltage frequency. The asynchronous motor is basically an ac transformer, with the line voltage applied to the primary (stator) and a voltage induced in the secondary (rotor). These motors are divided into two basic construction types: the squirrel-cage rotor and the wire-wound rotor. Wire-wound induction motors used to be the only induction motor that could have its speed varied, but with the advent of variable-frequency controls, speed can now be varied in the formerly constant-speed squirrel-cage induction motor. The squirrel-cage induction motor is the workhorse of industry and most common among the induction motors. Figure 9-15 is a photograph of a horizontal squirrel-cage rotor induction motor driving a centrifugal pump. This motor and pump are part of a large air-conditioning system. The motor is rated at 60 hp with 460-V ac, 3-phase, 60-Hz line input and turns at 1770 rpm. This pump delivers 190 gallons per minute (gpm) of chilled water to large air-conditioning units.

There are numerous ways to draw induction motors depending on the type of drawing required. Figure 9-14 shows the basic way to draw induction motors in an elementary diagram and a starter wiring diagram. The windings of many induction motors can be connected to two voltages or dual voltages. Figure 9-16 shows the actual connection diagram of a 3-phase dual-voltage, wye-connected

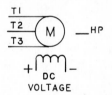

A) DC EXCITED SYNCHRONOUS MOTOR

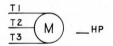

B) SQUIRREL CAGE ROTOR INDUCTION MOTOR

C) WOUND ROTOR INDUCTION MOTOR

FIG. 9-14 Elementary diagram symbols of polyphase ac motors: *(a)* dc-excited synchronous motor; *(b)* squirrel-cage rotor induction motor; *(c)* wire-wound rotor induction motor.

FIG. 9-15 Photograph of a 60-hp squirrel-cage rotor induction motor driving a centrifugal pump.

stator winding, induction motor that might be encountered by an electrician. For simplicity, the windings are often shown as straightline with labeling of only the connecting points. It is important to be able to draw motor winding connections, for this is how the motor is actually connected, and it might be necessary to communicate this information to an electrician.

Another drawing that an electrician might need is the motor outline or the motor end and side view. This information is needed for the physical installation of the motor, that is, the connection of the motor (driver) to the driven element (pump, fan, etc.), the mounting of the motor and driver element, and the checking of the necessary clearances and possible obstructions in the installation location. Figure 9-17 shows the end and side views of a dripproof induction motor.

The dimensions are given by letters which must be correlated to actual dimensions shown in the associated chart. This method of using a chart to give motor dimension is standard among motor manufacturers. In fact, the motor sizes for a given horsepower, speed, and enclosure are standard among motor manufacturers and are given a frame number to represent a standard size.

Single-phase ac motors receive the voltage a single-phase (e.g., 240- or 208-V ac) or line and neutral (120-V ac) power source. These are commonly

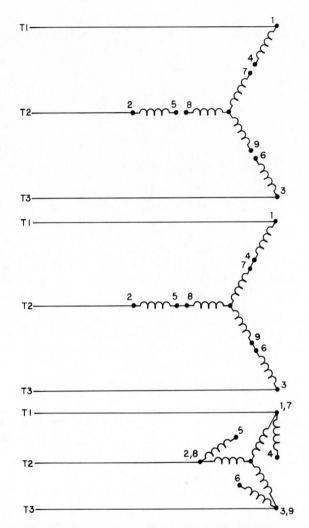

FIG. 9-16 Connection diagram of a three-phase, dual-voltage, wye-connected motor: *(a)* not connected; *(b)* high-voltage-connected; *(c)* low-voltage-connected.

called *fractional-horsepower* motors, although they are also available in the lower integral horsepower ratings. Generally, these motors are used for commercial and residential applications (furnaces, air conditioners, sump pumps, etc.) and are not found in large quantities in industry. Figure 9-18 shows the schematic symbols *for* single-phase motors (general use and shaded pole) and the schematic *of* four commonly found single-phase motors, namely, the split phase, the capacitor start, the permanent split capacitor, and the shaded pole.

Direct-current motors are used in a wide variety of industrial applications because their speed-torque relationship can be varied over a large range, typi-

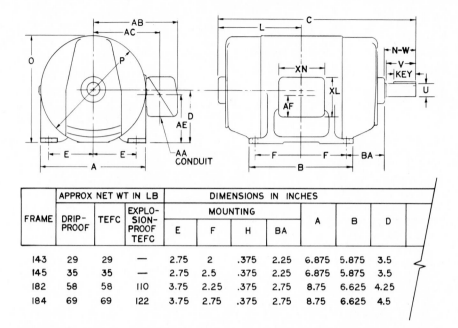

FRAME	APPROX NET WT IN LB			DIMENSIONS IN INCHES							
	DRIP-PROOF	TEFC	EXPLO-SION-PROOF TEFC	MOUNTING				A	B	D	
				E	F	H	BA				
143	29	29	—	2.75	2	.375	2.25	6.875	5.875	3.5	
145	35	35	—	2.75	2.5	.375	2.25	6.875	5.875	3.5	
182	58	58	110	3.75	2.25	.375	2.75	8.75	6.625	4.25	
184	69	69	122	3.75	2.75	.375	2.75	8.75	6.625	4.5	

FIG. 9-17 Side and end views of a squirrel-cage rotor and the associated dimension chart.

cally 8 to 1. Direct-current motors are applied where they can deliver three to five times their normal torque when starting. These motors have to be excited by either the same source voltage as the armature source voltage or separate excitation voltage; they are classified by the type of excitation of field they have. Figure 9-19 shows the symbols for several types of dc motor.

Definite and special-purpose motors are gearmotors, clutch and brake motors, torque motors, and timing motors. These motors are for special applications and are rarely encountered by drafters; therefore, they are not discussed in this text.

9-5 Programmable Controllers

In the previous edition of this text, there was a section on solid-state logic control which described various solid-state logic devices and PC boards and their application to industrial controls. Since that time, however, programmable controllers have essentially captured this market, forcing many manufacturers of industrial solid-state control to eliminate or minimize production of these devices.

The programmable controller is a programmable solid-state replacement for electromechanical "relay-type" controls or solid-state logic controls. Printed circuits were first manufactured in the late 1960s to meet the requirements of the automobile industry, which wanted to reduce the time and ex-

A. SCHEMATIC SYMBOL FOR A
SINGLE PHASE MOTOR (GENERAL)

B. SCHEMATIC SYMBOL FOR A
SHADED POLE INDUCTION MOTOR

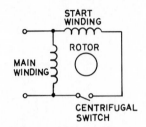

C. SCHEMATIC OF SPLIT PHASE
INDUCTION MOTOR

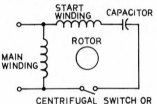

D. SCHEMATIC OF CAPACITOR
START INDUCTION MOTOR

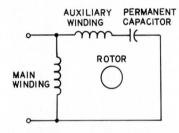

E. SCHEMATIC OF PERMANENT SPLIT
CAPACITOR INDUCTION MOTOR

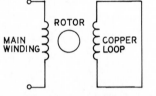

F. SCHEMATIC OF SHADED-POLE
INDUCTION MOTOR

FIG. 9-18 Symbols *for* a single phase motor: *(a)* general use; *(b)* shaded pole, and schematics *of* single-phase motors; *(c)* split phase; *(d)* capacitor start; *(e)* permanent split capacitor; *(f)* shaded pole.

pense of building and testing new relay control panels between model years. Thus the programmable controller contains the relays in programmable logic; therefore, new control sequences require only reprogramming, and not modifying existing or building new relay control panels.

The heart of a programmable controller is the processor, usually a microprocessor, but it could be keyboard-programmable solid-state logic or hardwired solid-state logic. A typical programmable controller consists of four components—the processor, the power supply, the input-output (I/O) modules, and the program console. Figure 9-20 is a block diagram that shows the arrangement of these components.

Programmable controllers used to have only digital input-outputs (I/O) or two I/O states (e.g., on-off, up-down, open-close, etc.), but today many programmable controllers have analog I/O or continuously varying I/O (e.g., temperature, pressure, 0 to 10 V dc, 4 to 20 mA, etc.). Today many programmable

A. SHUNT–WOUND DC MOTOR

B. SERIES WOUND DC MOTOR

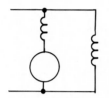

C. COMPOUND DC MOTOR

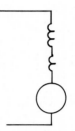

D. SERIES-WOUND WITH
COMMUTATING FIELD
DC MOTOR

E. SHUNT WOUND WITH
COMMUTATING FIELD
DC MOTOR

F. SEPARATELY EXCITED DC MOTOR
WITH COMMUTATING FIELD

FIG. 9-19 The symbols for several types of dc motor: *(a)* shunt-wound dc motor; *(b)* series-wound dc motor; *(c)* compound dc motor; *(d)* series-wound with commutating field dc motor; *(e)* shunt-wound with commutating field dc motor; *(f)* separately excited dc motor with commutating field.

controllers can perform proportional, integral, and derivative (PID) control, operate on local-area networks (LANs) and interface with host computers. In fact, some programmable controllers rival small computers for certain applications.

Figure 9-21 shows a programmable controller mounted within an industrial control cabinet. The processor and power supply are mounted in the upper left. Below the processor-power supply are three rows of I/O. On each I/O row the top half is the I/O logic modules and on the bottom, the I/O terminals where I/O signals are received and transmitted. The I/O wiring is connected to these terminals and is routed through the wireway (running vertically along the left side) to the remote I/O devices.

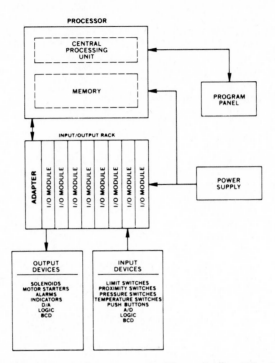

FIG. 9-20 **Block diagram of a programmable controller.** (Allen-Bradley Company.)

Because the programmable controller replaces relay ladder-type logic, it is actually programmed in a relay ladder-type language. Many devices used for programming programmable controllers (CRT, keyboard panel) have relay symbols on the entry buttons. Some programmable controllers have the ability to be programmed in a Boolean-algebra-type language or other assembly-type languages. Many programmable controllers can be interfaced with a computer, and some have the ability to print out their stored "logic" in a ladder diagram. Presently, there are no drawing standards governing the use of programmable controllers in control work. The drawings associated with an industrial control job involving programmable controllers should include an "elementary," a sequence of operations, a panel layout and interconnection diagram, a bill of materials, and an electrical layout. The drawings discussed in the previous sections will not be discussed in this section. Because the relays of a programmable controller are contained within the processor and are not discrete devices, there are varied opinions on what schematic-type or logic-type drawings are required for engineering, construction, and maintenance of the control system. Therefore, the authors will show a number of typical examples of the elementary diagram or logic diagram.

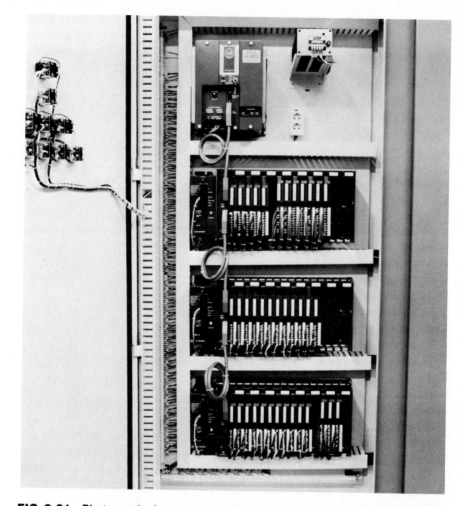

FIG. 9-21 Photograph of a programmable controller mounted within an industrial cabinet. (Allen-Bradley Company.)

Figures 9-22, 9-23, and 9-24 show a method of drawing a programmable controller in an industrial control system. The system shown is the controls of Load Station B of a large shipping conveyor system in an industrial plant. Essentially, the Load Station B controls allow an operator to transfer cartons from conveyor A to conveyor B, load conveyor B, and transfer cartons from conveyor B to the main conveyor. All is done by controlling the motors of conveyors A and B and the raise solenoid of conveyor B. There are many interlocks to permit logical and safe operation and lockouts to permit testing the system. There is even a timer (1 TMR) to stop the system until a carton stops shimmying.

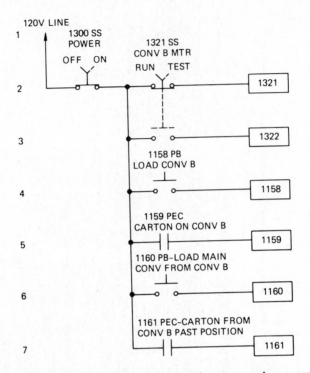

FIG. 9-22 Input diagram of a conveyor control system using a programmable controller.

Figure 9-22 is the input diagram showing the connections from the input device to the input modules of the programmable controller. This diagram is similar to a ladder diagram, but only one side of the power source is shown. Connections to external devices, such as photocells (e.g., 1159PEC), are not shown, but their associated contacts that control a programmable controller input are shown. The inputs to the programmable controller are shown as rectangles and are identified with a four-digit number.

Figure 9-23 is a relay-ladder logic representation of the internal programmable controller logic used. This is the logic used to program the programmable controller. The programmable controller "outputs" are shown as circles and are identified with a three-digit number. These outputs are either internal (400-series number), driving other internal relays, or external (100-series number), driving output devices, such as solenoids and motors. This logic diagram presents a couple of new twists in the standard relay-ladder diagram. For example, contact 492 on line 2 is actually connected to the left side of the ladder and is thus in parallel with contact 1158 on line 1.

Figure 9-24 is the output diagram showing the connections from the output modules of the programmable controller to the output devices. Again, this diagram is similar to a ladder diagram, except that only the neutral of the power

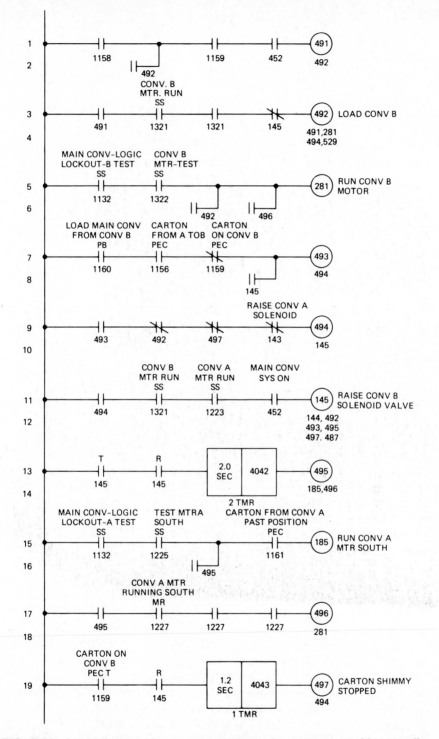

FIG. 9-23 Logic diagram of a conveyor system using a programmable controller.

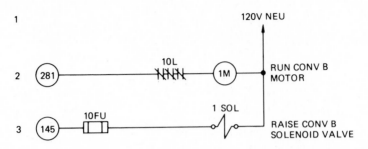

FIG. 9-24 Output diagram of a conveyor control system using a programmable controller.

source is shown. The programmable controller logic of this system controls two devices, namely, the B conveyor motor and raise solenoid.

Figures 9-25, 9-26, and 9-27 show another method of drawing a programmable controller control system. This system marks codes on relay cans. Essentially, it works by a revolving feeder initially loading seven relays in a magazine, which has been placed in the marker chute. The actuation of the initial start pushbutton (6PB) causes the advance solenoid (1SOL) to advance the magazine and the advance-up solenoid (4SOL) to engage the magazine and move it seven positions. Then, the escapement solenoid (3SOL) and the in-place solenoid (2SOL) allow the feeder to load and set another relay in the magazine. After 17 relays have been loaded and advanced, the marker clutch (1CLUTCH) is energized and starts marking the relays as they continue down the chute until all relays are marked.

Figure 9-25 is basically a ladder-type elementary diagram, showing all controls exterior to the programmable controller, the power connections, and the connections to the programmable controller input module and from the programmable controller output modules. The inputs to the programmable controller are identified with an X and a number (e.g., X2 is Start Print Operation) and the outputs from the programmable controller with a Y and a number (e.g., Y1 is an advance solenoid). The remainder of the circuit is self-explanatory.

Figure 9-26 is a logic *representation* of the programmable controller logic used for the control system. Note the word *representation*—the circuit is not the actual logic circuit inside the programmable controller, but one engineer's way of logically depicting how the control system should operate. It should be emphasized that this diagram could just as well have been done using relay symbols, as in our previous example. However, it is easier to optimize using logic-function symbols. The programmable controller inputs and outputs are the interface between this diagram and the preceding diagram. The control relay numbers are fictitious, but invaluable in helping the engineer depict the logic and enter the logic program into the programmable controller. The squares shown with titles in them are actual program subroutines designed to

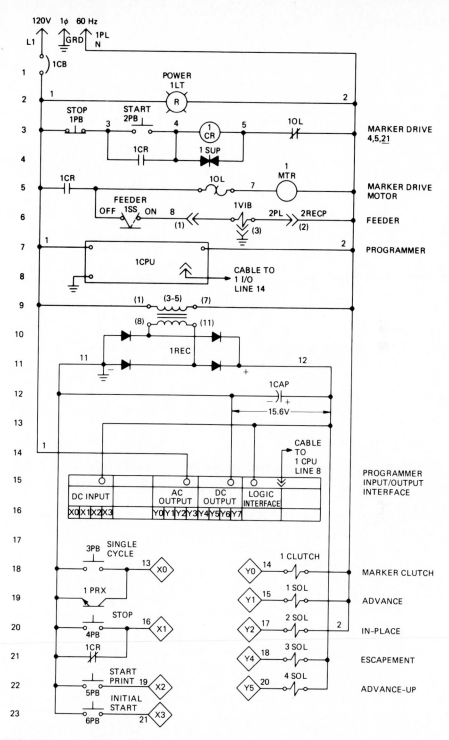

FIG. 9-25 Elementary diagram of a marking system using a programmable controller.

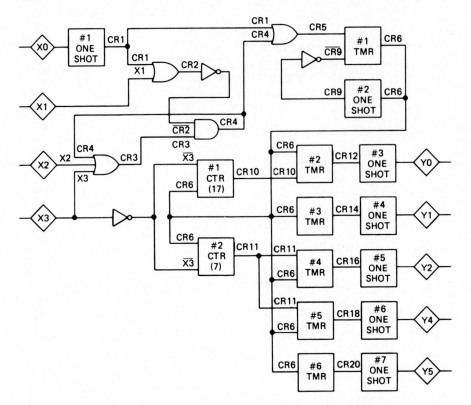

FIG. 9-26 Logic diagram of a marking system using a programmable controller.

operate the control system. For example, the box containing the no. 1 counter is actually a program used for counting.

Figure 9-27 is the actual program that must be entered into the programmable controller for the no. 1 counter program. The program works in this way: the outputs of X3 are inverted (set and reset), CR6 (counts) set, the timer counted up to 17 counts, at which time its output (CR10) is entered in #2TMR.

STORAGE LOCATION	DATA	I/O DEVICE
41	STR	CR6
42	STR/NOT	X3
43	CNTR	—
44	—	17
45	—	—
46	OUT	CR10

FIG. 9-27 One of the programs for the programmable controller used in the marking system.

9-6 Computer Control

Computer control in industry has become a reality from the computerizing of the process itself to the computerizing of shipping and warehousing operations. With the invention of the microprocessor, and thus the microcomputer, many engineers are finding it more economical to use a computer for an industrial control job. Computers are generally used for jobs whose requirements include at least one of the following: many control points, data processing, fast scanning, or fast control speeds. For the most part, computers, unlike programmable controllers, cannot be placed in the harsh ambient conditions of many manufacturing facilities without providing some type of climatic correction (air conditioning, filtering, etc.).

As mentioned at the outset of this chapter, as industrial control circuits become more complex, the drawings required become simpler. Such is the case with computer control. Other than the individual computer internal drawings or software (programming) flowcharts, the drawings particularly required for a computer control system are a system flow diagram and interface drawings. Many other drawings may be used in computer control, such as layouts, wiring diagrams, stock lists, etc.; however, these drawings have been covered in previous sections and need not be described in detail here.

Figure 9-28 shows a flow diagram of a computer numerical control (CNC) system. The concept of using digital computers to do numerical control work

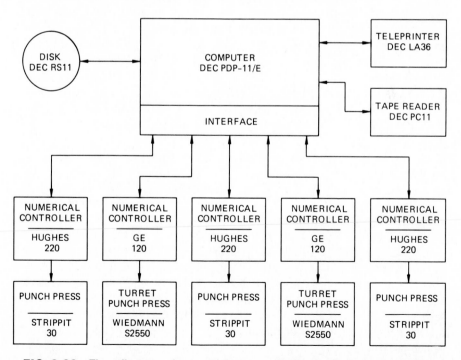

FIG. 9-28 Flow diagram of a computer-controlled numerical control system.

by replacing the tape readers is relatively new. In the system shown in this diagram, the computer was added; the existing tape readers were left to replace the computer in case of computer failure. The numerical control programs are read into the computer and stored on the disk. The teleprinter provides online access to the system and thus enables the operator to monitor or change programs. The computer interface is the electronics that lets the computer communicate with the machine controls, which in turn operate the numerically controlled punch press. The arrangement of flow-type diagrams should be as simple as possible to clearly present the "big" picture of the overall system. Generally, control flow should go from left to right or top to bottom. Symbols should be simple (rectangles, circles, etc.) and should be identified.

Computer interfaces can be relatively simple or complex, custom-engineered or stock-manufactured items, depending on the computer, the control equipment, and the type of control desired. Thus, computer interface drawings can also be relatively simple or extremely complex. In fact, the authors have seen computer interface drawings over 10 ft long, although they could probably have been made smaller. The interface drawings usually consist of logic diagrams and logic-module layouts similar to those found in Chap. 6. However, in complex interfaces, the individual logic circuits often are not shown on the logic diagram—it shows only the external connections, the module outline, and certain internal circuits represented by rectangles.

Figure 9-29 is an interface diagram showing an address selector. An address is a unique number that a control device is assigned by a designer so that it may receive and transmit its data from and to the computer. Inside the rectangle representing the address selector, there are a number of squares, each of which represents two or more logic functions. This address selector has the ability to select four addresses, namely, 764200, 764202, 764204, and 764206 (octal numbers). On the left-hand side of the diagram there are a number of circles with "B" inside, signifying the data bus. From the data bus, connections are run to the module card. The numbers associated with each bus connection tell where the bus is found in the module rack (e.g., B_8-S_1 refers to the module in row B, column 8, with connection at pin S_1). Adjacent to the address selector are its terminals, which are numbered. Inside the address-selector module and next to the terminals are alphanumeric combinations, which are acronyms for the computer bus functions (e.g., A_{17} is address line 17, CO is the data transfer mode). On the right side of the diagram there are connections with arrowheads going to other interface logic modules. For example, SELO selects address 764200 and, in combination with OUT HIGH or OUT LOW, selects the high or the low byte. The number/letter combination on the bottom of the address selector represents the location in the module layout (A_7) and the manufacturer's model number (M105, Digital Equipment address selector). As can be seen, computer interfaces can become complicated; however, like all drawings, a clear and logical layout of the diagram will provide better readability and understanding.

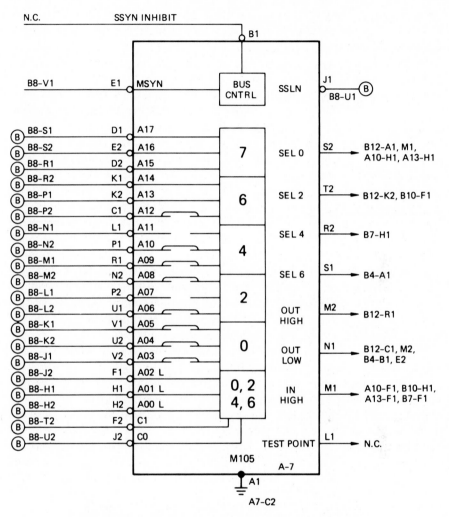

FIG. 9-29 Interface diagram of an address selector.

Computer control of industrial processes is definitely the future trend, and thus its associated drawings will continue to be with us in some form.

9-7 Robotics

As was discussed in Chap. 2, robotics is one of the main areas of computer-aided manufacturing (CAM). In many areas of manufacturing, such as automobile, robots are performing an increasing number of repetitive chores which were once done by humans. Today, far from being an R2D2 or C3PO of *Star Wars* fame, robots are, in this purest definition from the Robot Institute of America, "a re-programmable, multifunction manipulator designed to move material, parts, tools or specialized devices through various programmed motions for the performance of a variety of tasks."

Historically, one could say that the remote manipulator developed for handling radioactive materials in the 1940s is the predecessor of the modern robot. Robots have acknowledged productivity improvements in the areas of spot welding, tool loading, die casting operations, painting material handling of heavy or sharp objects, and repetitive tedious assembly-line operations.

Robots today consist of three major components: a manipulator, the controller, and the power supply. The manipulator is a series of mechanical linkages and joints which move in many directions to do the work required of a robot. Within the linkages, feedback devices are installed to sense the positions of these linkages and joints and transmit the positions back to the controller. These devices may be analog or digital and could include potentiometers, encoders, or limit switches. The controller has three basic functions: (1) to start and stop motion of the manipulator in the desired sequences and locations, (2) to store position and motion data in memory, and (3) to communicate to the "outside world." Figure 9-30 is a block diagram of a robot showing these major components and subcomponents. Robot controllers range from simple-step sequencers to stand-alone microcomputers and minicomputers. The function of the power supply is to provide the necessary power for the robot to do its job. This power supply can be hydraulic or electric or even a combination of both.

The arrangements of the mechanical manipulators are very diverse among robots, and so is the terminology used to describe the mechanical motions and components. An accepted way to describe robots is in terms of the coordinate systems in which they operate: (1) cylindrical, (2) spherical, and (3) jointed-spherical. A cylindrical coordinate robot has a vertical column mounted on a rotating base and a horizontal arm mounted on the vertical column. The column rotates; the arm moves in and out and up and down on the vertical

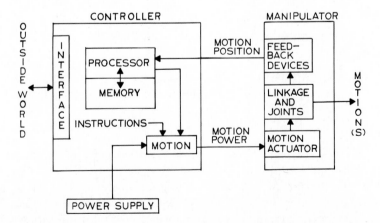

FIG. 9-30 Block diagram of a robot and its major components.

column. In a jointed-spherical coordinate robot, the configuration is similar to a tank turret. The arm moves in and out, pivots in a vertical plane, and rotates in a horizontal plane about the base. The jointed-spherical system has a base, an upper arm, and a forearm, with an "elbow" joint connecting the forearm to the upper arm and a shoulder joint connecting the upper arm to the base. These movements, as you may check by using your body, create movement that forms a sphere.

As mentioned in Chap. 2, the movements are also called *degrees of freedom* and are sometimes classified thus. All the coordinate systems mentioned previously have 3 degrees of freedom. If a wrist is added to the arm, it can provide three additional degrees of freedom; roll, rotation in a plane perpendicular to the end of the arm; pitch, rotation in a vertical plane through the arm; and yaw, rotation in a horizontal plane through the arm. Try these movements with your own wrist. Additional degrees of freedom may be added by putting the robot on tracks and moving it in and out or in a circle.

Figure 9-31 is a photograph of a robot and, a controller. The robot is a jointed-spherical or jointed-arm robot having 5 degrees of freedom; trunk, shoulder, elbow, and wrist (two). It can carry 10 kg (22 lb) and can be used for welding, sandblasting, and materials handling. It is electrically powered and uses either a point-to-point or linear/circular interpolation positioning method.

Figure 9-32 is a schematic of the main drive PC board of a teaching robot. This is one of many PC boards of this robot. This robot is designed to teach people the fundamentals of robotics. This particular circuit takes the logic signals from the on-board computer and converts them to the actual motor

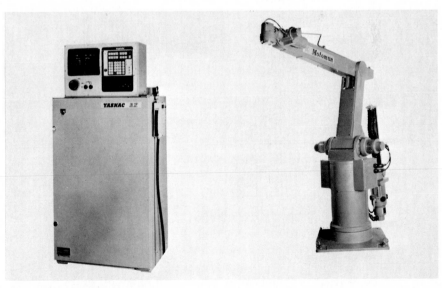

FIG. 9-31 Photograph of a 5-degree-of-freedom robot and its controller. (Yaskawa Electric America, Inc.)

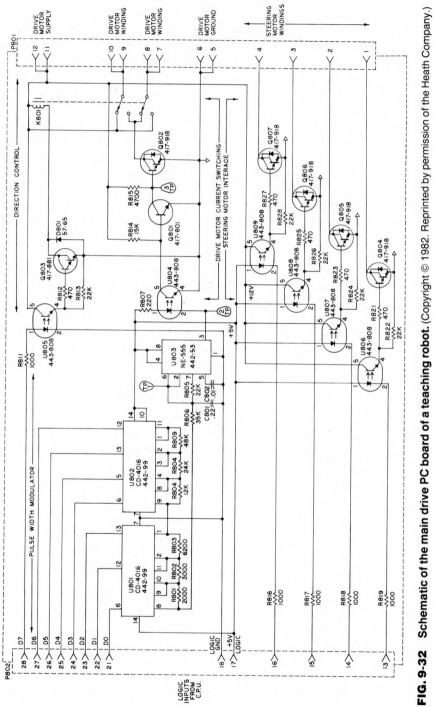

FIG. 9-32 **Schematic of the main drive PC board of a teaching robot.** (Copyright © 1982. Reprinted by permission of the Heath Company.)

drive signals. This schematic is typical of schematics associated with robots or for any other well-designed piece of electronic equipment. The inputs are on the left of the schematic and the outputs at the right. Note how the connector, which is mechanically in the same housing, is broken up and linked to preserve the input-output organization of the schematic. The input circuits are isolated from the amplifier by means of optoisolators. The output circuits are driven by Darlington amplifiers.

The future looks extremely bright for robots. The ever-improving price/performance ratio of computing power plus the incorporation of the advanced electronic technology are making possible the creation of robots that are capable of acquiring knowledge rivaling that of humans for many tasks. These new capabilities include the tactile (touch) and visual senses, closed-loop control, offline programming, and on-screen simulation. These features will give human operators greater control over the robots and will give robots the ability to sense and adjust to their working environment. Presently most robots are used for heavy-duty tasks such as welding, painting, and materials handling. In the future, the greatest potential for the use of robots is in the area of assembly-type operation. It is projected by the Robot Institute of America that robot production could increase from 18,600 in 1983 to 78,000 in 1990 or 420% increase in 7 years. An increase of 32,000 jobs in robot manufacturing, supplies, engineers, and technicians and users is also projected.

9-8 Instrumentation

Many industrial control systems have associated instrumentation which either measures or controls, or does both to, the process variables. As the complexity of industrial processes has increased, the complexity of instruments and instrument systems has also increased; thus, the complexity of associated drawings has also increased.

Besides the usual drawings associated with most industrial control (see chapter introduction), there is one particular drawing unique to the instrumentation field, namely, the balloon drawing. The balloon drawing is associated with the Instrument Society of America Standard ISA-S5.1, titled "Instrumentation Symbols and Identification," which has been adopted by ANSI as Y32.20-1975. The balloon drawing is essentially an instrumentation flow or logic diagram. Unlike most of the drawings encountered previously in this text, the balloon drawing can involve mechanical (hydraulic, pneumatic) signals and supplies (gas, air, steam, etc.), as well as electrical signals (current, voltage) and service voltage). However, this text will cover primarily electrical and electromechanical devices. The balloon is approximately a $\frac{7}{16}$-in.-diameter circle, used to represent an instrument or instrument tagging. An instrument is defined as a device used directly or indirectly to measure or control, or do both to, a variable. The term includes control valves, relief valves, annunciators, pushbuttons, etc. Tagging is the process of identifying other instrument symbols with a code. Figure 9-33 shows the balloon tag of an indicating voltmeter and will be used to demonstrate how identification is done in accordance with the ISA Standards.

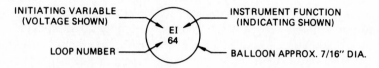

FIG. 9-33 Balloon diagram of an indecating voltage.

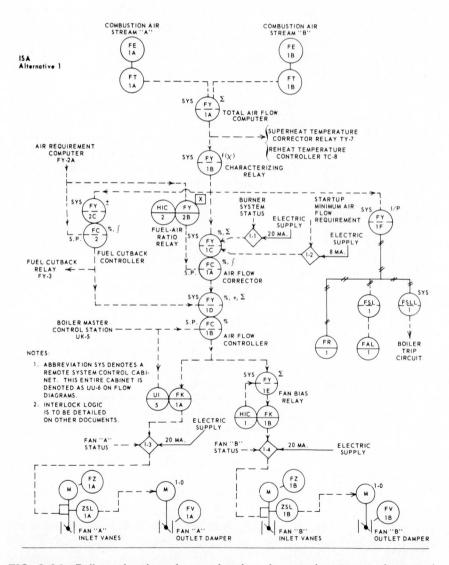

FIG. 9-34 Balloon drawing of a combustion air control system — alternate 1.
(Instrument Society of America.)

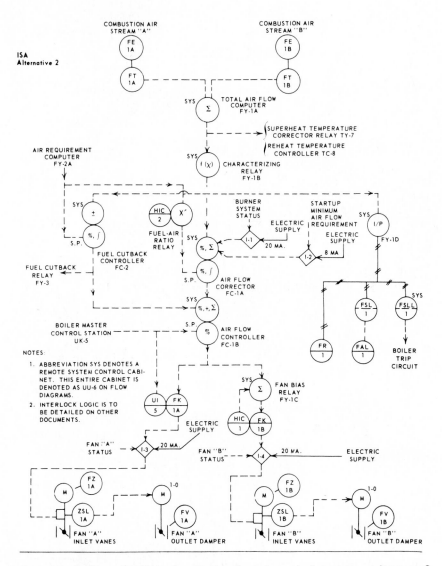

FIG. 9-35 **Balloon drawing of a combustion air control system — alternate 2.**
(Instrument Society of America.)

The first letter represents the initiating or measured variable — in this case, voltage. The second letter represents the instrument function — in this case, indication. The numbers below the letters are the loop numbers. A loop is a combination of one or more interconnected instruments arranged to measure or control a process, or both. Table 9-2 lists the instrument-function code letters and the associated measured or controlled letter. The standard should be consulted before using these codes because there are numerous qualifiers. The way of identifying relays with function symbols is shown in Table 9-3.

TABLE 9-2 Meanings of Identification Letters

This table applies only to the functional identification of instruments. Numbers in table refer to notes in the standard.

	FIRST LETTER		SUCCEEDING LETTERS (3)		
	MEASURED OR INITIATING VARIABLE (4)	MODIFIER	READOUT OR PASSIVE FUNCTION	OUTPUT FUNCTION	MODIFIER
A	Analysis (5)		Alarm		
B	Burner flame		User's choice (1)	User's choice (1)	User's choice (1)
C	Conductivity (electrical)			Control (13)	
D	Density (mass) or specific gravity	Differential (4)			
E	Voltage (eMF)		Primary element		
F	Flow rate	Ratio (fraction) (4)			
G	Gaging (dimensional)		Glass (9)		
H	Hand (manually initiated)				High (7, 15, 16)
I	Current (electrical)		Indicate (10)		
J	Power	Scan (7)			
K	Time or time schedule			Control station	
L	Level		Light (pilot) (11)		Low (7, 15, 16)

Letter	Measured or initiating variable	Modifier	Readout or passive function	Output function	Middle or intermediate (7, 15)
M	Moisture or humidity				User's choice
N (1)	User's choice		User's choice	User's choice	
O	User's choice (1)		Orifice (restriction)		
P	Pressure or vacuum		Point (test connection)		
Q	Quantity or event	Integrate of totalize (4)			
R	Radioactivity		Record or print		
S	Speed or frequency	Safety (8)		Switch (13)	
T	Temperature			Transmit	
U	Multivariable (6)		Multifunction (12)	Multifunction (12)	Multifunction (12)
V	Viscosity			Valve, damper, or louver (13)	
W	Weight or force		Well		
X (2)	Unclassified		Unclassified	Unclassified	Unclassified
Y	User's choice (1)			Relay or compute (13, 14)	
Z	Position			Drive, actuate, or unclassified final control element	

TABLE 9-3 Function Designations for Relays

The function designations associated with relays may be used as follows, individually or in combination. The use of a box enclosing a symbol is optional; the box is intended to avoid confusion by setting off the symbol from other markings on a diagram.

SYMBOL	FUNCTION
1. 1-0 or ON-OFF	Automatically connect, disconnect, or transfer one or more circuits provided that this is not the first such device in a loop $8 \cdot 2$
2. Σ or ADD	Add or totalize (add and subtract)†
3. Δ or DIFF.	Subtract†
4. $\pm$ $+$ $\boxed{-}$	Bias*
5. AVG.	Average
6. % or 1 : 3 or 2 : 1 (typical)	Gain or attenuate (input : output)*
7. $\boxed{\times}$	Multiply†
8. ÷	Divide†
9. $\boxed{\sqrt{}}$ or SQ. RT.	Extract square root
10. x^n or $x^{1/n}$	Raise to power
11. $f(x)$	Characterize
12. 1 : 1	Boost
13. $\boxed{>}$ or HIGHEST (MEASURED VARIABLE)	High-select. Select highest (higher) measured variable (not signal, unless so noted).
14. $\boxed{<}$ or LOWEST (MEASURED VARIABLE)	Low-select. Select lowest (lower) measured variable (not signal, unless so noted).
15. REV.	Reverse
16.	Convert
a. E/P or P/I (typical)	For input/output sequences of the following:

DESIGNATION	SIGNAL
E	Voltage
H	Hydraulic
I	Current (electrical)
O	Electromagnetic or sonic
P	Pneumatic
R	Resistance (electrical)

b. A/D or D/A	For input/output sequences of the following:

A	Analog
D	Digital

17. ∫	Integrate (time integral)
18. D or d/dt	Derivative or rate
19. 1/D	Inverse derivative
20. As required	Unclassified

* Used for single-input relay.
† Used for relay with two or more inputs.

Figure 9-34 shows a partial balloon drawing of a combustion air control system using the ISA instrument functional identifiers. Figure 9-35 shows an alternative and equally acceptable way of doing the same drawing using a combination of ISA instrument functional identifiers and functional designation symbols. Here, most of the balloons represent instruments with two instrument tags, shown at the bottom. The logic sequence of measuring to controlling flows from top to bottom; however, this orientation is optional. Instruments are arranged in terms of logic flow and will not necessarily correspond to signal-correction sequence. Signal lines may enter the balloons at any angle, and symbols may be drawn at any orientation. However, drawing neatness and organization leads to better understanding. Electrical signals are shown as dashed lines, and pneumatic signals are shown as solid lines with double crosshatch (//) interspaced. Directional arrowheads are added to signal lines when clarification of the flow of intelligence is needed. Explanatory notes are sometimes added adjacent to the symbols in order to clarify the function of an instrument or process device.

Individual instruments, like instrumentation systems, have become exceedingly complex in recent years. Thus, the drawings associated with instruments have also become complex. Many instruments are a mixture of sophisticated electronics, electromagnetic devices, and motors. Figure 9-36 shows a three-pen recorder, and Fig. 9-37 shows a block diagram of this recorder. This

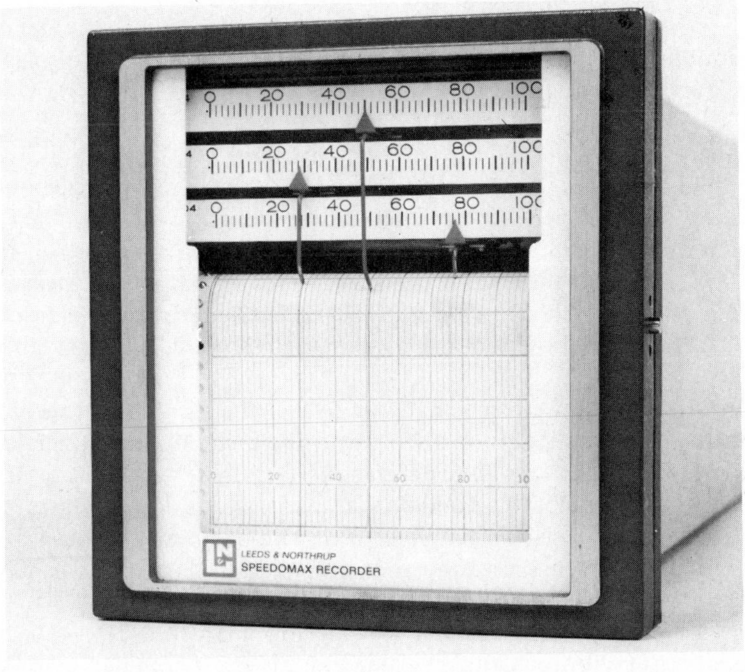

FIG. 9-36 **Photograph of a three-pen recorder.** (Leeds and Northrup Company.)

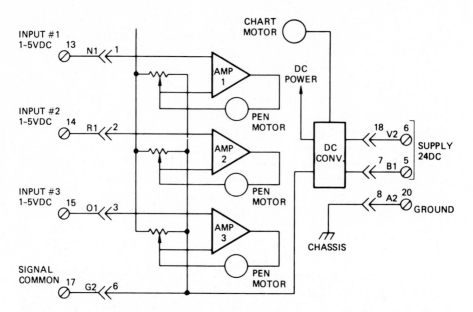

FIG. 9-37 **Block diagram of a three-pen recorder.** (Leeds and Northrup Company.)

instrument may be used in measuring or measuring and controlling a variable, such as voltage or temperature. The basic parts of this recorder, as shown in the block diagram, are: (1) the measuring circuit (shown on the left side), (2) the detector-amplifier, (3) the pen or balancing motor, and (4) the display or recording and indicating devices. The power input to the recorder is shown on the right side of the diagram.

The recorder works in the following manner. The measuring circuit for each input compares the incoming voltage signal to a known voltage, and the difference or error signal is detected and amplified by the amplifier. After amplification, the error signal is applied to the balancing motor, which, in turn, adjusts its position to nullify the error signal. Then, the position of the balancing or pen motor is indicated by an associated marker and recorded on the chart by an associated pen. This position is calibrated to the value of the variable measured.

Figure 9-38 shows a schematic of one of the recorder's three detector amplifiers. This amplifier is on a separate printed-circuit card, which is represented by the dashed lines. The detection and preliminary amplification is done in the integrated circuit (IC$_1$), and final amplification is done in the transistorized (Q_1 and Q_2) push-pull amplifier. The amplified error signal drives the balancing motor. Note the symbol for a ground in the lower right-hand side of the drawing.

Figure 9-39 is a block diagram showing how this recorder might be used in a closed-loop control system in which completely automatic control is required. The system consists of (1) a detector to produce a dc voltage in propor-

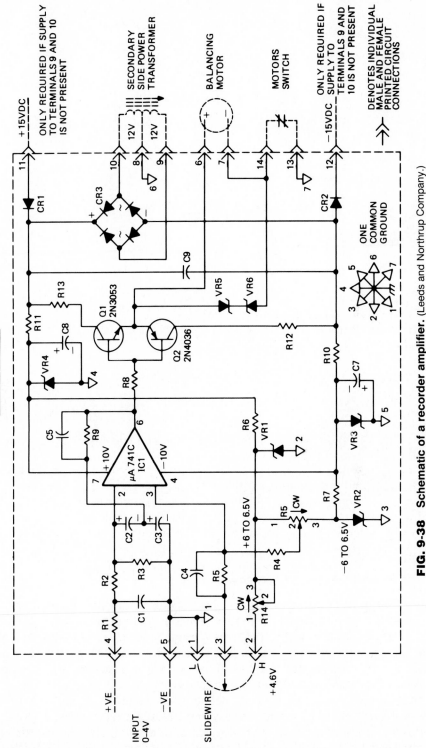

FIG. 9-38 Schematic of a recorder amplifier. (Leeds and Northrup Company.)

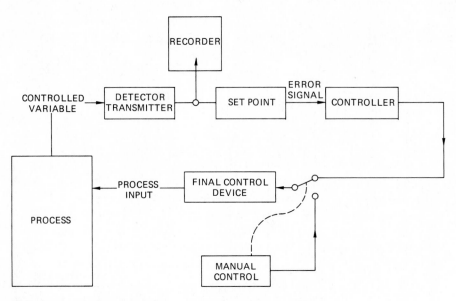

FIG. 9-39 **Block diagram of an automatic control system.** (Leeds and Northrup Company.)

tion to the controlled variable, (2) a set-point unit to provide a control set point to which the process is to be controlled (and to compare the voltage from the detector with the set-point voltage to provide an error voltage), (3) a controller to provide an output related to the detector and error voltage, (4) a final control device, such as a valve drive motor or SCR, etc., and (5) a recorder to provide the process operator with information about how well the process is being controlled and to show trends with time of the process variable. The two- and three-pen recorders may be used to provide operator information from two or three control loops. When used with an associated input selector unit, each pen may be switched into any one of several different control loops selected by the process operator.

9-9 Power Semiconductors

In the last two decades, the thyristor family of power semiconductors has taken and is taking over some of the control functions traditionally assigned to electromechanical devices. A thyristor is a semiconductor switch whose on-off action depends upon positive feedback. Basic functions being taken over include conversion (ac to dc), rectification (ac to dc), inversion (dc to ac), and cycloconversion (higher-frequency ac to lower-frequency ac). Thyristors are found in a wide range of controls, including solid-state relays, motor starters, motor-speed controllers, and furnace controllers. Thyristors offer a designer longer life, more precision, and faster control, and have no moving parts compared with electromechanical devices; however, thyristors are less efficient and consequently radiate more heat. But the future for the thyristor is extremely rosy: projections show them being used in applications ranging from common household appliances to large power-handling applications.

The best known and most representative of the thyristor family is the SCR or silicon controlled rectifier (sometimes called the *semiconductor controlled rectifier*). The SCR is a reverse blocking triode; it operates in only one direction. For operation during the entire cycle, two SCRs (back to back) are required. SCRs can be bought either separately for independent design or in prepackaged control systems.

The SCR controls used in industry are generally complex, with many components. The drawings encountered include: elementary layouts, wiring diagrams, stock lists, etc.; only the elementary diagram will be discussed in this section because of its differences. The only standards associated with SCR control drawings are the ANSI Standards on symbology and diagrams. The authors have found that most companies have unique ways of drawing SCR elementary control diagrams, but generally the elementary diagrams involving SCRs are hybrids of electronics schematics and electromechanical elementary diagrams.

Generally, dc motors are more adaptable to speed control than ac motors; however, most industrial plants do not have a dc source. Therefore, ac must be converted to dc. Figure 9-40 is a partial schematic of a dc, SCR variable-speed

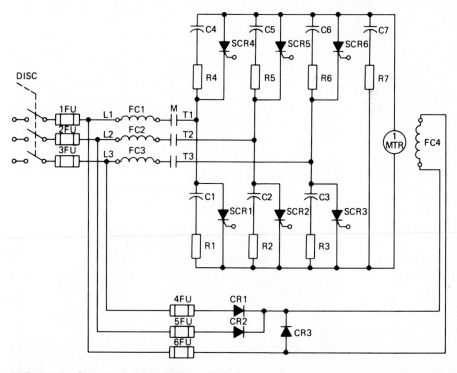

FIG. 9-40 Schematic of an SCR variable-speed drive for dc motors. (General Electric Company.)

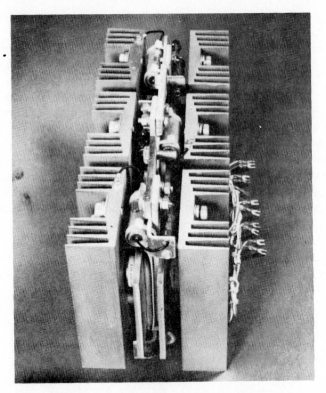

FIG. 9-41 **Photograph of an SCR module.** (General Electric Company.)

FIG. 9-42 **Photograph of a dc SCR variable-speed drive.** (Allen-Bradley Company.)

drive. This type of unit has virtually replaced the shunt-field-rheostat and armature-circuit-resistance methods of speed control. This speed controller operates in the following manner: three-phase ac power enters the fuse, is fed through the line reactor (FC) and the line contactor (M), and enters the SCR power-conversion module, where six SCRs, working independently but in a predetermined sequence, rectify the ac to an adjustable dc voltage. The dc power is fed through a shunt resistor (RT) to the motor armature. The speed of the motor is proportional to the dc voltage applied. The dc voltage is adjusted by changing the positive trigger voltages applied to the gates of the SCRs, which changes their conduction period and, thus, the amount of dc voltage. The trigger circuit, power supply, and miscellaneous control and interface circuits are not shown here owing to their complexity.

This schematic of a speed controller has the power circuit going from left to right. Generally, in SCR control schematics, the power circuit goes in this direction or from top to bottom.

Figure 9-41 shows an SCR power module for this speed controller. Note the large heat sinks to radiate and convect the heat generated by the SCRs. The capacitor and resistor that shunt each SCR may also be observed.

Figure 9-42 shows a dc speed controller. The SCRs are in the lower right-hand corner of the enclosure. Note the printed circuitry through the entire enclosure and the motor contactors in the upper half of the enclosure.

SUMMARY

The drawings in this chapter have been selected from among thousands of control systems that are employed in United States industry today. These drawings represent many different industrial control systems, from basic motor control to more sophisticated computer control. Different types of control may stand alone, or they may be integrated with other types of control. Formats have been suggested for doing each type of control drawing. Several drawings that are common to almost all types of control work include an elementary or logic diagram, a panel layout, a parts list, and assembly drawings. Symbology varies from control type to control type, and it even varies within control type. (There are six sets of logic symbology in use today. See Appendix B.) Some symbology used in this chapter in certain types of diagram is somewhat different in appearance from that in electrical diagrams earlier in the book. In some cases, the types of device used, such as relays and contacts, have their own special symbols. Sometimes new and special symbols, such as instrument balloon tags, have been introduced. Standard identification of components and the numbering of wires and terminal boards is extremely important in industrial control systems because of their complexity and vast number of components. And, finally, the trends of industrial controls have been presented, with their indications of more emphasis on programmable controllers and computers and less on electromechanical devices in the future.

QUESTIONS

9-1. Name five different drawings generally encountered in all types of control work.

9-2. Name four functions of motor control.

9-3. Describe the load circuit in a motor controller; in the control circuit.

9-4. What is the main difference between Figs. 9-2 and 9-4?

9-5. What is three-wire control? Two-wire control?

9-6. What is the difference between a contactor and a motor starter? Between a circuit breaker and a line switch?

9-7. Name two drawings strictly associated with motor control work.

9-8. What standards are generally used when doing electromechanical control drawings? Is the use of these standards mandatory?

9-9. For what words do the following abbreviations stand: PB, PS, CR, LS, OL, TR?

9-10. Why is a ladder diagram called a ladder diagram?

9-11. What is the purpose of a sequence of operations?

9-12. Describe a panel layout and its function; an interconnection diagram.

9-13. What information appears on the bill of materials (stock list)?

9-14. Name three types of motor and what they are used for.

9-15. Name four types of single-phase ac motor.

9-16. What is a winding connector diagram? What is its purpose?

9-17. What type of motor is called the "workhorse of industry"?

9-18. What are the two main differences between an elementary diagram (schematic) and a logic diagram?

9-19. Name the four basic components of a programmable controller.

9-20. Name the two types of symbology generally used in programmable controller elementary diagrams.

9-21. Name the control system requirements that practically dictate the use of a computer.

9-22. What are the major components of a robot?

9-23. What are the three coordinate system designs that a robot may be designed to work with?

9-24. What does "degrees of freedom" mean in description of a robot?

9-25. How should control flow be drawn on flow diagrams?

9-26. Describe the information that is found in a balloon tag.

9-27. What is the different aspect of balloon diagrams (not balloons) compared with any other type of drawing discussed in the text?

9-28. Which of the alternative methods of drawing balloon diagrams would you prefer, and why?

9-29. What is a thyristor? An SCR?

9-30. What basic functions do thyristors perform?

PROBLEMS

9-1. Redraw Fig. 9-2, using three-wire control with a Hands-Off-Auto (HOA) selector switch; a pressure (drop-in-pressure-to-operate) switch to be used in the automatic mode. Use $8\frac{1}{2} \times 11$ paper.

9-2. Redraw Fig. 9-2, using three-wire control with a HOA selector switch. A temperature (rise-in-temperature-to-operate) switch to be used in the automatic mode. Replace line switch and fuses with a circuit breaker. Use $8\frac{1}{2} \times 11$ paper.

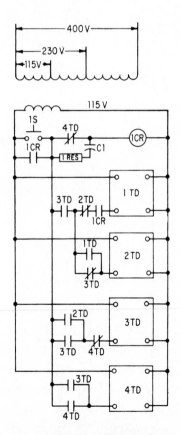

FIG. 9-43 (Prob. 9-3.) A ladder diagram for a sequence timer.

9-3. Make a drawing of the ladder diagram for the sequence timer shown in Fig. 9-43. The following changes might be in order: Add a stop pushbutton switch and selector switch to the appropriate part of the diagram. Add overload protection for the delay-period circuits and in other places you deem advisable. Use $8\frac{1}{2} \times 11$ or 11×17 paper.

9-4. Adjustable-speed drives for dc motors are of four types: (a) motor generator, (2) electron tube, (3) magnetic amplifier, and (4) silicon-controlled rectifier. The tube and magnetic amplifier are obsolete. The motor generator is becoming obsolete. Figure 9-44 shows in diagrammatic form the two types that are still popular. Draw these two diagrams and label all parts. (3-ϕ means 3-phase.) Use $8\frac{1}{2} \times 11$ paper.

9-5. Draw the elementary diagram of an automatic control system that has three 10-hp pumps, which come on with pressure drops to 85, 72, and 60 psi, respectively. Use a HOA selector switch to provide a hand operation for each pump. Use $8\frac{1}{2} \times 11$ or 11×17 paper.

9-6. In Prob. 9-5, we don't want to run one pump more than another; therefore, we want to modify the circuit by installing a manually operated pump-sequence control switch. When it is in position 1, 2, or 3, the pump

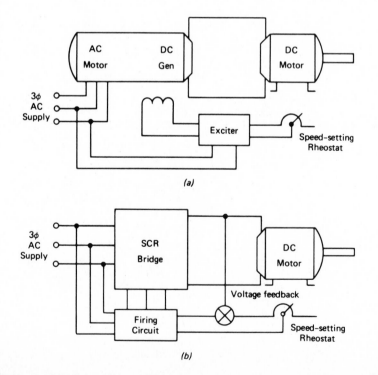

(a)

(b)

FIG. 9-44 (Prob. 9-4.) Two types of adjustable speed drive: *(a)* motor-generator; *(b)* controlled rectifier drive.

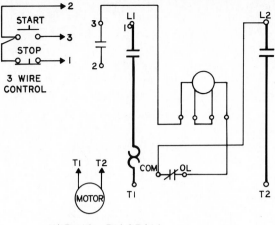

WIRING DIAGRAM

FIG. 9-45   (Prob. 9-7.) A wiring diagram of a single-phase motor and three-wire control.

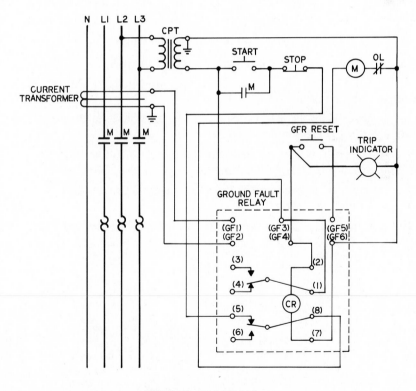

WIRING DIAGRAM

FIG. 9-46   (Prob. 9-8.) Wiring diagram of a motor circuit with a ground fault interrupter.

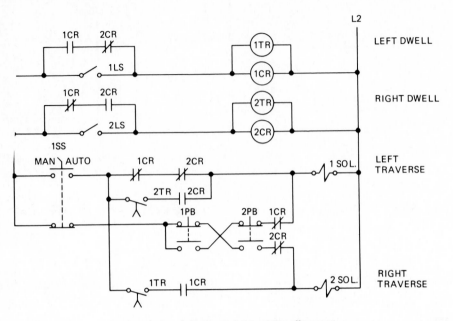

FIG. 9-47 (Prob. 9-9.) Ladder diagram.

sequence would be 1-2-3, 2-3-1, or 3-1-2, respectively. Use a one-pole selector switch and three relays. Use 11 × 17 paper.

9-7. Figure 9-45 is the wiring diagram of a single-phase motor and three-wire control. Draw the elementary diagram.

9-8. Figure 9-46 is a wiring diagram of a motor circuit with a ground fault interrupter. Draw the elementary diagram.

9-9. Figure 9-47 is an example of a ladder diagram. Convert to a PC diagram as shown in Figs. 9-25 and 9-26.

9-10. Redraw Fig. 9-8 as a programmable-controller-based control system,

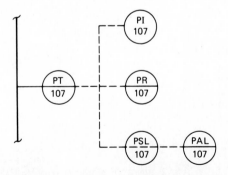

FIG. 9-48 (Prob. 9-13.) Balloon drawing.

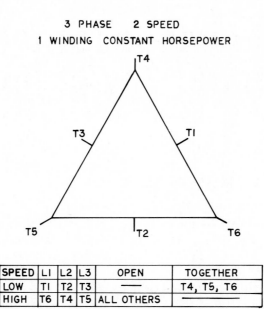

3 PHASE   2 SPEED
1 WINDING CONSTANT HORSEPOWER

SPEED	LI	L2	L3	OPEN	TOGETHER
LOW	TI	T2	T3	—	T4, T5, T6
HIGH	T6	T4	T5	ALL OTHERS	—

FIG. 9-49 (Prob. 9-14.) A connection diagram of three-phase, two-speed, single-winding constant-horsepower motor.

using logic symbols, but not symbols for programs. Use Figs. 9-25 and 9-26 as a guide. Use 11×17 paper.

9-11. A computer-based control system numerically controls a rotary drill press and a multiple-turret boring machine. In addition, the peripherals are a CRT terminal and a line printer. Draw the flow diagram, using Fig. 9-28 as a guide. Use $8\frac{1}{2} \times 11$ paper.

9-12. Draw a balloon tag of an indication ammeter on loop 101. A level alarm on loop 37.

9-13. Figure 9-48 is a balloon drawing of a pressure transmitter feeding a pressure indicator, a pressure recorder, and a low-pressure switch, which in turn feeds a low-pressure alarm. Redraw this circuit. Use $8\frac{1}{2} \times 11$ paper.

9-14. Figure 9-49 is a connection diagram of a three-phase, two-speed, one-winding, constant-horsepower motor. Draw it and connect it for low-speed operation.

10

Drawings for the Electric Power Field

10-1 Types of Electric Power Generation

An electric power generating station is shown in Fig. 10-1. The station has a capacity of 1100 MW,[1] comprised of two 550-MW units that burn pulverized lignite. The units are equipped with air quality control systems which include electrostatic precipitators and wet flue gas scrubbers for sulfur dioxide removal.

A large boiler for each unit provides steam at about 1000°F. This steam is piped to the high-pressure turbine, which is at the left of the unit shown in Fig. 10-2. The pressure of the steam turns the blades of the high-pressure turbine. This turbine and the steam turn the blades of the intermediate turbine. By means of crossover piping the exhaust from the intermediate turbine reaches the two double-flow tandem compound low-pressure turbine shells and causes those turbines to rotate. A large shaft from the low-pressure turbines turns the generator, which generates ac electricity that goes to a switch yard nearby. Here it is converted to 400-kV dc power for transmission over a distance of 400 mi at a voltage of 400 kV dc.

Another way to produce electricity is by conversion of solar energy. The photograph in Fig. 10-3 shows Solar One, a large central receiver solar electric power generating plant. It is a joint undertaking by the U.S. Department of Energy and several California utilities[2] and agencies near Barstow, California. Its purpose is to establish technical feasibility of such plants; gather data on operation, maintenance, etc.; collect environmental impact information; and develop or stimulate commercial and utility acceptance of such systems. It is designed to deliver at least 10 MW of electric power to the utility grid for a 4-hr period on the shortest day of the year and a 7.8-hr period on the longest.

As shown in Fig. 10-4, the Solar One pilot plant consists of seven systems: (1) the collector system, which focuses incoming solar radiation; (2) the receiver system, which absorbs the redirected energy and converts feedwater to steam; (3) the thermal storage system, which minimizes energy transients and stores thermal energy for use during transient or nonsunshine periods; (4) the power

[1] A megawatt is 1,000,000 W of power. It is also 1000 kW, a useful comparison because electricity is usually measured in kilowatt-hours (kWh).

[2] By *utility*, we mean a company or municipal agency that produces or distributes electric power (electricity).

376

FIG. 10-1   Coal-Creek (North Dakota) Station electric generating plant. Two 550-MW units and their boilers and stacks are shown close together in the right center. Cooling towers are at the left, and the main switch yard is in the foreground. (Black & Veatch, Engineers-Architects.)

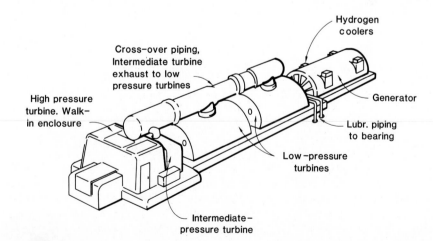

FIG. 10-2   Sketch of a large steam turbine-generator. This unit is called a four-flow tandem compound turbine generator. (General Electric and Black & Veatch.)

FIG. 10-3 The Solar One central-receiver electric power generating facility. The total height of the tower and receiver is 241 ft., and 1818 collectors (heliostats) are spaced around the tower. Reflection of the sun can be seen on those beyond the tower and also on the receiver, which is the target of reflections from the heliostats. (McDonnell-Douglas Astronautics Company and U.S. Department of Energy.)

generation system, which converts the collected thermal power into electric power delivered to the busbar; (5) the plant control system, which controls the plant and gathers data; (6) the plant support system, which includes piping, wiring, and other facilities and equipment; and (7) the beam characterization system, which calibrates the collector system equipment.

The collector system has 1818 individually controlled tracking heliostats, each of which has 12 mirror panels 120 in. long by 43 in. wide. Each heliostat is driven by two motors which are controlled by a central computer system. The receiver system comprises a tower-mounted unit that has 24 energy-absorbing panels that form a cylinder 45 ft high and 23 ft in diameter. Each panel has 70 vertical tubes, $\frac{1}{2}$ in. in outer diameter. Hot (950°F) steam at 1465 psi is produced here and delivered to either a turbine generator, the storage system, or both simultaneously. The turbine and generator operate similarly to those shown in Fig. 10-2. The turbine and generator operate at 3600 rpm and deliver between 10.5 and 13.8 kV to the main station. Figures 6-2 and 6-18 show flow diagrams for central receiver stations or plants.

There are other methods of providing electricity. Hydroelectric power plants produce cheap electricity, but most of the good sites have already been identified and put in use for this purpose. Wind farms with hundreds of windmills (wind generators) in certain windy locations have been constructed. Photovoltaic plants are quite promising for the future because the production of solar (photovoltaic) cells continues to rise and the price continues to drop. Also,

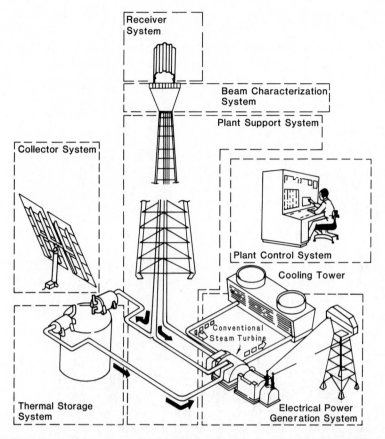

FIG. 10-4 The seven systems of the Solar One pilot plant are shown in pictorial form. Arrows indicate that steam generated in the receiver system can be sent to either the power generation system or the thermal storage system, which provides a steady-state source of steam for the turbine. (McDonnell-Douglas Astronautics Company.)

small photovoltaic units are proving to be good investments and reliable power sources for remote locations. Plants producing as much as 16 MW have been built in the western United States. A large photovoltaic plant resembles a central receiver plant without the tower.

Drawings that are made for electric power installations, whether they are for electric utilities, industry, or large commercial buildings, are as complex as those made in various electronics areas. In fact, a complete set of drawings for a typical electric generating plant just about covers the entire field of electrical drawing. Included in such a set of drawings would be one-line circuit diagrams, three-line diagrams, control and elementary schematics, logic diagrams, general arrangements, and connection diagrams, as well as plans, elevations, sections, and details of structures and equipment. In general, the ANSI Standard

Y14.15, "Electrical and Electronic Diagrams," presents a good format and reference for single-line, schematic-type, and connection diagrams, and one should consult it, if beginning to work in this field. Electrical design and associated drawings should take into account the electrical safety requirements required by the "National Electrical Code," which is discussed at length in Chap. 11.

10-2 The One- (Single-) Line Diagram

One of the first drawings made in a set of drawings for the design of a large project is a one-line diagram of the entire plant or system. Such a drawing shows by means of single lines and symbols the major equipment, switching devices, and connecting circuits of a plant or system.

Figure 10-5 is a good example of a one-line diagram. The single line running down through the figure actually represents three lines in this three-phase system. The power path can easily be traced from the shielded aluminum cable, steel reinforced (ACSR) downward past a grounded lightning arrester, through a GOAB (trade name) disconnect switch, fused disconnect switches, and a step-down transformer. It continues on down through an oil circuit breaker and auxiliary equipment. However, part of the electricity is sent through potential transformers, where it is stepped down to a small voltage for metering. In addition, current transformers near the OCB are used for current-measuring meters, none of which is shown in the drawing, although each is referred to in a note. From here, the power goes through a disconnect switch, past another lightning arrester, and out through a 12.47-kV FCWD (type F Copperweld) line.

Salient features which should be observed in the making of one-line drawings, as shown in Fig. 10-5, are:

1. Use of standard symbols, abbreviations, and designations
2. Highest voltage lines placed at the top or left of the drawing, if possible, with successively lower voltages placed downward or to the right
3. Main circuits drawn in the most direct and logical sequence
4. Lines between symbols drawn either vertically or horizontally, with minimum crossing of lines
5. Generous spacing to avoid crowding symbols and *notes*
6. When pertinent, information included on rating of equipment, feeder lines and transformers, vector relationship of equipment, ground or neutral connection of power circuits, protective measures, types of switch and circuit breaker, and instrumentation and metering

If the above points are observed in the making of a one-line diagram, the drawing will impart a clear picture of the system so represented. The one-line

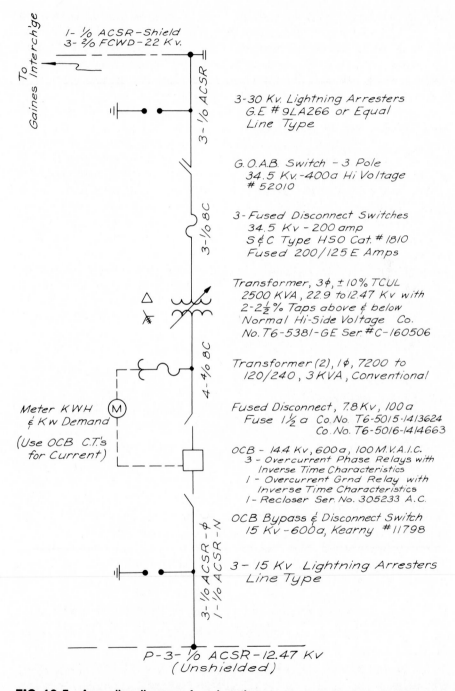

To Gaines Interchge

1- 1/0 ACSR–Shield
3- 2/0 FCWD–22 Kv.

3- 1/0 ACSR

3- 1/0 BC

3- 1/0 BC

4- 4/0 BC

3-30 Kv. Lightning Arresters
G.E. # 9LA266 or Equal
Line Type

G.O.A.B. Switch – 3 Pole
34.5 Kv.–400a Hi Voltage
52010

3- Fused Disconnect Switches
34.5 Kv – 200 amp
S & C Type HSO Cat. # 1810
Fused 200/125 E Amps

Transformer, 3φ, ±10% TCUL
2500 KVA, 22.9 to 12.47 Kv with
2-2½% Taps above & below
Normal Hi-Side Voltage   Co.
No. T6-5381-GE Ser. # C-160506

Transformer (2), 1φ, 7200 to
120/240, 3 KVA, Conventional

Meter KWH
& Kw Demand

(Use OCB C.T.'s
for Current)

Fused Disconnect, 7.8 Kv, 100a
Fuse 1½ a Co. No. T6-5015-1413624
Co. No. T6-5016-1414663

OCB – 14.4 Kv, 600a, 100 M.V.A.I.C.
3 - Overcurrent Phase Relays with
Inverse Time Characteristics
1 - Overcurrent Grnd Relay with
Inverse Time Characteristics
1 - Recloser Ser. No. 305233 A.C.

OCB Bypass & Disconnect Switch
15 Kv –600a, Kearny # 11798

3 - 1/0 ACSR-φ
1 - 1/0 ACSR-N

3 – 15 Kv Lightning Arresters
Line Type

P-3- 1/0 ACSR–12.47 Kv
(Unshielded)

FIG. 10-5   **A one-line diagram of a substation.** (Southwestern Public Service Company.)

drawing, which is used for many purposes, is a distinctive drawing peculiar to the electrical field and as such has made a great contribution to the industry. Although old, Fig. 10-5 is still a good example of a one-line diagram. A diagram drawn today might have higher voltages, such as 161 or 138 kV at the top, more meters and switches along the way, and more bushing-type transformers. Wire sizes would be correspondingly larger than those shown.

Approximately one-half of a one-line diagram for a part of a large electronics manufacturing plant is shown in Fig. 10-6. This diagram shows the electric energy coming from the power company after having been reduced to 12.47 kV. A small substation on plant property reduces this supply to 4.16 kV by means of an Askarel-filled (A-F) transformer. The 600-MCM conductor has an area equal to 600,000 circular mils. A circular mil is a unit used in specifying the cross-sectional area of round conductors. It is equal to the area of a circle whose diameter is 1 mil (0.001 in.). (A wire size table is given in Appendix D.)

The path of the energy can be traced to the motors at the bottom. These motors turn compressors and chillers for the plant air conditioning. To facilitate reading the diagram, a legend showing the abbreviations was printed on the diagram. It is reproduced below:

A	Ammeter	49	Overcurrent-overload relay
AS	Ammeter switch	50/51	Overcurrent (instantaneous and time) relay
V	Voltmeter		
VS	Voltmeter switch	51N	Overcurrent (ground) relay
TB	Test block	64	Ground-fault relay
27/59	Under/over voltage relay	67	Directional overcurrent relay
46	Phase-failure relay	87	Differential-protective relay

To the right of and below most of the symbols is a number. This number indicates how many of the devices are represented by the symbol; A_3, for instance, means that there will be three devices (one on each line) at that location.

In accordance with ANS Y14.15, "Electrical and Electronics Diagrams," the primary conductors have been shown with heavy lines. Some other details recommended by this standard are:

1. Winding connection symbols should be shown for all power equipment.

2. Neutral and ground connections should be shown for all grounded equipment.

3. Rating and type of load, when available, should be shown for each feeder circuit.

4. Current transformers should show the ratio of transformers and the ampere ratio. (Example: $3 - 600/5$.)

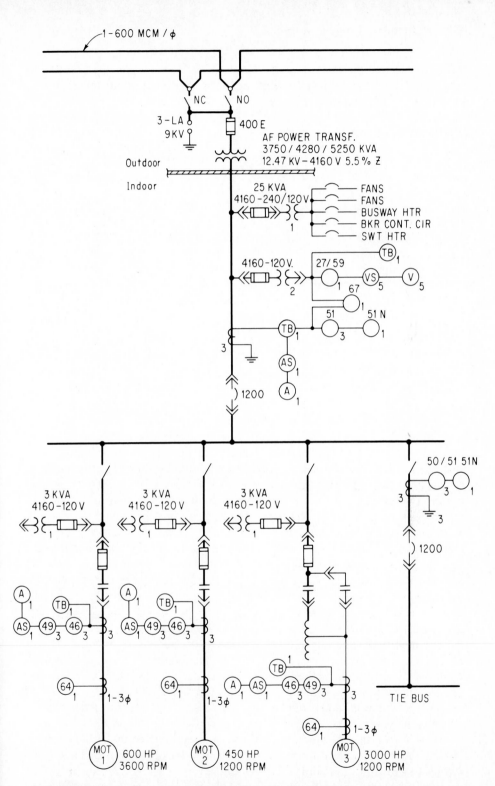

FIG. 10-6 Part of a one-line diagram for an industrial plant. (AT&T Technologies.)

5. Industrial control single-line diagrams may omit equipment ratings when they are used as standard drawings applying to more than one rating.

10-3 Three-Line Diagrams

A schematic-type drawing used in the electric power field is the three-line diagram, which is an expansion of the one-line diagram showing all three phases. Its primary purpose is to graphically illustrate the electrical connections and functions of the metering, relaying, and control power circuitry. It does not show the control of power equipment (e.g., circuit breakers) by this or other circuitry. Three-line diagrams are usually done for generating stations, substations, or any electrical installation that has metering or relaying circuitry. Three-line diagrams are an invaluable aid for construction and maintenance of electrical facilities in that they provide:

1. A complete three-wire schematic, showing all devices, wiring, and connections and their conceptual operation

2. A clean, concise picture of the point-to-point connections

3. The correct way to connect polarity-sensitive devices, such as current and potential transformers

Figure 10-7 shows a three-line diagram of a main breaker and its associated relaying and metering in a 15-kV distribution center in a large industrial complex. The breaker is one of several breakers in an enclosed, indoor, metal-clad switchgear installation. Power conductors in a three-line diagram are generally drawn at the top from left to right or on the left side (as in this case) from top to bottom. The power conductors, like those in the single-line diagram, are shown in heavier lines than the relaying and metering conductors. Device ratings are shown next to the device, and, if it is multirated, the maximum rating and the connected rating should both be shown. Device-function abbreviations and/or numerical codes are shown (e.g., 52 is a circuit breaker). Where two or more devices with the same number function designation are present, they are distinguished with lettered or numbered suffixes (e.g., $87-1$ means device 87 in A-phase). Device terminal numbers are generally shown to the left or below the conductor, while wire numbers are shown to the right or above the conductor. Wire numbers are generally a series of numbers or characters, such as C_1, C_{11}, C_{12}, which provides easier troubleshooting and intelligent conceptual continuity. If two arrangements of the same devices are used, for example, of a current transformer (CT), one CT's wire numbers would be $1\,C_1$, $1\,C_{12}$, . . . , and the other CT's wire numbers would be $2\,C_1$, $2\,C_{12}$, . . . , in order to differentiate between the wirings.

Figure 10-8 shows a similar, but 5-kV class, air breaker and cubicle (cell). Notice how the breaker is on roller wheels and withdraws from load and line stabs (not shown) in the cubicle.

Three-line diagrams should be done with good organization and attention to detail and spacing in order to present a clear, accurate picture of the metering

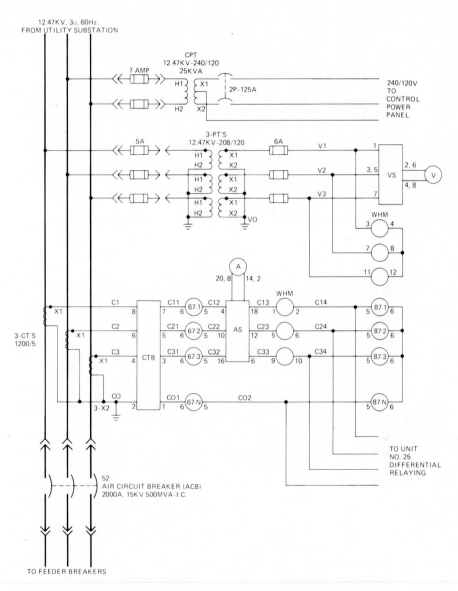

FIG. 10-7 Three-line diagram of 15-kV main breaker.

and relaying circuits. The three-line diagram plays an important part in presenting the total picture of a power system.

10-4 Logic Diagrams

Logic diagrams provide an effective means of communication among persons involved in design, construction, operation, and maintenance of electric generating stations. The five basic symbols are shown and explained in Fig. 10-9.

FIG. 10-8 **Photograph of 5-kV switchgear with breaker withdrawn.** (Courtesy of Westinghouse Electric Corporation.)

Also shown are typical equivalent electric circuits. Some supplementary logic symbols are shown in Fig. 10-10. These symbols convey necessary information used by the electrical control designer, and with their associated descriptive information (panel item, numbers, etc.), aid the overall project design coordination.

The symbols shown do not agree entirely with any one manufacturer's standard, but they are similar to the National Electrical Manufacturers' Association (NEMA) Standard. No particular device is represented by one of these symbols. This type of diagram, therefore, leaves the design of the circuit (including arrangement and selection of devices) up to the electrical control designer.

10-5 Relationship between Logic and Schematic Diagrams

Figure 10-11 shows the logic diagram(s) and schematic diagram of a system that provides correct operation and supervision of an auxiliary oil pump. The logic diagram is interpreted as follows:

> *If the control switch is in the start position or in the automatic position* and *the turbine hydraulic-oil pressure is low, the pump*

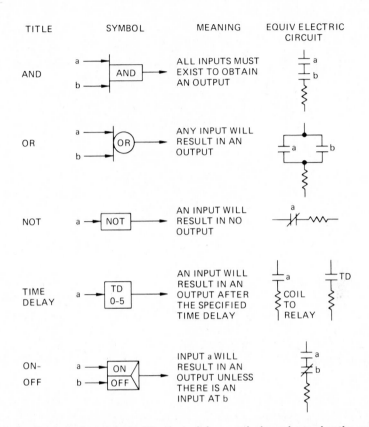

TITLE	SYMBOL	MEANING	EQUIV ELECTRIC CIRCUIT
AND	a → AND → b	ALL INPUTS MUST EXIST TO OBTAIN AN OUTPUT	a, b
OR	a → OR → b	ANY INPUT WILL RESULT IN AN OUTPUT	a, b
NOT	a → NOT →	AN INPUT WILL RESULT IN NO OUTPUT	a
TIME DELAY	a → TD 0-5 →	AN INPUT WILL RESULT IN AN OUTPUT AFTER THE SPECIFIED TIME DELAY	a, COIL TO RELAY, TD
ON- OFF	a → ON OFF → b	INPUT a WILL RESULT IN AN OUTPUT UNLESS THERE IS AN INPUT AT b	a, b

FIG. 10-9 Basic logic symbols. The time-delay symbol numbers give the range of adjustment desired in seconds.

starts. When the starting device is actuated or closed, a red light will go "on" on the turbine start-up panel (TSP) and on the control panel (CP). (The control switch is on the TSP, incidentally.)

If the control switch is in the stop position or if there is a motor overload, the pump stops. In this case a green light will be energized on the CP and the TSP.

If the motor had started and then tripped for some reason other than a control-switch action, an alarm would have sounded on the visual annunciator (item CP_{108}, window B_4) and a white light would have been energized on the TSP.

Certain conventional functions or actions are understood in the reading of these diagrams.

1. Unless otherwise indicated, control switches, pushbuttons, and selector switches are understood to be spring-return to normal.

2. Electrical interlock devices such as cell interlocks are not shown but are understood to be present in accordance with the clients' or office standards.

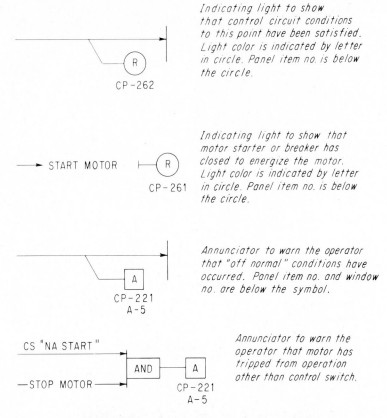

FIG. 10-10 Supplementary logic symbols.

3. "Motor trip" (as contrasted to "stop") is understood to mean that the motor has stopped for some reason other than a control-switch action.

The schematic diagram which applies to the logic diagram in Fig. 10-11 explains in detail all the items necessary for accurate electrical construction and operation, and also becomes a valuable maintenance tool. The control-switch development normally shown on this type of drawing has been deleted for simplicity.

It is possible, but not practical, for the system designer to dispense with schematic diagrams and to employ only logic diagrams in communicating with the fabricator of the control systems. However, according to one designer, the following reasons justify making of schematic diagrams for most systems:

1. To show a complete system with all the desired electrical features

2. To control the selection of alternatives in the circuitry and components to be used by the manufacturer

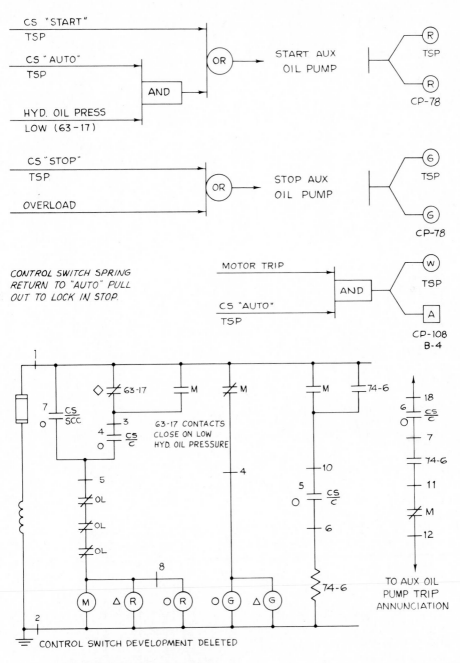

FIG. 10-11 **Example of a logic and a schematic diagram for an auxiliary oil-pump control system.** (Black & Veatch, Engineers-Architects.)

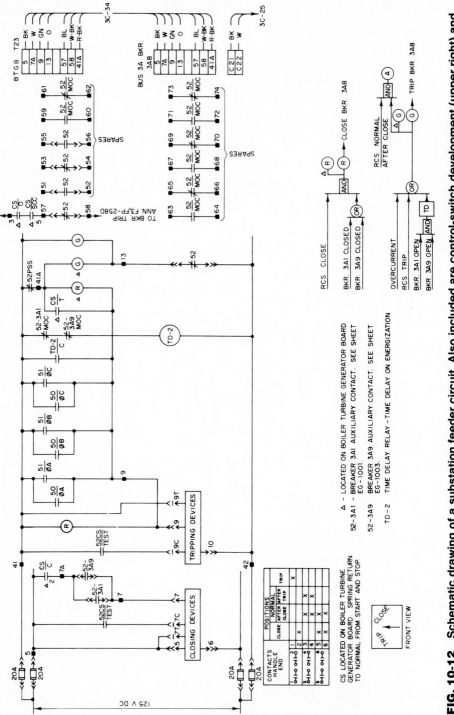

FIG. 10-12 Schematic drawing of a substation feeder circuit. Also included are control-switch development (upper right) and logic diagram (lower right).

3. To permit early determination and retain control of interconnecting cable requirements

Figure 10-12 is the control schematic of a substation feeder breaker which is used in a power generating station. The schematic shows the trip and close circuits that control the operation of the breaker. The trip and close devices which are on the breaker are represented on the diagram by rectangles bearing the labels "Tripping Devices" and "Closing Devices." These devices are usually solenoids that actuate spring-charged, mechanical operators on the breaker. The breaker and its associated contacts are represented by function number 52. The actual breaker, which is a power circuit device, is also not shown on the control schematic; however, the authors have seen control schematics on which it has been represented. The drawing also contains a logic diagram and control-switch development, which shows the switch position and the associated contact arrangement. Some of the abbreviations which may not be familiar to the reader are: CS, control switch; BKR, circuit breaker; BTGB, boiler turbine generator board; SCC, station control cabinet; PSS, permissive selector switch; RCS, remote-control switch; MOC, mechanically operated contact; and ϕA, phase A.

Many power generating stations and industrial plants use large (up to 20,000-hp) motors for pumps, such as condenser, tower, or boiler water; fans, such as boiler and tower; and auxiliary operations, such as coal pulverizers. The motors are characterized by their higher voltages (14 and 4 kV) and their larger and more complex controls. The schematic-type drawing used for the control circuitry is the elementary or ladder diagram, discussed in depth in Chap. 9. Figure 10-13 shows an elementary diagram of the motor controls for a small (350-hp) condenser water-pump motor. The contactor shown is of the removable 4-kV class. Many times, breakers are used in place of contactors for large-motor control. In addition to the normal fuse and overload protection, this motor also has phase-failure (PHFR) and ground-fault (GF) protection. Notice that current transformers are used in the overload and metering circuits. The legend explains the devices not familiar to the reader.

Schematic-type drawings, such as the control schematic and the elementary diagram, used in electric power work provide insight into the necessary functions of the controls along with the necessary electrical connections. As described earlier, the complexity of the controls required in electrical power systems sometimes dictates that logic diagrams be done to aid, but not replace, schematic-type drawings.

10-6 General Arrangement

The *general arrangement* drawing shows how the equipment will be physically arranged when it is manufactured and installed. It is an extremely important drawing in a complete set of electrical drawings because it is the only one that shows the actual equipment. General arrangements show equipment exteriors

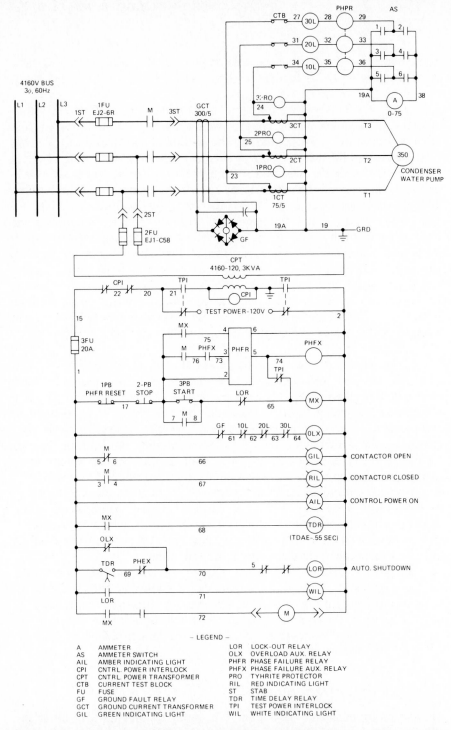

FIG. 10-13 Elementary diagram of a motor controller for a condenser water pump. (General Electric Company.)

and dimensions; the location of devices such as relays, meters, etc.; and pertinent details necessary for field installation, such as service entrances, mounting channels, etc. Such a drawing always contains an elevation and may also contain a plan (top) view, section views, details, and, sometimes, small, simple one-line or three-line diagrams. Because the general arrangement shows the equipment as it will physically exist, it is used by engineers to tell manufacturers how they would like the equipment fabricated and installed.

Figure 10-14 shows a general arrangement of a primary-selective, 15-kV-480Y/277-V, 100-kVA indoor secondary unit substation. This substation consists of three basic assemblies, namely the two primary (15-kV) switches, the transformer, and the low-voltage switchgear. The primary switches are primary-selective because they are interlocked so that only one switch may be closed at any time. The transformer is Askarel-filled (presently not used due to environmental regulations). The switchgear has main, tie, and several feeder breakers. This general arrangement contains an elevation, a plan view, and a simple single-line. This drawing is extensively dimensioned and shows some details. The single-line shows the key interlocking system that is used for personnel safety and proper operating procedures. Figure 10-15 shows an indoor substation similar to the one shown in the previous general arrangement.

Figure 10-16*a* shows two views of a large outdoor substation. The lower view is one of three sections, the location of which is identified in the top view with flaglike shapes and the letter A. Figure 10-16*b* shows a section of a newer substation. Most of the major elements of this substation are identified by "balloons" of various shapes and further described in a schedule on the right side of the sheet. For example, hexagons 1, 5, and 6 are (1) 161-kV circuit switcher, (5) 161- to 13.2-kV power transformer, and (6) 13.2-kV grounding transformer. Squares 2 and 4 are supporting stands for a 161-kV switcher and 161-kV bus, respectively. Circle 5 denotes an A-frame structure.

There are no accepted standards in drawing general arrangements; however, they should be done in such a manner as to give an accurate representation of the physical structure and how it will be installed. These drawings should be as simple as possible, without superfluous detail, neat, and good examples of orthographic projection as it applies to the electrical industry. Occasionally an isometric projection or some other type of pictorial view is used as a supplement to, or a substitute for, the views shown here.

10-7 Connection Diagrams

Connection diagrams are usually necessary and desirable, for electrical installations and their complexity directly depend on the complexity of the electrical installation itself. Connection or wiring diagrams show the physical wiring connections of control, metering, relaying, and auxiliary devices and do not generally involve the power conductors or devices. There are two basic types of wiring diagram: the diagrams associated with the internal wiring of pieces of equipment, commonly called *wiring diagrams,* and those diagrams associated

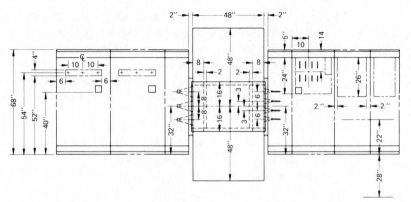

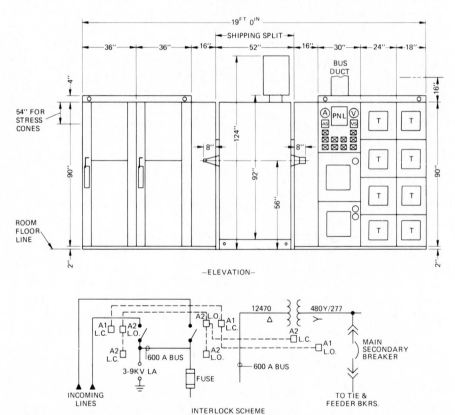

FIG. 10-14   General arrangement of a primary-selective, secondary unit substation. The primary switches are on the left, the transformer is in the middle, and the secondary switchgear is on the right. (Gould-Brown Boveri.)

FIG. 10-15 **Photograph of a secondary unit substation.** (Gould-Brown Boveri.)

with the external wiring connecting the individual pieces of equipment to one another, called *interconnection diagrams.*

Figure 10-17 shows the wiring diagram of a 15-kV air-circuit-breaker (ACB) cubicle that is used as a feeder breaker for a large industrial plant. It is one of several feeder breakers in a line-up (row) that make up a 15-kV distribution center. This wiring diagram shows the whole cubicle opened up from left to right, showing the rear view of the front door, the inside left side of the cubicle, the inside front of the cubicle, and the inside right side of the cubicle; it also shows a simplified three-line diagram, showing CT connections. All devices (relays, meters, etc.) are shown in their proportional sizes. This diagram uses highway-type connections, described previously in the text. The use of highway-type connections is predominant in the industry; however, the authors have seen wide variations. Note that the wires are numbered as described in the three-line-diagram section.

Figure 10-18 shows a partial interconnection diagram for a 15-kV switchgear line-up, which includes the breaker in the previous drawing. Again, highway-type connections are used. The interconnecting wires carry the same number all through each piece of equipment. The common or interconnecting wires associated with this figure involve ac control power (*ACN*, HB), dc control power (1, 2), and the 120-V metering voltage (P_1, P_2, and P_3).

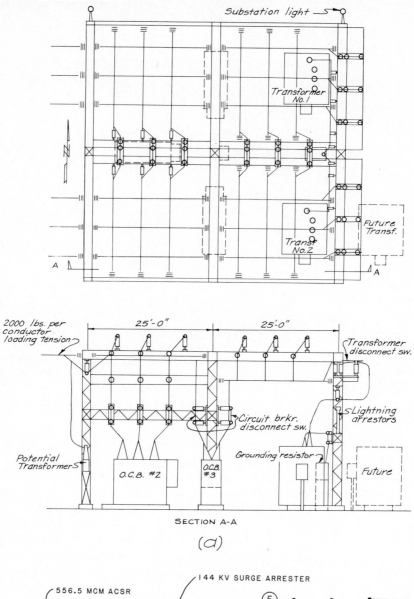

SECTION A-A

(a)

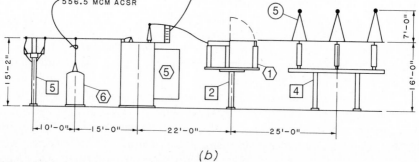

(b)

FIG. 10-16 General layouts: *(a)* top view and section of a substation; *(b)* elevation of a newer low-profile substation. (Black & Veatch, Engineers-Architects.)

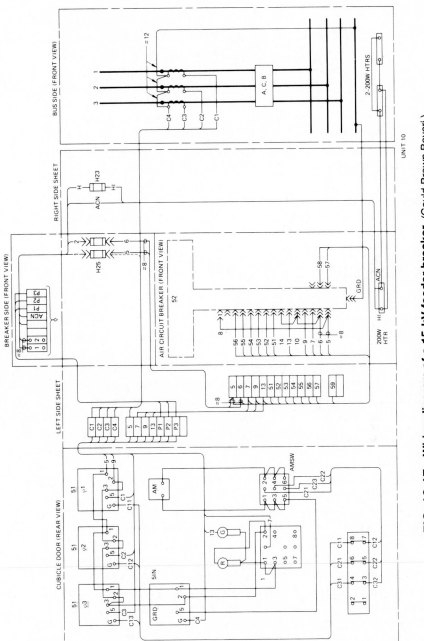

FIG. 10-17 Wiring diagram of a 15-kV feeder breaker. (Gould-Brown Boveri.)

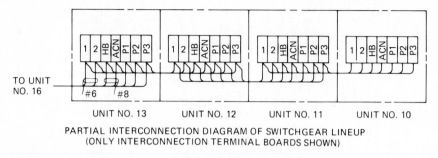

TO UNIT
NO. 16

#6 #8

UNIT NO. 13   UNIT NO. 12   UNIT NO. 11   UNIT NO. 10

PARTIAL INTERCONNECTION DIAGRAM OF SWITCHGEAR LINEUP
(ONLY INTERCONNECTION TERMINAL BOARDS SHOWN)

FIG. 10-18 Partial interconnection diagram of a line-up of 15-kV metal-clad switchgear, showing only interconnections. (Gould-Brown Boveri.)

Connection drawings, as these examples have illustrated, should show the best perspective and identification of the actual wiring. Routing of wiring must be shown exactly as it is to be done. Connection drawings are an important part of the electrical drawing package in that they show the actual wiring.

10-8 Power Distribution Plans

Power distribution plans or layouts are plan views showing the electrical distribution equipment (e.g., substation, bus ducts, etc.), the utilization equipment (e.g., pumps, furnaces, etc.), and the general routing of the electrical services between them. Because these plan views show equipment and are generally to scale, they are also used in the physical planning and installation of the equipment and services. Equipment is usually shown as an outline with little detailing. Distribution plans may also contain equipment ratings, exact service routings, service specifications (conductor information, raceway information), mechanical services, or other miscellaneous information. What should be included will depend on the firm for which you work. Experience has shown that if power layouts are done in areas that are congested or are rearranged quite often, they should be done in $\frac{1}{4}$- or $\frac{3}{8}$-in. scale; smaller scales have proved slightly more difficult to work with.

Figure 10-19 shows a power distribution plan of a mass-production machine shop. Here the electrical distribution equipment is 480- and 208/120-V, 3-ϕ enclosed bus ducts, which are mounted over the equipment near the ceiling. Bus ducts contain solid metal bars (3 or 4), conductors which are insulated from each other, and the enclosure housing. The bus duct receives its electric service from a substation (480 V) or a transformer (480 V – 208Y/120 V) connected to a 480-V bus duct. The bus ducts are plug-in. This allows bus-plug disconnects to be plugged into them much as plugs are plugged into common house receptacles. From the bus ducts, the electric service is run via the bus plug and conduit to the disconnect on the machine. This machine shop only requires 30-A bus plugs. Each service to the machines shows only the number of conductors, conductor size, conduit size, and phasing, if less than three-phase; voltages are obvious because the service source is shown. Conductor material (copper), conductor insulation (THWN), and conduit material (rigid metal-

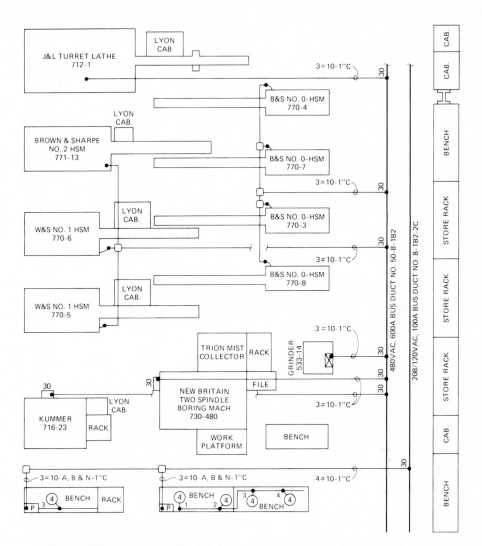

FIG. 10-19 **Power distribution plan of a mass-production machine shop.**

steel) are not shown because they are standard within this factory. This plan shows equipment names and numbers to locate it more accurately and differentiate it from similar equipment of the same model. Some of the symbols are standard—for example, a disconnect or motor controller—while other symbols are not—for example, a circled number 4, which is a 115-V, 15-A two-outlet receptacle, and a P, which is a 1-ϕ, 208/120-V, 4-pole power panel. The number next to the circled number is the panel circuit number.

Power distribution plans are an integral part of the electrical drawing package in that they are the only drawings that show service routings and equipment space occupancies.

10-9 Details and Specifications

No set of drawings covering a power installation of any size is complete without detail drawings. There may be several large sheets filled with small detail drawings. These may be roughly divided into two categories: (1) special and (2) standard details.

Special details, which cover the particular project being delineated, may include many different items, installations, or interconnections.

Examples of these drawings appear in Fig. 10-20. Two views showing the arrangement of trays are given (see Fig. 10-20a). These are rectangular metal ducts in which conduits are run throughout the structure. The conduits themselves are usually not shown, but their tray placement can be determined elsewhere, such as in the circuit routing designation in the specifications. Figure 10-20b shows a section of a *duct bank,* which is an underground encasement with six banks of 3-in. Korduct in which cables are run. Korduct is a thin-walled pipe used for encasement in concrete. Transite pipe is used for direct burial in the ground, as are ducts of fiber and plastic.

These are just three of the many details that are encountered in drawings for the electric power industry. Standard details, which may be used for more than one project, are usually drawn just once, but reproduced many times. Sometimes they are reduced to $8\frac{1}{2} \times 11$-in. size, then bound in the printed specifications.

The preparation of specifications is an important part of the engineer's job. The person who makes the electrical drawings can expect to do much work on the specifications. Specifications are usually done for installations, equipment, or material. Specifications generally consist of a legal portion, which engineers do not write; specific requirements; general provisions; and general requirements. The specific requirements include such things as scope of work, work by others, contract drawings, information to be submitted with bid, etc. The general provisions, which are usually written by cutting and pasting from previous specifications, contain specifications on such things as codes, interferences, quality of work, materials, tests, contractor's required facilities (office, telephone, etc.), trucking, toilet facilities, etc., or items of a general nature that need to be spelled out and clarified. The general requirements would contain the specifications for equipment and installation materials and methods required for the electrical installation itself. Beside requirements for an electrician's work, this portion of the specifications would spell out the work for other trades.

10-10 General Trends

The number of control circuits in generating-station construction is so large that drawing them on one-line diagrams has been largely abandoned in favor of logic diagrams. Similarly, the practice of drawing circuits on electrical plans has given way to showing raceway numbers (trays, conduits, etc.) only. All details are organized in the raceway and conduit schedules and by interconnection diagrams or schematics. Figure 10-20a shows tray details in elevation; similar drawings are used to show tray and raceway details in plan (top) views.

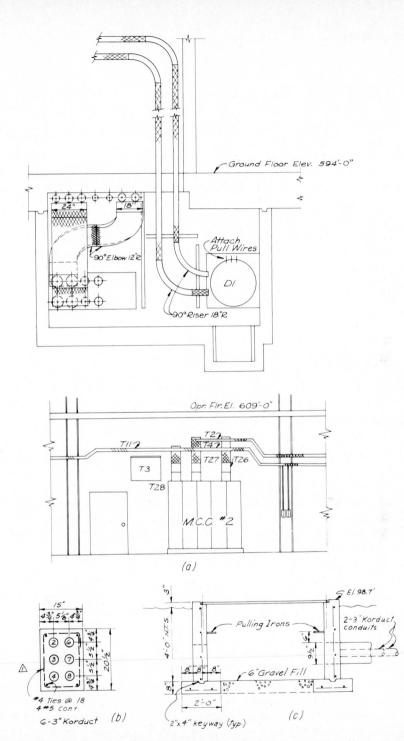

FIG. 10-20 Special details that are part of a set of drawings for a power-generating station: *(a)* tray details; *(b)* duct bank; *(c)* pull box. (Black & Veatch, Consulting Engineers.)

Another trend is that some manufacturers are furnishing internal wiring diagrams of their equipment that show only device terminals and wire numbers, and not the routing of the wiring. These diagrams, therefore, are less cluttered than the typical wiring diagrams; however, the value of this trend is debatable. Another trend is the use of computer-generated drawings by manufacturers and engineering firms. This trend is most apparent in bills of materials, standard details, control schematics, and wiring diagrams. These drawings are generally clear, very well organized, and of good quality.

As in electronics, there have been great advances in technology in the electrical field. Some of these advances are better insulating materials in components (transformers, breakers, wire, etc.); larger voltages in generation (260 kV), transmission (765 kV), and distribution (161 kV); larger and more efficient generating plants; and, of course, nuclear generating stations. But the largest and yet most subtle technical advance in the electric power field is the tremendous expansion in the use of semiconductor-controlled equipment. A partial list is as follows:

1. Static exciters with capabilities up to 3000 kW for electric generators

2. Solid-state protective devices, such as relays, which give more and better protective characteristics for the power-handling devices, such as breakers

3. SCRs for rectification and inversion, one of the prime reasons that dc transmission over long distances has become a practical, economic reality

4. Solid-state speed controls for motors, which have all but replaced their mechanical predecessors

However, the complete online computer control of power generating facilities represents the ultimate in this direction, and it has been accomplished by some utilities and industries. All these technical advances will have an effect on the number and complexity of drawings needed to show the complete picture of an electric power installation.

SUMMARY

The explanations and drawings in this chapter give a representative, but not complete, picture of the drawings used in the electric power field. It is felt that the drawings are typical enough and the coverage broad enough that the reader will have a reasonable knowledge of the drawings done in this area.

The one-line diagram is almost unique to the electric power field, showing in abbreviated form the course or energy flow of the power circuit and associated devices. Logic diagrams are an invaluable aid in developing an overall control scheme. Schematic-type drawings, such as the three-line, control, and elementary diagrams, have the functions of illustrating metering and relaying, as in power circuits, breaker control, and motor control, respectively. General arrangements show how the electric power equipment will be physically arranged when manufactured and installed. Connection-type diagrams show the physical wiring of individual pieces of equipment or the interconnection between pieces of equipment. Power distribution plans show the physical layouts of distribution and utilization equipment and the routing of electric services between them. Special and standard detail drawings are necessary and round out the set of drawings required for any large installation. Printed specifications, sometimes volumes of them, will accompany sets of electric power drawings. Engineers and drafters engaged in the preparation of such drawings may expect to participate in the preparation of these specifications.

QUESTIONS

10-1. Describe a one-line diagram and its function.

10-2. How can one indicate that more than one relay of the same type is used at one location on a one-line diagram?

10-3. At which end of a one-line diagram are the higher-voltage lines and equipment placed?

10-4. Name five different kinds of relay that might be shown on a one-line diagram for an electric substation.

10-5. Describe briefly what the following abbreviations stand for: CT, PT, BKR, OCB, FCWD, R, MCM, CS, AWG, ACB.

10-6. What is the purpose of a three-line diagram? Describe its arrangement.

10-7. How is more than one relay of the same type, connected to different lines via PTs or CTs, identified on a three-line diagram?

10-8. Describe how a logic diagram is used in electric design, and its interrelation to control schematics.

10-9. What are the five main functions used in logic diagrams for the electric power field?

10-10. What type(s) of circuit can be appropriately described by means of logic diagrams?

10-11. What is understood about pushbuttons and control switches in reading the logic diagrams shown in this chapter?

10-12. Why is it impractical to use only logic diagrams and not have control schematics?

10-13. Describe a control diagram. What is its function? What does the number 52 refer to?

10-14. What is the function of a general-arrangement drawing? What are the typical views found on the general arrangement?

10-15. What is the diameter of a 20-gage wire? Of a 500-MCM conductor?

10-16. What method of drawing connection or wiring diagrams is generally used in the electric power field?

10-17. Describe an electric power distribution plan. What is its primary function? What information is found on it?

10-18. What is meant by raceways in electric power construction? By duct banks, trays, and conduits?

10-19. Describe the two types of detail drawings presented in this chapter.

10-20. What are specifications? Do drawings accompany them?

10-21. What are some of the trends in this field? How do you feel they will affect the electrical drawings?

PROBLEMS

10-1. Figure 10-21 shows the plan (top) view of the grounding for a large solar energy generating station. Each space on the scales at top and left represents 250 ft. Let each space represent 3/4 in., using a 3/4 scale or 20 units on a 30 scale. Draw the layout on a sheet of $8\frac{1}{2} \times 11$ paper, using one of the two scales.

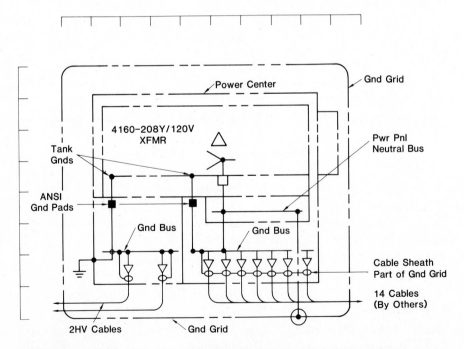

FIG. 10-21 (Prob. 10-1.) The grounding plan for a central receiver electric power generating plant.

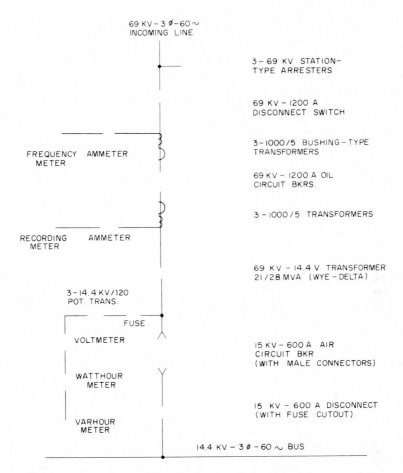

69 KV – 3 ∅ – 60 ~
INCOMING LINE

3 – 69 KV STATION–
TYPE ARRESTERS

69 KV – 1200 A
DISCONNECT SWITCH

3 –1000/5 BUSHING–TYPE
TRANSFORMERS

FREQUENCY   AMMETER
METER

69 KV – 1200 A OIL
CIRCUIT BKRS.

3 – 1000/5 TRANSFORMERS

RECORDING   AMMETER
METER

69 KV – 14.4 V TRANSFORMER
21/28 MVA (WYE – DELTA)

3 – 14.4 KV/120
POT. TRANS.

FUSE

VOLTMETER

15 KV – 600 A AIR
CIRCUIT BKR
(WITH MALE CONNECTORS)

WATTHOUR
METER

15 KV – 600 A DISCONNECT
(WITH FUSE CUTOUT)

VARHOUR
METER

14.4 KV – 3 ∅ – 60 ~ BUS

FIG. 10-22 **(Prob. 10-3.) Incomplete one-line diagram of substation.**

10-2. Starting at the top and going down, draw the symbols in the following list (a through j) on a vertical line (except where noted). Place identification of each symbol at the right of symbol on your *one-line drawing*.

  a. 69-kV feeder line (horizontal line)
  b. 1200-A 69-kV disconnect switch
  c. 69-kV arresters (Show leading away on a horizontal line.)
  d. 69-kV oil circuit breaker
  e. 69-/14.4-kV 21/28-MVA transformer, wye-delta
  f. 15-kV 600-A air circuit breaker with male and female connectors
  g. 5000-A 15-kV disconnect switch
  h. 1000/5 current transformers (Show leading horizontally to ammeter and
    watthour meter.)

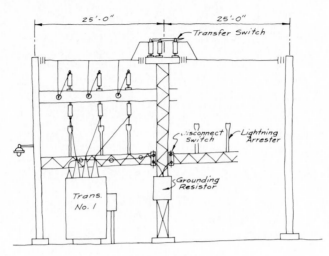

Section A-A

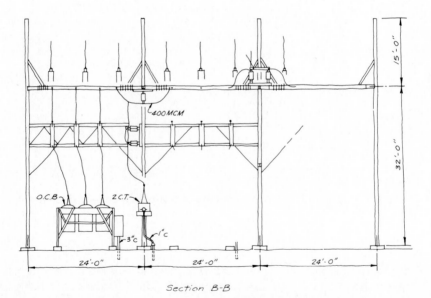

Section B-B

FIG. 10-23 (Prob. 10-4.) Sketches of substation sections. Section *A-A* (incomplete) through substation yard. Section *B-B* (incomplete) through substation yard.

    i. 15-kV/120-V potential transformers (Show leading horizontally to voltmeter.)

    j. Ground

10-3. Complete the one-line diagram of the substation shown in Fig. 10-22 by adding the missing symbols. Place suitable identification to the right of each symbol, except for meters and fuse.

10-4. Figure 10-23 shows two incomplete sketches of sections through two different substations. The right-hand part in section *A-A,* which contains transformer No. 2, is identical to the left-hand part shown. In section *B-B,* the left-hand span is complete, the center span is complete — except for the circuit-breaker arrangement, which is identical to that shown at the left — and the right-hand span is identical to the center span. Make a mechanical drawing of each of the sections, completing where necessary.

10-5. Redraw the floodlight detail shown in Fig. 10-24 to a size approximately two or three times that shown in the figure.

10-6. An installation has several motors and a control panel. Draw the following logic diagrams for part of this, as follows:

a. *Control Switch Automatic* and *Safety* must both be on. *Motor No. 1* will start if the preceding is in the on-state *or* if the manual control switch is on. When motor runs, a yellow light goes on at location 2, control panel.

b. *Motor No. 1* will stop if the *Stop Control Switch or* the *Overload, or* the *Motor Trip* is on. When the above happens, a green light goes on at location 2, control panel, and a buzzer sounds at the motor (MP_1).

c. For the *Auxiliary Motor* to start, both the *Manual Control Switch* and the *Safety* must be on. Also the *Visual* indicator must be on. Use $8\frac{1}{2} \times 11$ paper for each drawing.

10-7. Figure 10-25 shows a gas-burner valve functional diagram, which is a type of diagram used by some consulting engineers. Make a logic diagram for this valve and add it to the functional diagram shown.

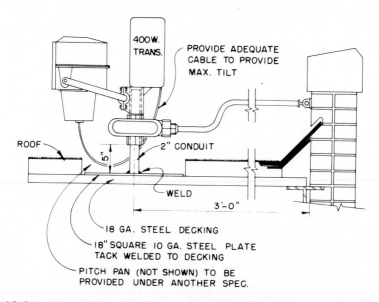

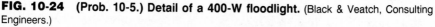

FIG. 10-24 **(Prob. 10-5.) Detail of a 400-W floodlight.** (Black & Veatch, Consulting Engineers.)

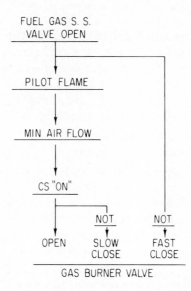

FUEL GAS S. S.
VALVE OPEN

PILOT FLAME

MIN AIR FLOW

CS "ON"

OPEN   NOT   NOT
          SLOW   FAST
          CLOSE   CLOSE

GAS BURNER VALVE

FIG. 10-25 (Prob. 10-7.) Functional diagram of a gas-burner valve.

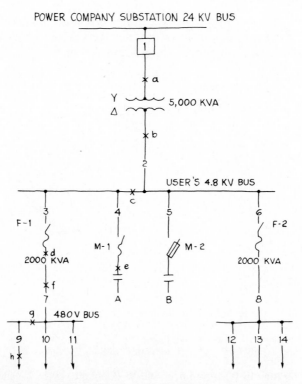

POWER COMPANY SUBSTATION 24 KV BUS

1

× a

Y 5,000 KVA
Δ

× b

2

USER'S 4.8 KV BUS

× c

3   4   5   6
F-1   F-2

× d   M-1   M-2
2000 KVA   2000 KVA

× f
7   × e   A   B   8

9   480 V BUS
×

9   10   11   12   13   14
h ×

FIG. 10-26 (Prob. 10-8.) A diagram of electric power service to a large plant.

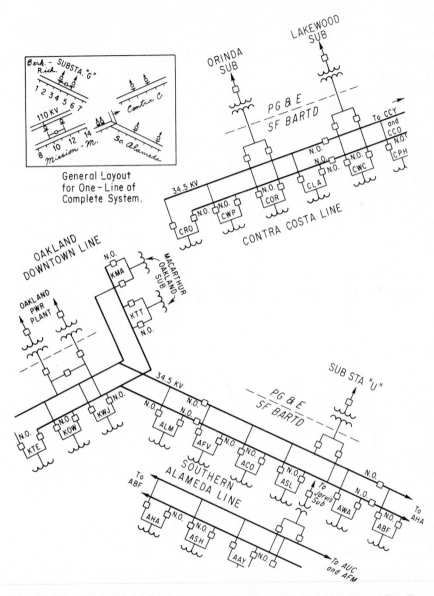

FIG. 10-27 **(Prob. 10-9.) Partial one-line diagram of electrification of San Francisco Bay Area Rapid Transit System.** (Parsons-Brinckerhoff-Tudor-Bechtel, Engineering Consultants.)

a. For *fast close* of valve, *Fuel Gas SS valve* must be open.
b. For *slow close* of valve, *CS* on, *Min Air Flow,* and *Pilot Flame* must *all* be in the On state.
c. For *opening* of *Gas Burner Valve,* the events of *both* a and b must all be in the On state. *Note:* Use NOT symbols for closing of the valve. When valve

is closed, a green light shows at 33–13. When open, a red light shows at 33–12. Use $8\frac{1}{2} \times 11$ or 11×17 paper for each drawing.

10-8. Draw a complete one-line diagram of the electric power distribution in an industrial plant, of which Fig. 10-26 is a diagrammatic sketch. Circuit breakers (oil) are at locations 1 through 14, and 500-hp motors are at A and B. Possible location of fault currents are at locations a through h. Use standard symbols throughout. Use $8\frac{1}{2} \times 11$ paper.

10-9. About 60 percent of the one-line diagram for the power distribution of the Bay Area Rapid Transit System (BART) is shown in Fig. 10-27. Make a complete one-line drawing of the system with a layout looking like that shown at the upper left. Stations on the Berkeley-Richmond line are (1) RRY, (2) RRI, (3) RCN, (4) RCP, (5) RNB, (6) RBE, and (7) RAS. Stations on the Mission-Market line are (8) MDC, (9) MBP, (10) MGP, (11) MTF, (12) MSS, (13) MPS, and (14) BTW. Configurations of these two lines are similar to the Contra Costa line. Their locations show in the left part of the General Layout inset. Include all lettering. Use 11×17 or 12×18 paper.

11

Electrical Drawing for Architectural Plans

An integral part of any set of drawings for the construction of a building is the wiring plan or layout. Several standards apply to this type of design and graphical presentation. Symbols for the drawings (other than those used previously in this text) are shown and explained in ANS Y32.9, "Graphical Electrical Symbols for Architectural Plans," Mil Std 15-3, "Electrical Wiring Symbols for Architectural and Electrical Layout Drawings," and in the *Residential Wiring Handbook* published by the Industry Committee on Interior Wiring Design.

The National Electrical Code (NEC), published by the National Fire Protection Association (NFPA) and the American National Standards Institute (ANSI), provides the minimum design criteria necessary to safeguard persons and property practically from the hazards arising from the use of electricity. The Code is voluntarily written by knowledgeable persons in all diverse groups associated with the electrical industry, including unions, manufacturers, inspection agencies, users, technical societies, contractors, utilities, insurance underwriters, and governmental agencies. Many of these organizations are represented by associations or societies. The Code is not intended as a design specification or instruction manual for *untrained* personnel.

The NEC covers electrical conductors and equipment installed within or on public and private buildings, structures, mobile homes, recreational vehicles, industrial substations, and other premises (yards, carnivals, parking lots, etc.). It also covers the conductors that connect the installations to a supply of electricity and other outside conductors. In general, the NEC does not cover installations in ship, water craft, railroads, aircraft, automobiles, or mines, nor does it cover communication equipment used by communication utilities or installations under the direct control of electric utilities. The NEC is purely advisory as far as NFPA and ANSI are concerned, but it is offered for use in law and for regulatory purposes.

Many political entities (e.g., cities and towns) and insurance underwriters have adopted the NEC in toto. Their primary reason for adopting the NEC is to minimize or eliminate the potential hazards associated with improper electrical installations, such as fire and electrical shock to personnel. Certain political entities often have building codes which are more restrictive than the NEC, and these must be met when doing electrical designs and installations within their jurisdiction.

11-1 National Electrical Code (NEC) Definitions and Contents

Because the NEC is such an important document, persons engaged in producing electrical drawings for architectural structures must be familiar with it and with other local codes. These persons should also be conversant with standard terminology and equipment. For the benefit of the reader, we give some of the definitions used in the code and a brief explanation of its contents, so that the rest of this chapter can be followed more easily. However, it should be remembered that the NEC is the standard for the minimum provisions associated with electrical installations necessary for personnel and property safety; it is *not* a drawing standard.

Some of the definitions used in the NEC are a little strange compared with their everyday use; however, they should be learned because they are peculiar and essential to the proper use of the Code. Some of the NEC definitions that are more applicable to the information contained in this chapter are as follows:

Accessible:

As applied to wiring methods: Capable of being removed or exposed without damaging the building structure or finish, or not permanently closed in by the structure or finish of the building. (See "Concealed" and "Exposed.")

As applied to equipment: Admitting close approach; not guarded by locked doors, elevation or other effective means. (See "Accessible, readily.")

Accessible, readily: (Readily accessible.) Capable of being reached quickly for operation, renewal, or inspections, without requiring those to whom ready access is requisite to climb over or remove obstacles or to resort to the use of portable ladders, chairs, etc. (See "Accessible.")

Ampacity: The current in amperes that a conductor can carry continuously under the conditions of use without exceeding its temperature rating.

Appliance: Utilization equipment, generally other than industrial, normally built in standardized sizes or types, which is installed or connected as a unit to perform one or more functions, such as clothes washing, air conditioning, food mixing, deep frying, etc.

Attachment Plug (Plug Cap): A device which, when inserted into a receptacle, establishes connection between the conductors of the attached flexible cord and the conductors connected permanently to the receptacle.

Branch Circuit: The circuit conductors between the final overcurrent device protecting the circuit and the outlet(s).

Appliance: A branch circuit supplying energy to one or more outlets to which appliances are to be connected. Such circuits have no permanently connected lighting fixture not a part of an appliance.

General-purpose: A branch circuit that supplies a number of outlets for lighting and appliances.

Individual: A branch circuit that supplies only one utilization equipment.

Multiwire: A branch circuit consisting of two or more ungrounded conductors which have a potential difference between them, and an identified grounded conductor which has equal potential difference between it and each ungrounded conductor of the circuit and which is connected to the neutral conductor of the system.

Building: A structure which stands alone or which is cut off from adjoining structures by fire wall with all openings therein protected by approved fire doors.

Cabinet: An enclosure designed for either surface or flush mounting and provided with a frame, mat, or trim in which a swinging door or doors are or may be hung.

Circuit Breaker: A device designed to open and close a circuit by nonautomatic means and to open the circuit automatically on a predetermined overcurrent without injury to itself when properly applied within its rating.

Concealed: Rendered inaccessible by the structure or finish of the building. Wires in concealed raceways are considered concealed, even though they may become accessible by withdrawing them. (See "Accessible, As applied to wiring methods.")

Conductor:

Bare: A conductor having no covering or electrical insulation whatsoever. (See "Conductor, Covered.")

Covered: A conductor encased within material of composition or thickness that is not recognized by this Code as electrical insulation. (See "Conductor, Bare.")

Insulated: A conductor encased within material of composition and thickness that is recognized by this Code as electrical insulation.

Dead Front: Without live parts exposed to a person on the operating side of the equipment.

Device: A unit of an electrical system which is intended to carry, but not utilize, electric energy.

Disconnecting Means: A device, or group of devices, or other means by which the conductors of a circuit can be disconnected from their source of supply.

Enclosure: The case or housing of apparatus, or the fence or walls surrounding an installation to prevent personnel from accidentally contacting energized parts or to protect the equipment from physical damage.

Exposed:

As applied to live parts: Capable of being inadvertently touched or approached nearer than a safe distance by a person. It is applied to parts not suitably guarded, isolated, or insulated. (See "Accessible" and "Concealed.")

Feeder: All circuit conductors between the service equipment or the generator switchboard of an isolated plant and the final branch-circuit overcurrent device.

Grounding Conductor, Equipment: The conductor used to connect the non-current-carrying metal parts of equipment, raceways, and other enclosures to the system grounded conductor and/or the grounding electrode conductor at the service equipment or at the source of a separately derived system.

Ground-Fault Circuit-Interrupter: A device intended for the protection of personnel that functions to deenergize a circuit or portion thereof within an established period of time when a current to ground exceeds some predetermined value that is less than that required for operation of the overcurrent protective device of the supply circuit.

Lighting Outlet: An outlet intended for the direct connection of a lampholder, a lighting fixture, or a pendant cord terminating in a lampholder.

Location:

Damp: Partially protected locations under canopies, marquees, roofed open porches, and similar locations, and interior locations subject to moderate degrees of moisture, such as some basements, some barns, and some cold-storage warehouses.

Dry: A location not normally subject to dampness or wetness. A location classified as dry may be temporarily subject to dampness or wetness, as in the case of a building under construction.

Wet: Installations underground or in concrete slabs or masonry in direct contact with the earth; locations subject to saturation with water or other liquids, such as vehicle washing areas; and locations exposed to weather and unprotected.

Outlet: A point on the wiring system at which current is taken to supply utilization equipment.

Panelboard: A single panel or group of panel units designed for assembly in the form of a single panel, including busses, automatic overcurrent devices, and with-or-without switches for the control of light, heat, or power circuits; designed to be placed in a cabinet or cutout box, placed in or against a wall or partition and accessible only from the front. (See "Switchboard.")

Raceway: An enclosed channel designed expressly for holding wires, cables, or busbars, with additional functions as permitted in this Code. (Raceways may consist of metal or insulating material, and the term includes rigid metal conduit, rigid nonmetallic conduit, intermediate metal conduit, liquid tight flexible metal conduit, flexible metallic tubing, flexible metal conduit, electrical nonmetallic tubing, electrical metallic tubing, underfloor raceways, cellular concrete floor raceways, cellular metal floor raceways, surface raceways, wireways, and busways.)

Receptacle: A receptacle is a contact device installed at the outlet for the connection of a single-attachment plug. (A single receptacle is a single-contact device with no other contact device on the same yoke. A multiple receptacle is a single device containing two or more receptacles.)

Receptacle Outlet: An outlet where one or more receptacles are installed.

Remote-Control Circuit: Any electric circuit that controls any other circuit through a relay or an equivalent device.

Service: The conductors and equipment for delivering energy from the electricity supply to the wiring system of the premises served.

Service-Entrance Conductors:

Overhead system: The service conductors between the terminals of the service equipment and a point usually outside the building, clear of building walls, where joined by tap or splice to the service drop.

Underground system: The service conductors between the terminals of the service equipment and the point of connection to the service lateral. (Where service equipment is located outside the building walls, there may be no service-entrance conductors, or they may be entirely outside the building.)

Service Equipment: The necessary equipment, usually consisting of a circuit breaker or switch and fuses, and their accessories, located near the point of entrance of supply conductors to a building, other structure, or otherwise defined area, and intended to constitute the main control and means of cutoff of the supply.

Switchboard: A large single panel, frame, or assembly of panels on which are mounted, on the face or back or both, switches, overcurrent and other protective devices, buses, and usually instruments. Switchboards are generally accessible from the rear as well as from the front and are not intended to be installed in cabinets. (See "Panelboard.")

Utilization Equipment: Equipment which utilizes electric energy for mechanical, chemical, heating, lighting, or similar purposes.

Voltage (of a Circuit): The greatest root-mean-square (effective) difference of potential between any two conductors of the circuit concerned.

Briefly, the NEC contains standards on installation, application, construction, materials, and equipment associated with the electrical industry. Standards are found in the following areas:

1. Wiring design and protection, which include circuits (branch, feeder, etc.), protective devices (fuses, circuit breakers, surge arresters, etc.), and grounding of all types.

2. Wiring methods and materials, which include cable, raceways, busways, wireways, boxes, fittings, panelboards, switchboards, etc., of all types.

3. Equipment for general use, such as flexible cords, lighting fixtures, appliances, heating–ventilating–air-conditioning equipment, motors, motor controllers, generators, transformers, capacitors, resistors, reactors, and batteries.

4. Equipment and methods associated with special occupancies, such as places where fire or explosion hazards may exist (garages, bulk-storage plants, aircraft hangars), health facilities, theaters, studios, manufactured buildings, mobile homes and parks, recreational vehicles, and marinas or boatyards.

5. Special occupancies such as hazardous areas, theaters, places of assembly, manufactured buildings, agricultural buildings, mobile homes, recreational vehicles, and marinas or boatyards.

6. Special equipment such as electric signs, cranes, hoists, elevators, escalators, electric welders, sound-recording equipment, data-processing equipment, x-rays, induction-dielectric heating equipment, metal-working tools, irrigation equipment, and swimming pools.

7. Special electrical conditions, such as emergency systems; systems over 600 V; installations under 50 V; remote-control, signaling, and limited-power circuits; standby power-generation equipment; and fire-protective signaling systems.

8. Communications systems such as telephone, telegraph, central alarm stations, radio and TV receiving and transmitting equipment, and CATV systems.

The preceding has been a relatively brief description of the terminology and contents of the NEC. If questions should arise, the Code should be consulted. It is available from NFPA or ANSI.

11-2 Simplified and True Wiring Diagrams

A true wiring diagram shows every wire and its connection in a system, or circuit. Such a diagram is shown in Fig. 11-1a, in which four ceiling light-fixture outlets are depicted, two of which are connected to, and controlled by, individual single-pole single-throw switches. A simplified arrangement of this branch is shown at the right in the same figure. Here, approved symbols have been used for the light outlets, the switches, and the wire run, which may be of nonmetallic sheathed cables, armored cables, or any approved method of running conductors between outlets. The two parallel dashes across the wire runs indicate that a two-wire conductor is to be used. Actually, according to the standards, when a two-wire run is to be installed, the dashes may be omitted. If the conductor is to be composed of more than two wires, dashes indicating the number of wires must be provided on the drawing.

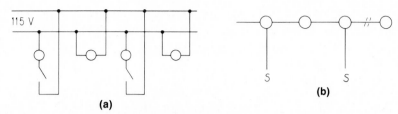

FIG. 11-1 Wiring diagrams of light-fixture outlets on a circuit: *(a)* true wiring diagram; *(b)* simplified, or installation, diagram.

11-3 Wiring Symbols on a Simple Floor Plan

The architect usually shows the location of lights, convenience and special-purpose outlets, and the desired switching arrangements on a floor plan. For small, simple structures, the required symbols and wiring arrangements may be drawn on the same floor plan (Fig. 11-2) that shows all information necessary for the erection of the building. For larger or more complicated structures, complete wiring details will probably be drawn on separate floor plans, called *electrical*

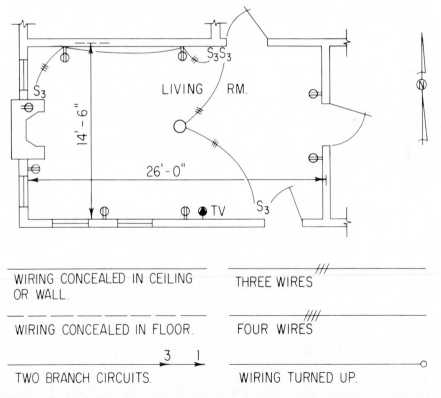

FIG. 11-2 Floor plan of a room with fixtures, outlets, and switches. Standard wiring symbols are below.

layouts or *electrical plans.* In either case, the simplified type of diagram, such as that shown in Fig. 11-1*b*, will be used. This wiring layout will be drawn by an architect, engineer, or drafter who is familiar with the engineering and building-code requirements.

The living-room plan (Fig. 11-2) shows two circuits: (1) the three-way switching arrangement for the ceiling outlet and (2) a similar arrangement for the two convenience outlets on the north wall. The outlet symbols, including the special-purpose outlet (indicated for TV antenna), are taken from ANS Y32.9. Also in accordance with the standard, the wire symbols for the switch-to-ceiling-light runs are drawn with a medium-weight solid line, indicating that the wires are to be concealed in the walls or ceiling above. Where no perpendicular dashes are shown across the wires, the conductors must have two wires. With the addition of a little more information about fixtures, the floor plan, of which the living-room plan in Fig. 11-2 is a part, will yield enough information for the satisfactory installation of the complete electrical system.

11-4 Separate Electrical Plans

A plan for the electrical system of a small business building appears in Fig. 11-3. This drawing was one of several, including plans and details for heating, air conditioning, and plumbing, which appeared on a single sheet.

This electrical plan shows the location of three separate distribution panelboards and the proposed location of a future one. Panel C provides power service — mainly for the motors which run the mechanical equipment; panels A and B supply electricity for lighting and the other electrical needs of offices 102, 103, 104 and the vestibule. Each branch circuit is documented with an arrow pointing in the general direction of the panel and a designation such as A-2. This means, for example, that the nine fluorescent light fixtures in office 102 are on the same circuit, No. 2, which is fed at panel A. A separate telephone circuit, enclosed in conduit, is also shown.

A number of symbols that either do not appear in or differ from those shown in ANS Y32.9 are drawn in this electrical plan. The wall bracket outlets have the four prongs which are still widely used,[1] but which are not shown in ANS Y32.9. A long-and-short-dashed line is used for circuit legs, regardless of whether the wires are run in the ceiling above, or the floor below, and the short-dash symbol is used for the switch leg. Confusion in the interpretation of these symbols is avoided by preparation of a legend.

11-5 Fixture Schedule and Legend

Figure 11-4 shows a legend and fixture schedule that accompanies the electrical plan in Fig. 11-3. Inclusion of such schedules and legends is the customary

---

[1] One explanation for the continued popularity of the four prongs is that many persons feel that the plain circular symbol listed in ANS Y32.9 may be easily confused with other circular symbols which may appear on drawings.

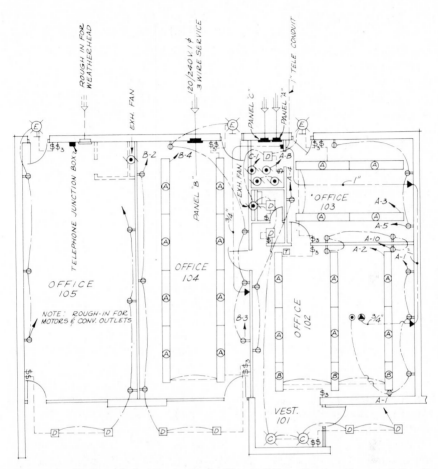

FIG. 11-3 Electrical plan for a small office building. (Brasher, Spencer & Goyette, Architects-Engineers.)

practice of architects and consulting engineers who prepare electrical layouts and details for the construction of buildings. The installation of the electrical system is facilitated by the inclusion of a letter designation at each fixture symbol and cross-referenced designations in an accompanying schedule. The exact form of the schedules has not been standardized. A "remarks" column has been omitted from the original schedule from which Fig. 11-4 was taken in order to conserve space.

A drawing should show intent, and in the most concise manner. Time, money, and argument will be saved if the proper information is placed in the legend. Too often, the person in the field does not see the written specifications, but does see the drawings and the legend. If the symbol for a floor convenience outlet appears, for example, the worker knows what the symbol means, but not which of the 50 available combinations is required, unless it is specifically

ELECTRICAL FIXTURE SCHEDULE					
MARK	MFG.	CAT. NO.	MOUNTING	WATTS	LAMP	FINISH
A	FELCO	4039	RECESS	3-95	430 MA	STANDARD
B	FELCO	4033	RECESS	3-38	430 MA	STANDARD
C	LITECRAFT	2305	WALL BRACKET	100	I.F.	SATIN ALUM.
D	PRESCOLITE	488-6600	RECESS	100	I.F.	STANDARD
E	PRESCOLITE	WE-2	WALL BRACKET	100	I.F.	SATIN CHROME

LEGEND :

☐   CEILING LIGHT OUTLET - LETTER DENOTES FIXTURE

▭ₒ   FLUORESCENT LIGHT OUTLET "   "   "

⊢◯   WALL BRACKET OUTLET   "   "   "

⊖   DUPLEX CONV. OUTLET

⊙   FLOOR CONV. OUTLET

⦸   MOTOR OUTLET

◀   TELEPHONE OUTLET

⬤   FLOOR TELEPHONE OUTLET

⊢T   THERMOSTAT

ⱷ   SINGLE POLE SWITCH

ⱷ³   THREE WAY SWITCH

— — —   SWITCH LEG

— · —   CIRCUIT LEG

— ·· —   EMPTY CONDUITS

FIG. 11-4 Fixture schedule and legends for electrical plan of small office building. (Brasher, Spencer & Goyette, Architects-Engineers.)

stated. Now, if the legend reads "FLOOR CONVENIENCE OUTLET— Frank Adam FB-3," the worker will know exactly what device to install.

11-6 Example of Electrical Layout

Another electrical layout is shown in Fig. 11-5; this depicts the second floor of a three-story office building. This floor arrangement might be suitable for a drafting or design room. The wiring symbols conform reasonably to the American Standard code. Arrows indicate the circuit and number; half arrows indicate partial circuits.

Included with this electrical plan were a legend, lighting-fixture schedule, panel schedule, disconnect-switch schedule, telephone-circuit riser diagram, and electrical riser diagram. Of these, only the electrical riser diagram is shown in this chapter. Before it is discussed, a word about how energy is distributed throughout a structure seems to be advisable.

As electric energy is brought into a building, it is usually first passed through a meter. From here it is brought into a main load center. In a small building or residence this load center consists of a fuse box or circuit breaker to

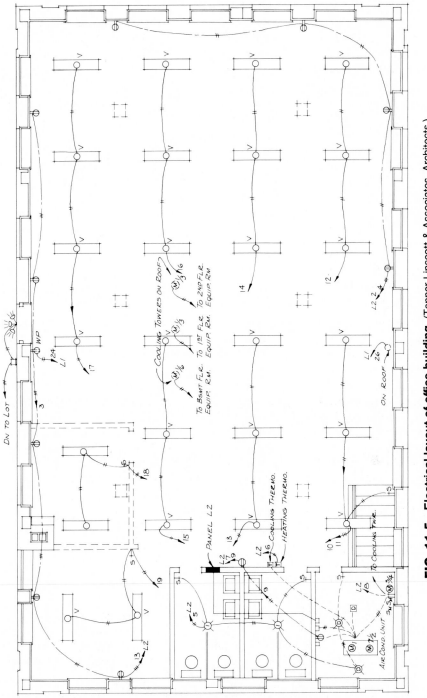

FIG. 11-5   Electrical layout of office building. (Tanner-Linscott & Associates, Architects.)

which each branch circuit is connected. Through these branch circuits energy is fed to each outlet, lamp, or appliance. In a larger structure the main load center may be a large panelboard, or switchboard, with circuit breakers, disconnect switches, and other controlling devices. From this main load center, electric energy is fed through large conductors called *feeders* to branch load centers, often panelboards, or *panels* as they are sometimes called. From these panels energy is delivered through branch circuits to each outlet, fixture, appliance, or motor. Such factors as location of panelboards, voltage and copper losses, etc., determine the method used in connecting the branch and main load centers. Many different interconnecting layouts are used. More than one panelboard may be necessary on the same floor of a large building in order that excessively long runs of branch circuits be avoided. Sometimes separate panels are used for lighting circuits only, and others for motors and machines only. Often, panelboards are used for combinations of types of circuit.

11-7 Electrical Riser Diagram

A riser diagram shows how electric energy is distributed through a building from the time it enters the building until it arrives at the branch load centers. It is drawn as an elevation and usually is not to scale. An electrical riser diagram for the building whose second-floor electrical plan is shown in Fig. 11-5 appears in Fig. 11-6. This diagram shows the electrical service passing through four meters, then through four disconnect switches. From switches S_1, S_2, and S_3 the current goes to lighting panels L_2, L_3, and L_B. From the latter two panelboards, energy goes to the lighting and convenience outlets in the ground floor and to mechanical equipment for the ground and first floors. Current travels from panelboard L_3 through a subfeeder to L_1, which supplies all the electrical service on the first floor, except for air-conditioning equipment. Panel L_2 supplies electric energy to most of the circuits on the second floor. However, the electrical layout for this floor (see Fig. 11-5) indicates that outside lighting and mechanical equipment on the roof are connected to panelboards on lower floors.

The panelboard schedule which accompanies the riser diagram includes the following information about each panel:

1. The number of branch circuits to be served
2. Current-handling capability (amperage)
3. Name of manufacturer and catalogue number of panel

Similar information is given in the disconnect-switch schedule.

11-8 Electrical Drawings for Large Buildings

The next series of drawings illustrates the type of drawing required for large buildings. The standards are not as much help in showing such a complicated

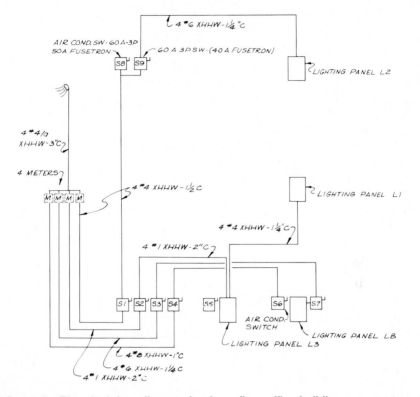

FIG. 11-6 Electrical riser diagram for three-floor office building. (J. R. DeRigne & Associates, Consulting Engineers.)

system; hence liberal use is made of orthographic projection, pictorial drawing, schematic diagrams, notes, and specifications.[1] A very brief description of the general electrical layout of a large office building will be given by referring to the drawings, which in some instances have been simplified from the original for clarity.

The underground cables of the power and light company enter the building at the upper right, as shown in Fig. 11-7. They immediately are brought into a main service entrance, which is shown in some detail in Fig. 11-8. In the main service entrance are current and potential transformers (used for measuring current and potential), fuses, and switches, as shown in the one-line diagram. From the main service entrance the electrical service is fed to two 1000-kVA transformers (see Fig. 11-7), which step the voltage down to a four-phase state that provides 208-V three-phase service and 120-V single-phase service. From these transformers current is brought through 3000-A 4-ϕ (4-phase) busses to the main switchboard, shown in pictorial form in Fig. 11-9. From this switch-

---

[1] "The National Electrical Code" (ANSI-C1) contains information that is most helpful in the planning of electrical systems for large buildings.

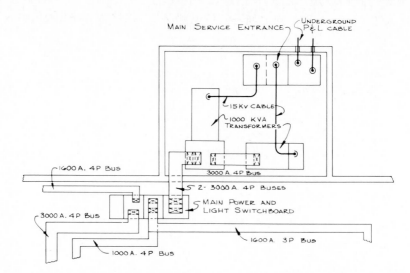

FIG. 11-7 Part of basement plan of large office building, showing details of electrical service. (W. L. Cassell, Mechanical Engineer, and Tanner-Linscott & Associates, Architects.)

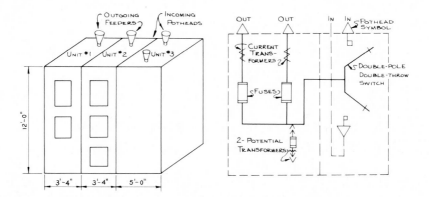

FIG. 11-8 Pictorial and one-line diagrams of main service entrance of large office building.

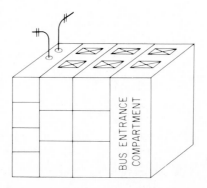

FIG. 11-9 Pictorial drawing of main power and light switchboard of a large office building.

gear, electric power is distributed by busses to various parts of the basement, as shown in Fig. 11-7. Actually the large copper or aluminum busbars themselves are not shown. Just the bus ducts are drawn — much the same as heating and air-conditioning ducts are drawn.

Service to upper floors is fed through the large 3000-A 4-ϕ bus — which is shown in both Figs. 11-7 and 11-10 (the riser diagram) — to branch load centers

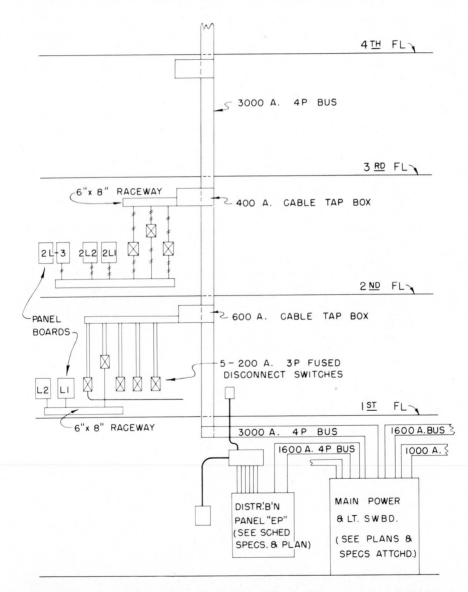

FIG. 11-10 **Part of riser diagram for a large office building.** (W. L. Cassell, Mechanical Engineer.)

on each floor. At each branch load center are a cable tap box, disconnect switches, and panelboards. Here, use is made of metal rectangular raceways (also called "trays"), in which connecting cables are placed. From the panelboards, service is distributed throughout the floors, as explained in the following two paragraphs.

Figure 11-11 shows a part of the electrical plan for the first floor. Shown in somewhat slight detail is the load center with its cable tap box and panels. Of special note is the underfloor duct system, shown by means of dotted lines. In these ducts, sometimes called *raceways,* are placed the wires for telephone service, 120-V single-phase electrical service, and 208-V three-phase service if desired. Outlets may be placed at any point along each of these ducts, and electrical service provided at each of these points. The ducts are covered with $2\frac{1}{2}$ in. of concrete, on which additional flooring material is often laid. The outlets are usually installed after the latter has been laid, but before the conductor cable has been pushed or drawn through the raceways. However, additional outlets may be emplaced along these ducts after the building has been completed and occupied, although new wiring may have to be poked through the ducts.

Figure 11-12 is a photograph of a raceway system being installed. Note that the raceways are often divided into more than one compartment. It is also possible to have a complete undergrid or cellular floor system. In this type of construction wires can be laid 12 in. apart all the way across the room in both directions. This provides a high degree of flexibility because a telephone, minicomputer, or intercom connection can be placed anywhere in the room.

Details of various electrical hookups and appurtenances are often desirable or necessary. They may take several forms and have different amounts of information. Two detail drawings are shown in Fig. 11-13.

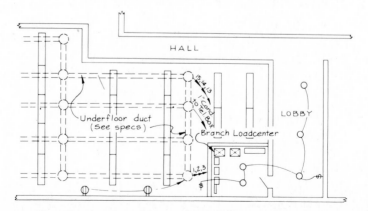

FIG. 11-11 Simplified electrical plan of first floor of a large office building, showing underfloor ducts.

FIG. 11-12 Installation of an underfloor duct system. (Walker-Parkersburg.)

11-9 Coordination and Organization of Drawings

There are usually areas on every drawing which, from the electrical contractor's viewpoint, should be made more comprehensive to help in preparing bids. Also the person doing the installing would benefit by having more information about correct or desired procedure, in many situations. However, it is possible to put too much information on the drawing — in trying to answer every question that may arise — and, as a result, confuse everybody from the contractor to the final inspector. Yet when special treatment is indicated, pertinent details should be supplied in order to present a logical and definite manner for the installation.

Many offices which prepare electrical and mechanical drawings for structures have improved their coordination with all crafts doing the work on a job. Such coordination (both before work starts and during the work) avoids conflicts between various mechanical items on the project. Last-minute changes in equipment by manufacturers cause some conflicts and cannot be fully pre-

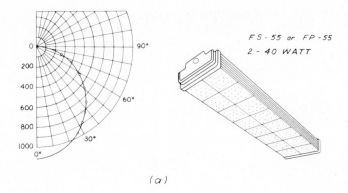

(a)

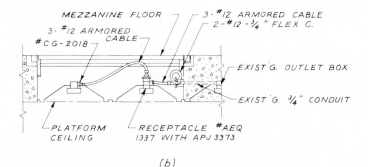

(b)

FIG. 11-13 Architectural details: *(a)* part of luminaire specification; *(b)* cabling detail. (San Francisco Bay Area Rapid Transit District.)

dicted in advance. However, obvious conflicts can be avoided by proper coordination while the drawings are being made.

If a consulting engineer is doing the wiring and mechanical design for an architect, their offices should also coordinate. Such things as space for raceways, ducts, and air-conditioning equipment must be provided.

11-10 Load Computation

With the information supplied in the National Electric Code (NEC), it is possible to compute the number of branch circuits and feeders required in many types of structure. Recently the NEC applied the term *volt-amperes* to general-usage areas. Volt-amperes is the apparent power that goes through a circuit. It is mathematically composed of watts (real power) and vars (reactive) (actually the square root of the sum of the squares). The ratio of watts divided by volt-amperes is called the power factor (pf). Therefore, if the pf equals 1 or 100 percent, the quantity of watts equals the quantity of volt-amperes. Since Tables 11-1 through 11-3 use either volt-amperes or watts at 100 percent power factor, they are equivalent. Let us assume that a residential dwelling has an area of 1490 sq

TABLE 11-1   General Lighting Loads by Occupancy

TYPE OF OCCUPANCY	UNIT LOAD PER SQUARE FOOT (VOLT-AMPERES)
Auditoriums	1
Banks	5
Dwellings (other than hotels)	3
Industrial buildings	2
Office buildings	5

TABLE 11-2   Calculation of Feeder Loads by Occupancy

TYPE OF OCCUPANCY	PORTION OF LIGHTING LOAD TO WHICH *DEMAND FACTOR* APPLIES (VOLT-AMPERES)	FEEDER DEMAND FACTOR (PERCENT)
Dwellings (other than hotels)	First 3000 or less at	100
	Next 3001 to 120,000 at	35
	Remainder over 120,000 at	25
Warehouses (storage)	First 12,500 or less at	100
	Remainder over 12,500 at	50
Others (except hospitals, hotels, and apartment buildings)	Total wattage	100

TABLE 11-3   Demand Loads for Household Electric Ranges, Ovens, and Counter-Mounted Cooking Units over 1.75-kW Rating

NUMBER OF APPLIANCES	MAXIMUM DEMAND (NOT OVER 12-kW RATING)	DEMAND FACTORS	
		LESS THAN 3.5-kW RATING	3.5 TO 8.75-kW RATING
1	8 kW (8000 W)	80%	80%
2	11 kW	75%	65%
3	14 kW	70%	55%

ft, exclusive of basement, attic and garage, and a 12-kW kitchen range. Using Tables 11-1 through 11-3, we compute the general lighting load and other requirements as follows:

1. General lighting load
1490 sq ft at 3 VA/sq ft = 4470 VA (see Table 11-1)

2. Minimum number of branch circuits required
  a. General lighting load (based on 115 V): 4470 ÷ 115 = 39.0 A. This can be handled by three 15-A two-wire circuits or two 20-A two-wire circuits. (The NEC recommends one circuit for each 500 sq ft.)
  b. Small-appliance load: (Sec. 220-3 states that two or more 20-A branch circuits shall be installed to take care of small appliances in the kitchen, laundry, dining room, and breakfast room.)

3. Minimum-size feeders required:

COMPUTED LOAD	VOLT-AMPERES
General lighting	4470
Small appliances (computed at 1500 VA for each circuit in accordance with Sec. 220-3)	3000
Total (without range)	7470
3000 W at 100% (Table 11-2)	3000
7470 − 3000 = 4470 VA at 35%	1565
Net computed (without range)	4565
Range load (see Table 11-3)	8000
Total net computed	12565

If 115/230-V three-wire system feeders are used, the current load will be 12,565 ÷ 230, or 55 A. But, because the computed load exceeds 10 kVA, service conductors shall be 100 A [Sec. 230–41(b)(2)]. Feeder size can be selected from one of several tables in the NEC, which gives ratings for types of copper, aluminum, and copper-clad aluminum conductor. The feeders might be selected as No. 3 AWG copper conductor with rubber-type insulation, such as RHW or RUH, or with a thermoplastic-type insulation, such as THW or THWN; or the feeder might be selected as No. 1 AWG aluminum or copper-clad aluminum with the same types of insulation. As mentioned earlier, city building codes may have other requirements not specified in the National Code. Kansas City requires that all wiring be run in conduit for buildings (including houses) that happen to be in the Class A fire district. In other areas, conduits must be used in duplexes and multiple-residency buildings of two stories or more.

SUMMARY

Depicting electrical requirements for residential buildings, offices, and other structures requires a combination of graphical treatment, knowledge of building codes, and written specifications. For small structures, such as houses, it is usually sufficient to show electrical outlets, fixtures, and switches right on the house plans and to indicate what switching arrangements are required. In general, the rest of the circuiting arrangement is left to the constructor. But in larger buildings, it is necessary to depict all the circuit arrangements, including panelboard locations, in order to comply with safety requirements, to allow plenty of flexibility for future equipment and expansion, and to minimize expense of installing electrical work. National drawing standards ANS Y32.9 and Mil Std 15–3 (which are identical) are followed by most engineers who prepare electrical drawings. In depicting electrical requirements for buildings, it is necessary or desirable to make wiring plans, riser diagrams, and schedules for fixtures, raceways, conductors, etc. Symbol legends are sometimes used, especially if nonstandard symbols are used in drawings. Pictorial views and orthographic views of certain features and equipment are often quite helpful. Coordination by the person making the drawings with the architect, craftspeople, and equipment manufacturers is highly desirable because it will minimize conflicts and unnecessary installation expenses. Local building codes must be followed by the designer and constructor. When the local code does not cover a situation, the NEC must be consulted. If there is no local electrical code, the NEC should be used.

QUESTIONS

11-1. What authority, or authorities, shows the symbols to be used in making a residential floor plan that shows the electrical arrangement?

11-2. Why are riser diagrams usually not drawn to scale?

11-3. Why do we not use bus ducts in residential structures?

11-4. Sketch two different symbols which might be used to show a TV outlet.

11-5. What column headings would you use for a fixture schedule for a small building?

11-6. Why is it not customary to show two dashes across a circuit line for a two-wire conductor, when three dashes are customarily used to show a three-wire circuit?

11-7. Can you use the same graphical symbol for an electric-range outlet and for a dishwasher outlet?

11-8. Name three different schedules pertaining to the electrical system of a large building that might be found among the drawings for that building.

11-9. Name or describe two situations in which lines for the conductors themselves are not shown on drawings describing the electrical layout of a large building.

11-10. Does ANS Y32.2 cover the graphical portrayal of the electrical system of a residence?

11-11. Briefly describe three methods of providing underfloor electrical service to an office building whereby electric current can be supplied at evenly spaced intervals throughout or across a room.

11-12. Where does a branch circuit begin? Name two kinds of branch circuit.

11-13. Where are overcurrent devices located?

11-14. What are two reasons for limiting the number of wires in a raceway?

11-15. What are the terminal points of a feeder (or feeder circuit)?

11-16. What is one difference between a switchboard and a panelboard?

11-17. What is the smallest rating a panelboard may have? Or to put it differently, what determines the minimum rating of a panelboard?

11-18. Under what conditions does the NEC say that raceways should not be installed?

11-19. What devices or systems can be used for grounding electrodes? (Name four.)

11-20. What is the minimum rating of a receptacle?

11-21. What can be done to avoid conflicts?

11-22. Why do some architects (or consulting engineers doing electrical design for architects) find it necessary or advisable to add legends to their drawings?

11-23. What sort of information would be contained in a legend for an electrical drawing?

PROBLEMS

Many of the problems are divided into two parts. Part a will be mainly a drawing requirement, and part b will involve computational requirements. In some cases, the instructor may require both parts; in other cases, only a or b. Computations do not need to be put on drawing paper. However, they should be neat, orderly, and correct.

11-1. The sketch of the floor plan of Fig. 11-14 shows all electrical outlets and fixtures and the relationships between the fixtures and switches of an apartment.

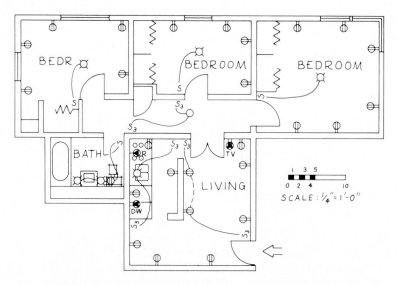

FIG. 11-14 (Prob. 11-1.) Plan of an apartment unit.

a. Make a mechanical drawing of this plan using currently approved symbols for wire runs, fixtures, and outlets. Wiring for duplex convenience outlets does not have to be shown except where they are controlled by a switch. Use 11 × 17 or 12 × 18 paper.

b. How many general 20-A and appliance 20-A branch circuits are required? If the range load is 8000 W and 110/220 three-wire feeders are used, what will be the total current load for the first floor?

11-2. How many 15-A branch circuits will be required for the room shown in Fig. 11-2? How many 20-A branch circuits would be required for the office layout in Fig. 11-5? (The interior dimensions are 22 × 57 ft. Each fluorescent luminaire is 80 W.)

11-3. Figure 11-15 shows the floor plan of an apartment with all outlets, fixtures, and switches. Switches, for the most part, work overhead lights, but in the living room they operate two of the four outlets.

a. Make a drawing (mechanical or freehand) of this floor plan, adding doors and the necessary wiring. Show TV outlets in the living room and largest bedroom. Use standard symbols. Use 11 × 17 or 12 × 18 paper.

b. A feeder system of 115/230 V and three wires is used and the kitchen range is rated at 11 kW. How many 15-A branch circuits will be required for the apartment? What will be the total current load?

11-4. Figure 11-16 is the floor plan of a one-story residence.

a. Draw this plan to a scale of $\frac{1}{4}$ in. = 1 ft 0 in. Show all switching arrange-

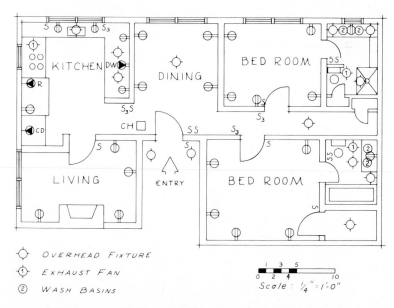

FIG. 11-15 (Prob. 11-3.) Floor plan of luxury apartment.

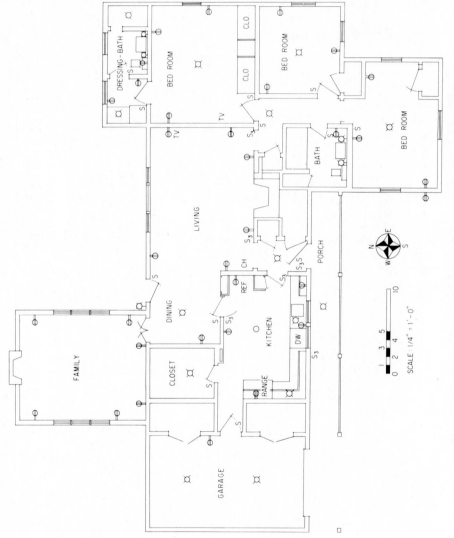

FIG. 11-16 (Prob. 11-4.) Floor plan of three-bedroom residence.

434

ments as you think they ought to be. At least two duplex outlets in the living room are to be controlled by switches at the north entry. Use standard symbols throughout the drawing.

b. The kitchen range has an 11-kW rating, and 220-V three-wire feeders are used. How many branch circuits (general and appliance) will be required? (What will the difference be if central air conditioning is installed?) What will be the total current load to be used for computing the size of the feeders?

11-5. Figure 11-17 is the first floor plan of a two-story house with all outlets and fixtures shown.

a. Draw this plan to a scale of $\frac{1}{4}$ in. = 1 ft 0 in., and show all switching arrangements as you believe they should be made. Use correct symbols as given in the Appendix. Optional: show circuits for the convenience outlets. Use 11 × 17 paper.

b. Assume the house has three bedrooms and a hall upstairs. For an 8-kW kitchen range and the other equipment shown, how many and what kind of branch circuits would be required for the house? What is the total energy requirement?

11-6. The lighting fixtures for a drafting-design room are shown in Fig. 11-18. If the fluorescent lights are rated at 80 W each, estimate the number of branch circuits needed. An electrical outlet should be put near each desk. Four desks are placed under each row of luminaires, making a total of 24 drafting

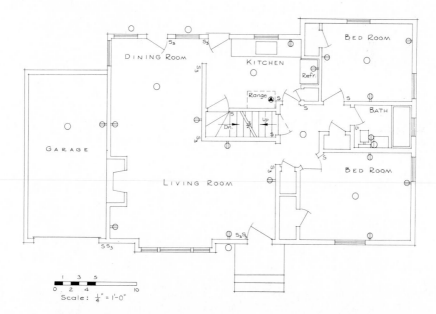

FIG. 11-17 (Prob. 11-5.) Floor plan of first story of residence.

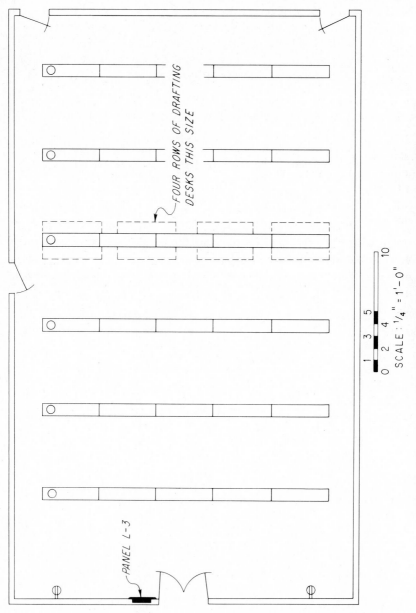

FOUR ROWS OF DRAFTING
DESKS THIS SIZE

PANEL L-3

SCALE: $\frac{1}{4}$" = 1'-0"

FIG. 11-18 (Probs. 11-6 through 11-8.) Floor plan of drafting-design room.

desks. Lights will be turned on and off at panel L_3. Other circuits go to panel L_1 on another floor. Make a scale drawing of this room, and show all circuits. Use 11×17 or 12×18 paper.

11-7.   Under each luminaire shown in Fig. 11-18 will be placed a designer's desk. It is desired to have provisions for electric erasers, telephones, and electric calculators at each desk. Make a detailed drawing that includes a system of underfloor raceways which will provide this electrical service to all desks. Assume that telephone conductors and 115-V conductors can be placed together in a raceway. Use 11×17 or 12×18 paper.

11-8.   Because the telephone company objected to having telephone lines in the same duct with other conductors, it was decided to install a cellular-type, underfloor wiring system for the drafting room in Fig. 11-18. Draw the floor plan to scale, showing a cellular system that will supply 115-V ac energy to outlets at each desk, and separate cellular ducts for intercom and telephone service to each desk. Overhead lighting may be included in the drawing. Use 11×17 or 12×18 paper.

11-9.   Figure 11-19 shows a section of a three-story building with a basement. Draw a riser diagram for this building as you believe it would be designed. Figure 11-6 will give you an idea of what conductor sizes to use. Use XHHW insulated wire; XHHW insulation is heat- and moisture-resistant and made of a

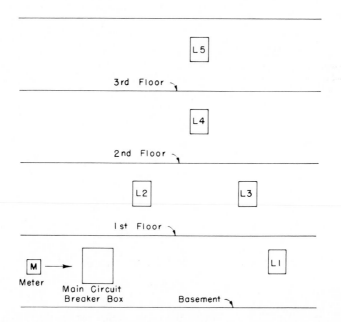

FIG. 11-19 (Prob. 11-9.) **Section elevation of three-story office building with electrical distribution equipment.**

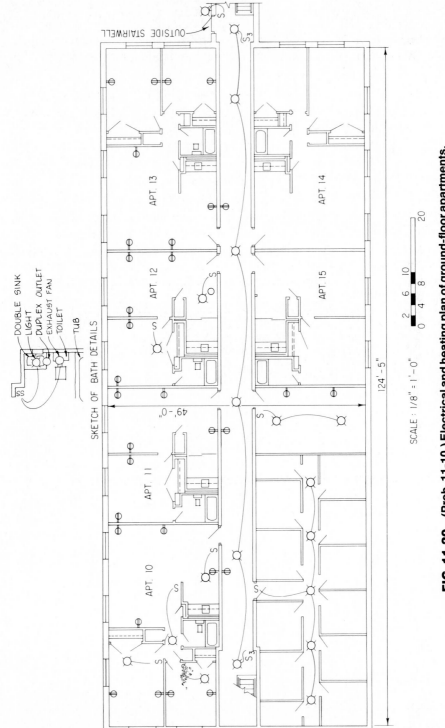

FIG. 11-20 (Prob. 11-10.) Electrical and heating plan of ground-floor apartments.

SCALE : 1/8" = 1'-0"

APT. 13

APT. 14

APT. 12

APT. 15

APT. 11

APT. 10

OUTSIDE STAIRWELL

124' - 5"

49' - 0"

SKETCH OF BATH DETAILS

DOUBLE SINK
LIGHT
DUPLEX OUTLET
EXHAUST FAN
TOILET
TUB

synthetic, cross-linked polyethylene. The current-carrying capabilities of XHHW insulated copper conductors in damp locations are as follows:

CONDUCTOR SIZE (AWG)	AMPACITY
14	15
12	20
10	30
8	45
6	65
4	85
3	100
2	115
1	130

11-10. The incomplete heating and electrical plan of one floor of an apartment building is shown in Fig. 11-20. Overall dimensions are also shown. Apartments 10, 13, and 14 are identical except that they are reversed. The same is true for apartments 11, 12, and 15.

a. Make a complete scale drawing of this plan, showing all circuiting and switching arrangements for all the rooms and the hall. Appliance loads are 8 kW for ranges, 2 kW for dishwashers, and 4.2 kW for washer-driers. These appliances will be located in the kitchen of each apartment. Their exact locations and outlets are not shown. (Use 12 × 18 paper for a $\frac{1}{8}$-in. to 1-ft 0-in. scale.)

b. If a copy of the NEC is available, compute the minimum number of branch circuits and the minimum size subfeeder for each apartment. Compute, also, the main feeder for this floor and the rating of the panel, if one panel is used.

11-11. Letter a fixture schedule to accompany an electrical plan. Such a fixture schedule might include the information shown in Prob. 1-7, Chap. 1.

12

Graphical
Representation of Data

The graph is one of the most effective tools of communication that any technically trained person can wield. With a graph, one can bring order to a collection
of data and present it in a picture form that tells a story. Also with a graph, one
can compare the performance of two or more related items or processes and
may be able to make certain computations not practicable by algebra, analytic
geometry, or calculus. In development and research the graph is used to determine the relationships of two or more variables, to compare laboratory data
with theory, and to determine if test data are accurate and reliable.

Because this information must be presented honestly, accurately, and as
clearly as possible, skill and judgment are required to make a good graph.
Therefore, an engineer or drafter should develop as much skill and knowledge
within this area as in any other area or type of technical drawing. The fundamental principles of graphical representation are:

1. Graphs should be truthful representations of the facts.

2. Graphs should be clear and easily read and understood.

3. Graphs should be so designed and constructed as to attract and hold the
attention.

Figure 12-1 is a good example of a well-drawn technical or engineering type
of graph.

12-1 General Concepts in Preparing Graphs

One of the most difficult problems in constructing a graph is choosing the
horizontal and vertical scales. As an example of this problem, Fig. 12-2 is
presented. In this figure the same "curve" is shown on four different graphs,
each having different arrangements of scales; thus, it may be difficult to recog-
nize that the same data are being presented. Graph *a* is the best of the four for
two reasons: (1) the data are presented more honestly, and (2) better use is made
of the available space than in any of the other three drawings. This problem of
the selection of scales can often be solved by using commercially prepared graph

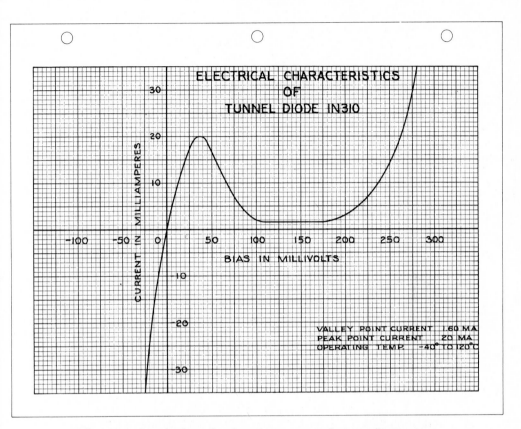

FIG. 12-1 A typical graph, plotted on rectangular coordinate paper.

paper and making wise use of the space available. However, it may not always be possible or desirable to use such paper.

Another problem in graph construction is deciding what type of graph to make. In other words, the problem might be, "What type of graph paper should be used?" There are many different types of graphs in use today, and many different kinds of graph paper. Most graph papers are printed on $8\frac{1}{2} \times 11$-in. paper, although larger sizes are available. Lines come in black, blue, green, orange, and red for many graph styles. Grids are available in rectangular, polar, and probability coordinates, to name several examples.

Most technical and engineering graphs are drawn on *rectangular coordinate* paper. (It is also called *arithmetic, rectilinear, cross-ruled,* and *square-grid* paper.) Typical spacings for this type of paper are 5, 10, and 20 lines (or spaces) to the inch and 10 to the centimeter. Other spacings such as 4, 6, and 8 lines to the inch are available, but are not in wide usage. The graph of Fig. 12-1 was made on paper that has 10 lines to the inch, as are several other examples in this chapter. If it is desired to make blueprints, Ozalid prints, or other similar types of reproduction, then thin, translucent graph paper can be used. If the graph

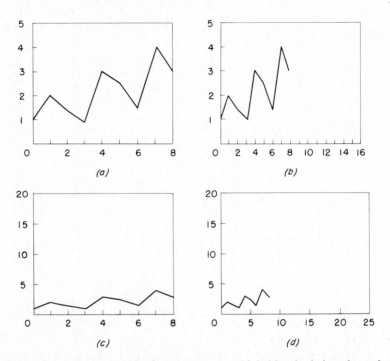

FIG. 12-2 Graphs having different scales on which identical data have been plotted.

paper has blue grid lines, these will not appear in the reproduction, but orange, red, or black lines will show up on a print. The graphs in this chapter which have the grid lines showing have been drawn on red-line graph paper.

In order to make good reproductions (by Ozalid, Thermofax, xerography, and other reproduction processes), lines and lettering must be made heavy and dark in order to "stand out" from the grid lines. Ink drawings give the best results, but dark, heavy pencil work will provide legible copies in some reproduction processes. In order to make lettering stand out even more, one can do the lettering on heavy, white paper, then paste the paper on the graph. This has been done with the title in Fig. 12-5 and in several other graphs in the chapter. This is effective, but not necessary, and will not reproduce on blueprints or Ozalid prints. Sometimes the thin grid lines on a graph can be used as guidelines. This has been done with the supporting data (lower right) of Fig. 12-1. This lettering should be the smallest lettering on the graph. The $\frac{1}{10}$-in. height between successive guidelines is ideal.

The tunnel-diode curve of this graph falls in two quadrants, making it necessary to arrange the vertical and horizontal scales so that negative as well as positive values can be plotted. By locating the zero point *(origin)* to the left of center, it was possible to draw the desired portion of the curve, show the scales and their captions, and provide space in the upper part for the title and in the lower right for the auxiliary or supporting data. The X and Y scale values could have been placed around the edges of the graph, but it is believed that they are more appropriately located close to their respective axes, as shown in Fig. 12-1.

12-2 Selection of Variables and Curve Fitting

Data to be plotted graphically are generally available in tabular form, with each point having two coordinates as follows:

POINT	COORDINATES	
1	0	0
2	10	2.6
3	20	3.8
4	30	5.3
5	40	7.8
6	50	10.1

One set of coordinate values must be plotted on the horizontal, or X axis (or *abscissa,* as it is often called), and the other set of coordinates must be plotted along the vertical, or Y axis (or *ordinate*). Standard practice is to plot the *independent* variable *horizontally* and the *dependent* variable *vertically.* The independent variable is that variable which the operator can control during a test, if one can be controlled. In some cases where a variable cannot be controlled, one variable is arbitrarily selected. *Time,* for instance, is generally considered to be the independent variable. A glance at the coordinates, appearing above, shows that the first set of coordinates progresses at even intervals of 10. It is obvious that the operator or observer was able to take readings at 10, 20, etc., either by controlling the variable or, if it were a natural phenomenon such as time, by taking readings at convenient intervals. Those coordinate values in the *left* column, then, represent the independent variable.

After the graph is laid out and the points are plotted, the problem of drawing, or "fitting," the final curve presents itself. Whether to draw a smooth curve or straight lines between successive points depends on several factors. Some of them are as follows:

1. Most physical phenomena are "continuous." This means that the curve showing the relationship between such variables should be smooth — with few inflections — and should pass through or near plotted points.

2. Data backed up by theory should be represented by a smooth curve. (However, if plotted points are not abundant, straight lines are often drawn between points, as with instrument calibration.)

3. Discrete, or discontinuous, data — representing a discontinuous variable having discrete increments — should be shown by joining successive points with straight lines.

4. Observed data not backed by theory or mathematical law should be represented by point-to-point straight lines, unless continuity can be definitely established.

Figure 12-1 includes a curve representing continuous data. Figure 12-2 has a curve which follows the discontinuous relationship of periodic observations. Figure 12-3*a* shows a theoretical curve and points taken from actual field data. Such points should be joined by a smooth curve. Figure 12-3*b* is a typical example of discontinuous data.

12-3 Curve Identification

Not infrequently, it is necessary to put more than one curve on a graph. The curves must be drawn or labeled so that they can be easily identified and distinguished from one another. There are three methods by which this can be done:

1. Clearly label each curve
2. Use a different type of line for each curve
3. Use different plotting symbols for each curve

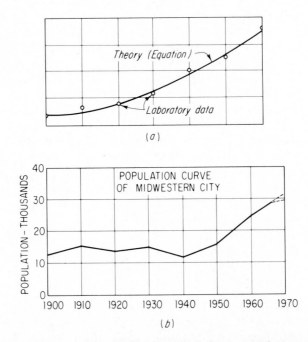

FIG. 12-3 Examples of curve fitting: *(a)* a situation in which a curve should be drawn through or near points plotted from laboratory data; *(b)* census figures, taken once every 10 years, which yield discontinuous, or discrete, data. Straight lines should be drawn from point to point.

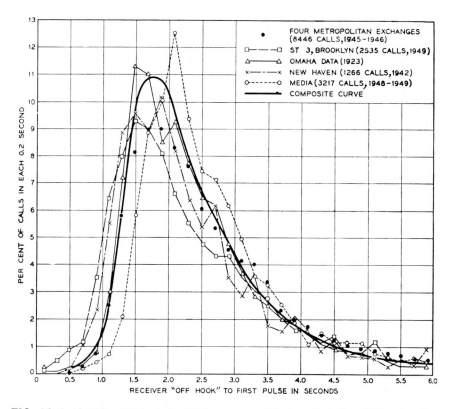

FIG. 12-4   A composite curve. The heavy curve is a composite of the other four curves. (Bell Telephone Laboratories.)

Often two or more of these methods are combined. For example, in Fig. 12-4 different lines and plotting symbols have been used, except for the *composite* curve, which is a sort of weighted average of the other curves. A similar identification system has been used in Fig. 12-5. The difference between the two figures is the method of labeling curves. In Fig. 12-4, a *legend* in the upper right corner identifies each line. In Fig. 12-5, each line is identified by means of a title, or caption, and a *leader* pointing to the curve itself. Both methods are widely used.

12-4 Zero-Point Location

In a great majority of cases, line graphs are drawn with the *origin* at the lower left corner. Figures 12-2 and 12-4 are in this category. In other cases, the vertical scale begins at zero, but the horizontal scale does not. Figure 12-3*b* is a good example. Most engineers feel that the *Y* scale should begin at zero; or conversely stated, they feel that starting the *Y* scale with a figure *not* equal to zero tends to distort the picture.

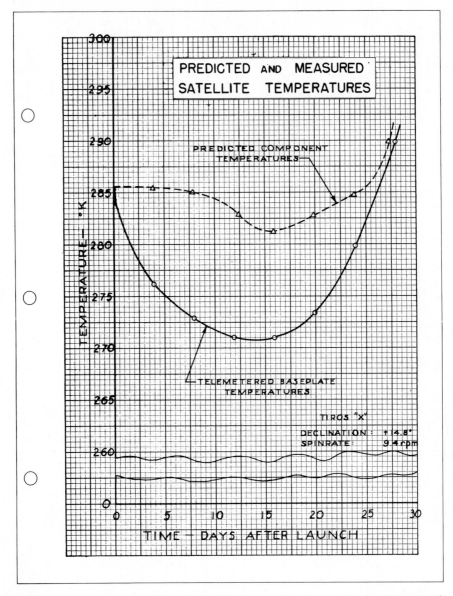

FIG. 12-5 A graph in which the vertical scale has been "broken" to permit greater vertical excursion.

Sometimes, however, the values to be plotted on the vertical scale are such that starting the scale at zero will produce a plotted curve that is too flat and inaccurate for working purposes. Figure 12-5 is an example of this problem. Note that the temperature plots of the weather satellite fall between 270° and 290°. By "breaking" the vertical scale between 0 and 260°, we were able to establish ordinate values which provided sufficient vertical latitude ("excursion") for accuracy and comparison of the two curve shapes. There are other techniques for accomplishing the same result, the point being to warn the reader that the vertical scale is not all there or that it starts at something other than zero.

12-5 Steps in Construction of an Engineering Graph

The following steps illustrate how a graph, as drawn in many engineering offices, is constructed. Such a graph is shown in Fig. 12-6.

1. Arrange the data in table form, preferably in a logical order, from the smallest figure to the largest. This will make plotting easier and faster, and will clearly show the upper and lower limits for each variable.

2. Determine which variable will be the independent variable and which the dependent, for the very practical reason that proper coordinate paper must be selected and a decision made whether the long dimension of the paper will be up and down or sideways on the drawing board. These will probably depend on which variable is which.

3. Select the graph paper that will best show the curve and that will accommodate the points to be plotted taking into account their extreme values. This may include a trial plotting of values along the horizontal and vertical axes. In the case of Fig. 12-6, we selected rectangular coordinate paper having fine lines spaced $\frac{1}{10}$ in. apart, and heavier lines $\frac{1}{2}$ in. apart.

4. Locate the zero point of each scale (origin of graph), allowing for margins within the grid portion of the paper at the bottom and left side. One-inch margins are common because they allow room for binding the graph at the left side and for scale values and their descriptions.

5. Letter in the values along and outside the two "baselines" provided in step 4. Standard practice is to use multiples of 1, 2, 5, or 10. In Fig. 12-6, multiples of 10 (each inch representing $10,000) were used on the vertical scale, and multiples of 2 (each half-inch representing 20 kW) were used on the vertical scale. Do not put these numbers too close together.

6. Letter in the descriptions of the scales close to the figures, as shown in Figs. 12-6 and 12-7. Vertical lettering should be used, and each description should be centered between the zero point and the figure at the other end of the baseline. Each description should tell what the scale values show, and what units the scale values represent. Typical standard abbreviations are: kW, A, Hz (cycles per second), and kHz (kilocycles per second).

7. Plot the points with a sharp pencil or pricking instrument. It is a good idea to circle these points immediately after plotting so that they will be easily seen when drawing the curve later. If the circles are to be shown permanently, as is so often done with experimental data, they should be hollow and from $\frac{1}{16}$ to $\frac{1}{10}$ in. in diameter.

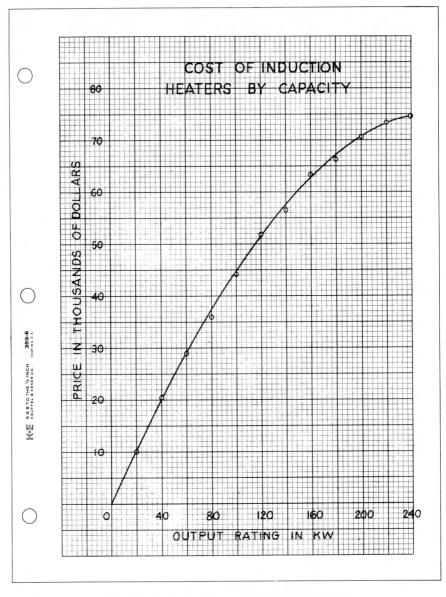

FIG. 12-6 Another typical engineering graph.

8. Draw in the curve. If the curve is not of the straight-line variety, many persons prefer to sketch it lightly freehand until the desired result is obtained, then to put it in with a heavy pencil line, using an irregular curve. If plotting symbols are used, the curve should not be drawn through the symbols. If a symbol is in the path of the curve, the latter should come up to and just touch each side of the symbol.

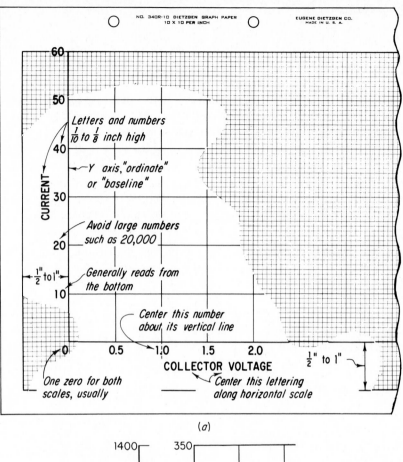

(a)

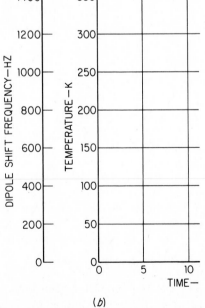

(b)

FIG. 12-7 *(a)* Layout details for the construction of a graph. *(b)* When two or more vertical scales are required.

9. Place the title on the graph. Titles — which are clear, yet as brief as possible — are commonly placed either above or below. This lettering should be vertical uppercase and larger than the numbers and letters describing the vertical and horizontal scales. Some firms require that the fine, preprinted grid lines be erased from around titles and borders drawn around the lettering.

10. Place supporting data, if desired or required, on the graph. Such data may include date of preparation, site of tests or observations, name of person or party making the tests or graph, source of data if not original, equations, and simple circuit diagrams. Such data are often placed in the lower right-hand area of the graph, but are sometimes placed elsewhere if circumstances require. (See Figs. 12-1 and 12-5.)

11. Complete the graph. This may include inking curve, borders, and lettering if inking is required. Sometimes letters and figures can be typed on graph paper with standard typewriters or varitypers if special ribbon is used.

A graph should be made interesting and clear through the use of different line weights. The curve should be the heaviest line on the graph, the border(s) or baselines the next heaviest, and grid lines (if not preprinted) the lightest.

12-6 Drawing a Smooth Curve

Drawing a curve generally involves two steps:

1. Sketching, freehand, a trial curve through (or near, as the case may be) the plotted points
2. Drawing the finished curve along the trial curve, using pencil or pen and a plastic curve or spline

Some of the plastic (often called *irregular* or *French*) curves and their usage are shown in Fig. 12-8. In 99 cases out of 100, it will not be possible to select a plastic curve that will match the plotted curve for its entire length. The next-best solution is to try a curve (if more than one is available) or that part of a curve (if only one is available) that will fit as much of the trial curve as is possible. (If the curve is to be inked later, it is a good idea to remember which parts of the curve were used at different locations along the final curve shape.)

Curve drawing is usually more satisfactory if the pencil is used on the edge of the curve that is away from the person who is drawing. This is somewhat analogous to using the upper (far) edge of a T square or horizontal bar.

12-7 Scales and Their Titles or Captions

There are good and bad ways to organize the numbers and descriptions along the baselines of a graph. Here are a few examples:

GOOD SELECTION OF NUMBERS

0	10	20	30	40	50
0	0.2	0.4	0.6	0.8	1.0
0	50	100	150	200	250

POOR SELECTION OF NUMBERS

0	10,000	20,000	30,000	40,000	50,000
0	30	60	90	120	150
0	7	14	21	28	35

GOOD TITLE OR CAPTION ORGANIZATION

RESISTANCE IN OHMS
RESISTANCE IN THOUSANDS OF OHMS
RESISTANCE—THOUSANDS OF OHMS
RESISTANCE IN MEGOHMS

SATISFACTORY BUT SOMETIMES CONFUSING

RESISTANCE—OHMS $\times 10^3$
PARTS PER $FT^3 \times 10^6$
TRIGGER CHARGE—10^{-10} COULOMBS
(The numbers shown are all positive.)

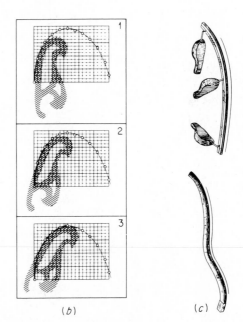

(*a*)   (*b*)   (*c*)

FIG. 12-8 **Irregular curves, splines, and use of the curve.** (Parts *a* and *c* from Frank Zozzora, *Engineering Drawing*, 2d ed., McGraw-Hill Book Company, New York, 1958; part *b* from Thomas E. French, Carl L. Svensen, Jay D. Helsel, Bryan Urbanick, *Mechanical Drawing*, 9th ed., McGraw-Hill Book Company, New York, 1980.)

12-8 Families of Curves

Figure 12-9 shows a *collector-characteristic* curve of a transistor which was obtained by varying the voltage and measuring collector current for several values of base current. This family of curves can be used for the determination of transistor performance in a common-emitter circuit and also for the calculation of other useful parameters. For example, it is often desirable to know the ratio of the direct collector current I_C to the direct base input current I_B. This *current gain* is typically around 49 or 50, calculated as follows:

$$\beta = \frac{I_C}{I_B} = \frac{0.981}{0.021} = 49$$

Other useful characteristics which can be obtained from such a table and other information supplied by a manufacturer are frequency cutoff, breakdown voltage, reach-through voltage, and storage time. When five or more curves are placed on a single graph, it is not very practical to draw a different type of line for each curve. Proper identification of each line is usually accomplished in the manner shown in Fig. 12-9.

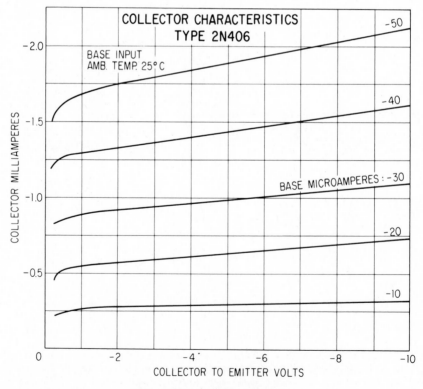

FIG. 12-9 A family of curves.

12-9 Graphs for Publication

When graphs are to appear in printed matter, such as books or technical journals, they are usually not drawn on commercial graph paper. ANS Y15.1, "Illustrations for Publication and Projection," is an appropriate guide for such cases. Covered are such points as minimum letter size and line weights before reduction to legible and clear size. A standard outline proportion of $6\frac{3}{4} \times 9$ in. ($\frac{3}{4}:1$) is recommended. No more coordinate lines should be drawn than are necessary to guide the eye. Because additional discussion is presupposed, supporting data and equations should be omitted. In the majority of published graphs, plotting symbols are not shown, although some do include these symbols. Figures 12-9, 12-14, and 12-17 are good examples of this type of graphical presentation.

In short, the published graph is a rather simple construction, uncluttered by minute data, yet very clear and legible. In order to have clarity and legibility, it might not include all data necessary to tell the whole story.

12-10 Line Graphs on Other Types of Graph Paper

As mentioned previously in this chapter, other kinds of commercially printed graph paper are available and used by many engineers and scientists for various reasons.

Figure 12-10 shows curves for cassette tapes over a range of frequencies from 30 to 12000 Hz. Like many curves in which frequency is one of the variables, it is plotted on semilogarithmic paper, in this case on a four-cycle paper. It would be impossible to plot points accurately over such a large range on linear (rectangular-coordinate or square-grid) paper. With the use of semilogarithmic paper with three, four, or five cycles, it is possible to plot accurately. When using such graph paper we do *not* convert the raw data to logarithms.

Another reason for using different kinds of graph paper is to achieve a pattern that yields a "straight-line curve." Sometimes, points that yield a curved pattern on one kind of paper yield a straight-line pattern on another kind of paper. The following will result if a straight-line curve can be drawn:

1. Future prediction (extrapolation) is easier if a straight line is used.

2. Two curves can be better compared if they are straight-line curves than if they are otherwise shaped.

3. The slope and equation of the line (and therefore of the data which produced the line) can be obtained graphically.

If the logarithmic scale is oriented vertically and the linear scale horizontally, a curve representing a constant *rate* of change will plot as a straight line. Such a curve would result if we were to plot the 2.8 percent annual increase in the rate of change of our output per worker-hour since World War II.

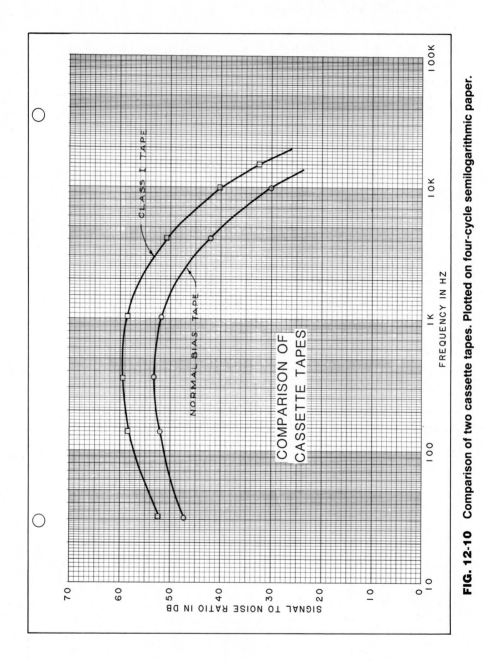

FIG. 12-10 Comparison of two cassette tapes. Plotted on four-cycle semilogarithmic paper.

12-11 The Logarithmic Scale

The logarithmic scale is a functional scale in which the distances are laid out to equal the function (logarithm), but the numbers at these distances are those of the variable. A logarithmic scale from 1 to 10 is called a *cycle.* The left edge of the scale is marked 1 (the logarithm of 1 is zero), and the right edge 10 (the logarithm of 10 is 1), and we have a *unit* scale from the log of 1 (zero) to the log of 10 (1). This unit scale is called a *cycle,* but we can scale intermediate distances like 1.5, 2, 3, 5, etc., and show those numbers at those points. A logarithmic scale from 10 to 100 or 100 to 1000, etc., is also called a *cycle,* or *modulus.*

It is impossible to have a zero showing on a logarithmic scale. If the scale includes the number 1 at either end, the zero is there *graphically* because the logarithm of 1 is zero. Also, the logarithm of zero is minus infinity. This would be impossible to plot graphically.

It is also worth noting that *interpolation* between marks on a logarithmic scale must be done *logarithmically.* When using logarithmic graph paper, one does not have to be concerned about this.

12-12 Equations of Straight-Line Plots

The following equations will apply in the situations indicated:

EQUATION	TYPE OF COORDINATE
$y = mx + b$	Rectangular
$y = \dfrac{x}{mx + b}$	Rectangular $\left(\dfrac{x}{y} \text{ values plotted against } x \text{ values}\right)$
$y = bx^m$	Logarithmic
$y = b(10)^{mx}$ or $y = b(e)^{mx}$	Semilogarithmic (logarithmic scale vertical)
$y = m \log x + b$	Semilogarithmic (logarithmic scale horizontal)

In these equations, m represents the *slope,* and b the *intercept.* We will show how to get these quantities in the next two examples.

12-13 Use of Logarithmic Paper

Logarithmic paper (both scales are arranged logarithmically) is used for reasons similar to semilogarithmic paper. If the range of plotted values is large in both the x and y directions, logarithmic paper, with the proper number of cycles, can be used very much as semilogarithmic paper was used to accommodate the large range of frequency values used in Fig. 12-10. As in the case of semilogarithmic graph paper, logarithmic paper is available with anywhere from one to five logarithmic cycles, and anything in between. The most-used papers have the same number of cycles in the horizontal and vertical directions, but papers are available with different combinations. Usually, however, the length of one horizontal cycle is the same as the length of one vertical cycle.

Another reason for plotting data on logarithmic paper is to get a straight-line pattern. It might be found by trial and error that a certain group of data provides a straight-line plot on such paper. Or theory or previous tests with similar data may indicate that a set of plotted points will yield a straight-line pattern. Figure 12-11 contains such a situation. Plotting of points and construction of the graph follow the same techniques, previously explained, for drawing a graph on rectangular coordinate paper. One difference is that the numbers around the edge are usually already printed on the sheet. The drafter often must

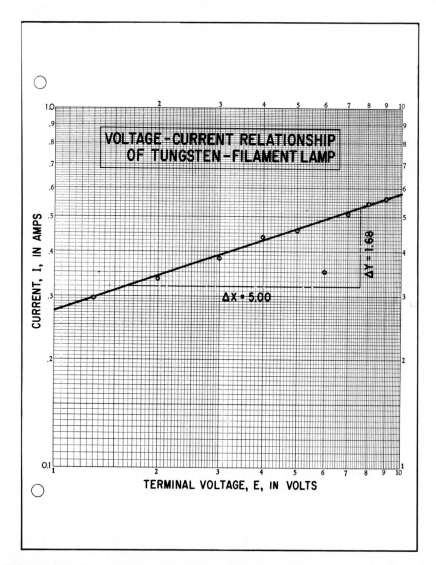

FIG. 12-11 Data plotted on logarithmic paper. These data yield a straight-line pattern ($\Delta Y/\Delta X$ represents the slope).

add zeroes or decimal points to the numbers of the scale.[1] Because these numbers are outside the grid, the scale caption must also be placed outside the grid. Another difference is that the curve itself should not be very thick, if one wants to obtain accuracy in measuring the slope and locating the intercept.

In Fig. 12-11, a right triangle with a 5-in. base has been drawn. Each leg of the triangle has been measured accurately with a decimal scale to obtain Δx and Δy values. (This is *construction* work, and it is sometimes erased and does not appear on the finished graph.) The important point about measuring the legs of the triangle on logarithmic paper (where the horizontal and vertical cycles are equal in length — which is usually the case) is that a linear scale be used and that the *same* scale be used to measure each leg. Instead of using a decimal scale, we could have used a 50 scale, 40 scale, quarter scale — or any *linear* scale. Using the logarithmic scale of the graph paper to measure the legs will *not* provide the correct slope.[2] The slope is obtained as follows:

$$m = \frac{\Delta y}{\Delta x} = \frac{1.68}{5.00} = 0.336 \text{ or } 0.34$$

The *intercept* is found by observing where the line intersects the Y scale where $x = 1$ (remember, the log of 1 is zero) and reading the value along the Y scale. The intercept in Fig. 12-11 is

$$b = 0.272$$

The equation for this line is, therefore,

$$y = 0.272x^{0.34}$$

or, using the abbreviations for the actual units plotted,

$$I = 0.27E^{0.34}$$

The third-decimal-place values have been dropped because there probably is not enough justification for this type of accuracy. Two-decimal accuracy is appropriate, however.

12-14 Use of Semilogarithmic Paper

Figure 12-12 illustrates a situation in which a series of plotted points provides a straight-line pattern, as shown by the line. As in the case of the logarithmic graph, previously cited, the scales had to be marked off around the outside edges of the grid because the log scale numbers had already been printed.

The Δy and Δx values shown were measured with the same *linear* scale. But because the two scales of a semilogarithmic graph are different, we have to

---

[1] Sometimes the printer does not leave enough room for this, or there is not much room for scale titles, presenting a rather awkward situation.

[2] If the horizontal and vertical cycles of the paper are not equal in length, one will have to select points on the line and work with their coordinates to get the slope:

$m = (\log Ay - \log By)/(\log Ax - \log Bx)$.

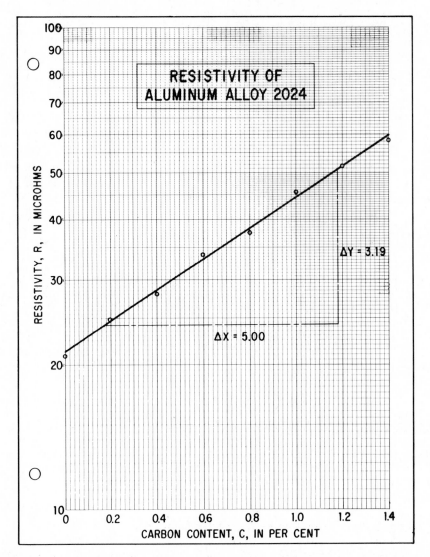

FIG. 12-12 Points that give a straight-line pattern when plotted on semilogarithmic paper.

do a little more work to get the correct slope. The additional work (beyond what was done in the case of the logarithmic graph) is to obtain values of *unity* for the X and Y scales. Unity for the X scale is 5.00 in. (the actual distance from $x = 0$ to $x = 1.0$). Unity for the Y scale is 10.00 in. (the distance from $\log 10 = 1$ to $\log 100 = 2$), in this case the length of the logarithmic cycle from 10 to 100. Now, we are ready to determine the slope.

$$m = \frac{\Delta y/\Delta x}{y \text{ unity length}/x \text{ unity length}} \quad \text{measured with the same linear scale}$$

$$m = \frac{3.19/5.00}{10.0/5.0} = 0.319 \text{ or } 0.32$$

The intercept is found by reading the value along the Y scale where the line intersects at $x = 0$:

$$b = 21.3$$

The equation is

$$y = 21.3 \ (10)^{0.32X}$$

or $\qquad\qquad\qquad\qquad\qquad R = 21.3 \ (10)^{0.32C}$

If it is preferred to use e instead of 10 in the equation, the slope must be divided by 0.434, which is $\log_{10} e$.

$$m = \frac{0.32}{0.434} = 0.74$$

and the equation becomes

$$R = 21.3 \ (e)^{0.74C}$$

If one cannot reconcile the graphical method just explained, one can check the slope by selecting two points on the curve and using their coordinates as follows:

$$m = \frac{\log Ay - \log By}{Ax - Bx}$$

We have done this by selecting points at $A = (1.16, 50)$ and $B = (0.46, 30)$ to get

$$m = \frac{1.699 - 1.476}{1.16 - 0.46} = 0.319 \text{ or } 0.32$$

There are other methods which can be used to obtain, or check, these equations. One method which requires more work but which is more accurate — and which can utilize a digital computer — is the method of least squares.

12-15 Polar Coordinates

Because of the directional characteristics of electrical devices such as lamps, antennas, and speakers, studies must be made of their performance in different directions. The results of these studies can be most appropriately shown on polar charts. These show two variables, one having a linear magnitude plotted on equally spaced concentric circles, and the other an angular quantity plotted radially with respect to a *pole* or origin. The plotted points are usually, but not always, joined to form a line or curve.

Figure 12-13 shows a polar plot of the power-radiation pattern of a half-wave dipole, as computed by an analog computer. Notice that the linear scale is

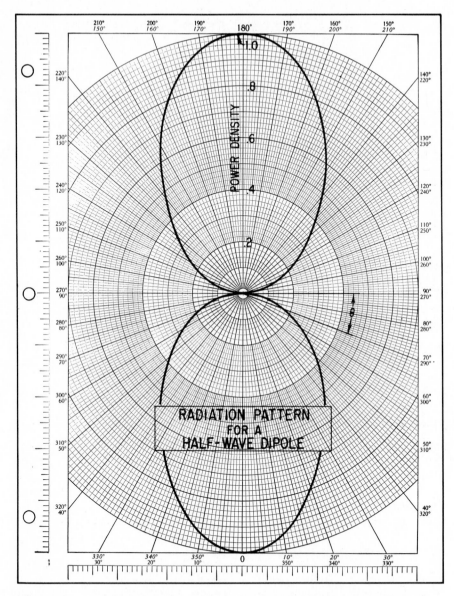

FIG. 12-13 A radiation pattern plotted on polar coordinate paper. (Zero at bottom.)

shown as units based on 1.0 being the maximum. Notice, also, that the zero is at the bottom of the graph. The curve has two large enclosures, called *lobes,* and points of zero magnitude, called *null points,* at 90 and 270°.

Figure 12-14 depicts the cross-talk coupling between two antennas at the same location. The two "envelopes" show the maximum cross-talk obtained

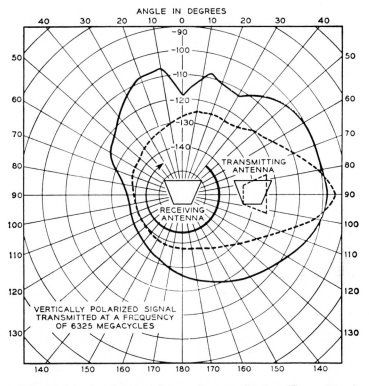

FIG. 12-14 Another graph on polar coordinates. (Zero at top.)

for two positions of the transmitting antenna when the receiving antenna is rotated clockwise. Notice that zero is at the top and that degrees increase both clockwise *and* counterclockwise from zero.

In other situations, it is sometimes desirable to plot zero at the right side of the graph and go counterclockwise, much as the mathematician plots values of trigonometric functions. It is possible to buy polar coordinate paper that has the zero at the top or at the bottom. Disk recording devices use polar charting, and, instead of radial graduations being in degrees, they are in hours or days.

12-16 Bar Charts

Some types of data do not lend themselves well to presentation in line graphs. They might be better suited for display in some other form, such as a bar chart. Also, data in graphical form must often be presented to clients or persons who do not have technical backgrounds. Such persons may find bar graphs, pie charts, and pictorial graphs easier to read.

Figure 12-15 is a bar chart in which the bars run up and down, for which reason it may also be called a *column chart*. Its construction is arrived at in much the same manner as a line graph that is plotted on rectangular coordinate

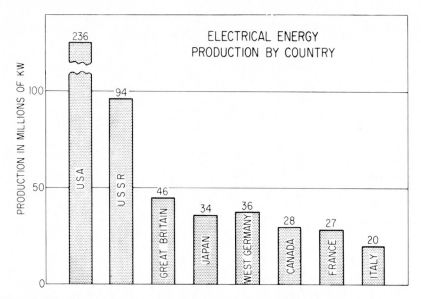

FIG. 12-15 A bar chart or graph. (Titles for bars are often placed below the bars.)

paper. General practice is as follows:

1. Several major horizontal grid lines should appear with their scale values.

2. Bars, or columns, should be shaded.

3. Widths between bars should be no wider than the bars themselves and may be less.

4. Bars are often arranged in ascending or descending order, but this is not a requirement.

To make an attractive bar chart, one should consider using preprinted appliqués for shading. Rather than make a scale which would accommodate the United States production of 236,000,000 kW, the authors decided to use a scale that would permit the bars of the other countries to be larger and "broke" the United States bar as shown. While not generally done, this is acceptable practice as long as the values are shown at the top of the bars.

Another type of bar chart is shown in Fig. 12-16. This is only one form of a horizontal bar graph. Another form has a zero line running up the middle, with bars representing positive values extending to the right and bars with negative values to the left. Bars may also be broken down individually into several parts. For example, if the information were available, we could show what percent of each country's electrical production was (1) hydro, (2) steam generating, and (3) nuclear, on the graph of Fig. 12-15.

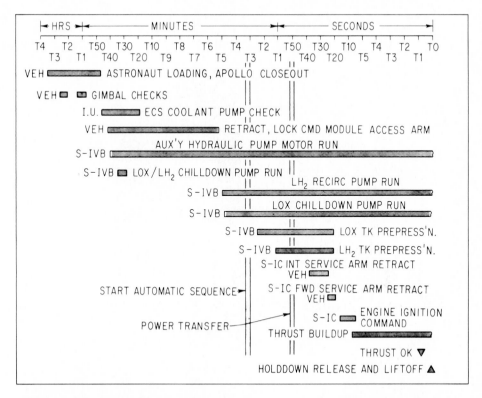

FIG. 12-16 Horizontal-bar chart showing typical prelaunch events for Saturn-Apollo space flight. (NASA.)

12-17 Pie Graph

A very popular type of chart, although held in low regard by statisticians, is the pie chart. It is most effective for displaying five to seven items that make up 100 percent. A well-designed pie graph has the largest item in the upper right sector, starting at 12 o'clock, followed by the next largest item, and so on in a sequence of decreasing size. The following principles of good construction apply:

1. The graph must be large enough to permit lettering within all but the very smallest sectors.

2. Arrangement of items should follow a sequence of decreasing sizes (Fig. 12-17).

3. Items should have a shading sequence of ever-increasing or -decreasing darkness.

4. The largest item should start at 12 o'clock and be on the "clockwise" side.

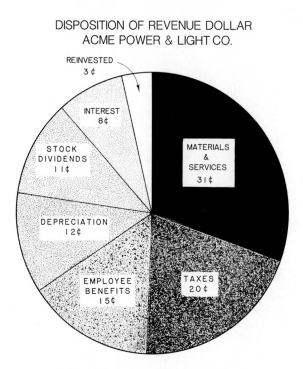

DISPOSITION OF REVENUE DOLLAR
ACME POWER & LIGHT CO.

FIG. 12-17 A pie graph. Notice that the material is arranged sequentially by size.

5. A title should be placed above or below the graph.

6. The percentage or value should show in each sector.

12-18 Pictorial Graphs

There is some evidence that graphs presented in pictorial form are remembered longer than those drawn in two dimensions. Many different ways of showing graphs in pictorial form have been used, including isometric, dimetric, oblique, and perspective construction. A three-variable pictorial graph has been shown in Fig. 12-18, which has been drawn in isometric projection. Making a graph like this takes quite a bit of time; for one thing, lettering is a little tricky. Therefore, one should be sure that the results will justify the extra effort. The material presented in Fig. 12-18 could have been presented in the form of the graph shown in Fig. 12-9, and vice versa.

Another pictorial drawing, shown in Fig. 12-19, is only partly a graph. This shows how ingenuity can be used to combine two different kinds of graphical representation. Other forms of pictorial representation of graphs have been used with success. Bar graphs have often been drawn pictorially. Oblique projection adds depth to the bars and makes possible the addition of more information in the third dimension.

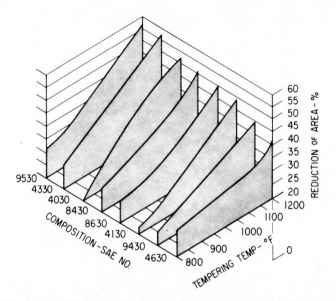

FIG. 12-18 A pictorial graph showing reduction of area of various steels after liquid quenching at various temperatures. (Three variables in isometric projection.)

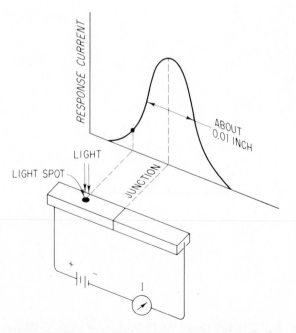

FIG. 12-19 A pictorial drawing (in dimetric projection) showing the response of a photodiode depending on where it is illuminated. The curve is a normal frequency-distribution curve.

12-19 Other Types of Graphic Presentation

The graphs shown in this chapter are of the type found most often (perhaps 95 percent of the time) in engineering, design, and research offices. Actually, there are a number of other ways to present material graphically, and some of these graphical forms can be used for calculation as well. Some of these graphs have been assigned names as follows:

1. Alignment charts
2. Concurrency charts
3. Conversion scales
4. Network diagrams
5. Holographs
6. Derivative curve ⎫
7. Integration   ⎬ graphical calculus

    Figure 12-20 is a type of conversion scale that shows the comparative economic advantages of reducing the weights of various electrical or electronics

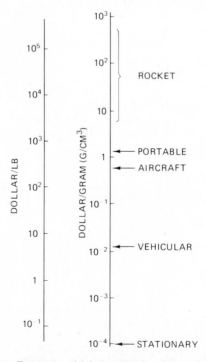

FIG. 12-20 **A line diagram, which is also a conversion diagram. This shows justifiable cost for removing by miniaturization one pound or gram of unnecessary weight.** (From Edward Keonjian, *Microelectronics*, McGraw-Hill Book Company, New York, 1963. Used by permission.)

gear. For the other types of graph, listed on the preceding page, the reader should refer to a good text on engineering graphics, graphics, or graphic science.

SUMMARY

Graphical construction is a very important phase of technical drawing. A careful survey of the problem is the first and most important step in successful graphic presentation. It should cover a careful consideration of items such as the following:

1. The purpose of the graph
2. The occasion for its use
3. The type of person who is to read it
4. The nature of the data
5. The medium of presentation to be used
6. The reproduction process to be used
7. The time available for preparation
8. Equipment and skill available

A graph should have a sufficiently professional appearance to inspire confidence in the facts presented. Layout and design are as important as quality of drafting. The type of data being portrayed graphically determines whether the curve should be smooth or a series of straight lines from point to point. Clarity and brevity are the two most important features of any graph. Some graphs can be used for purposes of computation. For example, equations such as $y = mx + b$ and $y = bx^m$ can be obtained if the curves appear as straight lines on the appropriate types of graph paper. Alignment charts, concurrency charts, and other graphical forms are useful in making certain computations.

QUESTIONS

12-1. What determines whether the plotted points on a graph should be connected with straight lines or be joined by a smooth curve?

12-2. What are two reasons why it is sometimes desirable to have data plotted in one straight line (by using different graph paper) instead of as a curve?

12-3. What determines whether data should be shown by means of a bar (or column) chart or by a pie chart?

12-4. Name four types of commercial graph paper that are available.

12-5. Along which axis is the dependent variable usually plotted?

12-6. In plotting the data from a laboratory experiment, how would you go about determining which variable is the independent one?

12-7. If it were necessary to draw two curves on one graph, how would you differentiate between these curves?

12-8. In what way would a graph that is made for a slide projector differ from a graph that is made to be part of an engineering report?

12-9. Name three purposes for which graphs are or may be used.

12-10. What are three different line spacings that may be purchased in commercial rectangular coordinate paper?

12-11. Why is there not a zero point on logarithmic graph paper?

12-12. What do we mean by a "cycle," as it pertains to commercial graph paper?

12-13. List three examples of supporting data that might appear on an engineering graph.

12-14. When using commercial graph paper, where would you place the title? The supporting data?

12-15. If three different weights of lines are to be used in drawing a graph, which parts of the graph will be depicted by which weights of line?

12-16. Name four shapes of plotting symbols that are commonly used in graph work. Which symbol is the most common?

12-17. In what numbers, or multiples thereof, do we usually lay out the baselines of a graph?

12-18. List four abbreviations that are often used to indicate typical units that are placed along a baseline.

12-19. If it is desired to show each point permanently by means of a circular plotting symbol, how large would you make the symbol?

12-20. What characteristic of a heater or antenna can best be shown by means of a polar graph?

12-21. In constructing a pie chart, what two sequences should be observed and followed?

12-22. What would be a reason for using graph paper with blue lines, as opposed to using red-line paper?

12-23. List two sequences of numbers placed along the axis of a rectangular coordinate graph that are considered to be poor sequences.

12-24. Why is it possible to draw a right triangle on a straight-line curve on logarithmic paper and use the linear lengths of the two legs to obtain the correct slope?

12-25. What are six factors that should be considered in the layout of a graph?

PROBLEMS

12-1. Prepare a pie chart for the 1990 forecast of the market for fiberoptic sensors. Follow the suggested arrangement of data given in Sec. 12-17. Use $8\frac{1}{2} \times 11$ paper.

TABLE A

Commercial	$120 Million
Military	92
Aerospace	68
Medical	45
Utility	35

12-2. Prepare a bar or pie chart for the increase in capacity (in megawatts) estimated by the Public Service Company (Table B).

TABLE B

	NET ADDITION	YEAR-END CAPACITY
1986	0.1.9	.'.5
1987	.1 /	: :
1988	.:	.' /
1989	.: .:	:.: .:
	.: .:	:. .:
	.: .:	:. .:
	.: .:	:. .:

12-3. Plot the hourly demand and normal-capability curves of the interconnected system of Mid-Western Power & Light Co. (Table C).

TABLE C

YEAR	MAXIMUM HOURLY DEMAND	NORMAL GENERATING CAPABILITY (THOUSANDS OF KILOWATTS)	YEAR	MAXIMUM HOURLY DEMAND	NORMAL GENERATING CAPABILITY (THOUSANDS OF KILOWATTS)
1977	256	263	1982	411	449
1978	265	263	1983	469	437
1979	318	325	1984	501	565
1980	352	364	1985	513	584
1981	381	364	1986	580	742

12-4. The 10-year history of generation data of East States Utilities Company is shown in Table D. Plot the kWh and Btu curves on one graph.

TABLE D

YEAR	kWh GENERATED (BILLIONS)	BTU REQUIRED (BILLIONS)	YEAR	kWh GENERATED (BILLIONS)	BTU REQUIRED (BILLIONS)
1981	1.32	19.2	1986	1.90	23.0
1982	1.43	20.4	1987	2.14	26.3
1983	1.54	20.4	1988	2.24	26.9
1984	1.71	22.2	1989	2.43	28.6
1985	1.80	23.2	1990	2.62	30.3

12-5. Make a bar chart of the data presented in Prob. 12-3.

12-6. Make a bar chart of the data presented in Prob. 12-4.

12-7. The following data (Table E) show the survival probability of a nonredundant system and a redundant system with $N = 10^5$ components. Plot both curves and identify each. Failure rate is $f = 10^{-7}$ per hour.

TABLE E

TIME (HOURS)	SURVIVAL PROBABILITY, P	
	NONREDUNDANT SYSTEM	REDUNDANT SYSTEM
0	1.0	1.0
200	0.2	1.0
500	0.05	1.0
1,000	0	1.0
2,000	0	0.89
4,000	0	0.85
8,000	0	0.66
12,000	0	0.44

12-8. The following data (Table F) include the capacitance per unit area and the breakdown voltages of SiO_2—silicon structures in microcircuits for different oxide thicknesses. Plot the two curves on the same graph, identify the curves, and provide a suitable title.

TABLE F

CURVE 1		CURVE 2	
CAPACITANCE (pFs/ SQ MIL)	OXIDE THICKNESS (Å)	BREAKDOWN VOLTAGE (mV)	OXIDE THICKNESS (Å)
0.41	500	0.055	500
0.22	1000	0.09	1000
0.14	1500	0.16	2000
0.105	2000	0.215	3000
0.07	3000		

12-9.   Plot the family of curves for the 2N1490 NPN transistor having a base input on a common-emitter circuit. Curves for four base currents are shown below (Table G).

TABLE G   Collector Characteristics

CURVE 1 25 BASE mA		CURVE 2 10 BASE mA	
COLLECTOR-TO-EMITTER VOLTAGES	COLLECTOR mA	COLLECTOR-TO-EMITTER VOLTAGES	COLLECTOR mA
0	0	0	0
0.5	160	0.5	160
1.0	300	1.0	250
1.5	395	1.5	315
2.0	450	2.0	375
2.5	500	2.5	420
		3.0	470

CURVE 3 3 BASE mA		CURVE 4 1 BASE mA	
COLLECTOR-TO-EMITTER VOLTAGES	COLLECTOR mA	COLLECTOR-TO-EMITTER VOLTAGES	COLLECTOR mA
0	0	0	0
0.5	110	0.5	90
1	190	1	112
2	270	2	125
3	320		Straight
4	352		to
		4.5	141

12-10. Table H provides data for four characteristic curves of a 2N2102 NPN transistor for an ambient temperature of 25°C. Plot the family of four curves for a common-emitter circuit having a base input. These are typical collector characteristics.

TABLE H

BASE CURRENT = 2 mA		BASE CURRENT = 8	
COLLECTOR-TO-EMITTER VOLTS	COLLECTOR mA	COLLECTOR-TO-EMITTER VOLTS	COLLECTOR mA
0	0	0	0
0.3	100	0.3	200
1	140	1	250
2	175	3	365
4	195	4	402
10	210	6	450
BASE CURRENT = 14		**BASE CURRENT = 22**	
COLLECTOR-TO-EMITTER VOLTS	COLLECTOR mA	COLLECTOR-TO-EMITTER VOLTS	COLLECTOR mA
0	0	0	0
0.3	200	0.3	200
0.6	300	0.6	300
1	330	1	375
2	402	2	482
4	502	2.2	500

12-11. On rectangular coordinate paper, plot the depreciation-cost curve and the maintenance-cost curve for electronic heating equipment, using the data given below (Table I). A suitable title would be "Costs of 60-Hz Electric Energy." Use different plotting symbols for each curve. Your instructor may want you to add the two curves graphically to get a total cost curve.

TABLE I

DEPRECIATION COST		MAINTENANCE COST	
COST/kWh (CENTS)	GENERATOR OUTPUT (kW)(INDEPENDENT VARIABLE)	COST/kWh (CENTS)	GENERATOR OUTPUT (kW)
6	300	4	200
4	1000	3	1000
3.6	2000	2.7	2000
3.2	8000	2.4	4000
3.0	16000	2.5	8000
		3.2	16000

12-12. Table J contains data on TM 999 board projects that have been subjected to reliability testing at 65°C.

TABLE J

BOARD OPERATING TIME (THOUSANDS OF HOURS)	NUMBER OF FAILURES (CUMULATIVE)
(-4)(Dynamic burn-in)	3
0 (Start of test)	3.6
2	4.1
4	5
6	6.3
8	7.5
10	8
12	8.25
20	9

A dynamic burn-in (to be plotted at the left of zero) significantly reduces failures. Plot the cumulative failures versus board operating time. The curve is straight between 12,000 and 20,000 h.

12-13. The annual requirements for artwork for three types of circuit manufacturing have been rising and will continue to do so. The data shown below (Table K) should be plotted on rectangular (arithmetic) graph paper to show past and estimated requirements through 1992.

TABLE K   Number of Artwork Layouts per Year (in Thousands)

YEAR	PRINTED WIRING BOARDS	THIN-FILM CIRCUITS	THICK-FILM CIRCUITS
1980	6.0	4.6	1.5
1982	9.0	6.5	3.0
1984	14.2	9.3	5.0

(Continued on next page)

TABLE K   *(Continued)*

YEAR	PRINTED WIRING BOARDS	THIN-FILM CIRCUITS	THICK-FILM CIRCUITS
1986	19.0	13.5	8.0
1988	23.0	20.6	11.6
1990	25.5	27.5	16.6
1992	29.0	33.0	22.5

12-14. On semilogarithmic paper (three cycles) make a graph for the data given below (Table L). An appropriate title might be "Input Resistance versus Load Resistance of Transistor Circuit."

TABLE L

LOAD RESISTANCE (OHMS, INDEPENDENT VARIABLE)	INPUT RESISTANCE (Ω)
10 kΩ	1020
50 kΩ	820
100 kΩ	700
300 kΩ	600
1 MΩ	560
10 MΩ	540

12-15. Construct a graph, using multicycle semilogarithmic paper, for the phase response of an *RC* amplifier. Let the phase shift, below, be the dependent variable (Table M).

TABLE M

RESPONSE CURVE WITHOUT FEEDBACK		RESPONSE CURVE WITH FEEDBACK	
FREQUENCY (UNITS AS SHOWN)	PHASE SHIFT (DEGREES)	FREQUENCY (UNITS AS SHOWN)	PHASE SHIFT (DEGREES)
		10 Hz	+130
10 Hz	+150	40 Hz	20
100 Hz	50	100 Hz	5
1 kHz	0	1 kHz	0
10 kHz	−20	10 kHz	0
100 kHz	−120	100 kHz	−20
1 MHz	−180	300 kHz	−100
		1 MHz	−150

12-16. Make a graph showing the two curves, one with and one without feedback, for the frequency response of the operational amplifier (Table N).

TABLE N

WITHOUT FEEDBACK		WITH FEEDBACK	
FREQUENCY (Hz)	GAIN (dB)	FREQUENCY (Hz)	GAIN (dB)
10	48	10	18
30	54	35	20
100	56	100	21
10^4	55	10^3	21
10^5	43	4×10^4	21
3×10^5	35	9×10^4	22
(300,000)		2×10^5	20
		3×10^5	15

12-17. On polar coordinate paper, plot the dipole radiation pattern using the data listed in Table O. There will be a major and two minor lobes. Do not number or circle points. This is a vertical plane pattern of a half-wave center-fed antenna. (Zero should be at bottom of chart.)

TABLE O

MAJOR LOBE			MINOR LOBE			MINOR LOBE		
POINT	ANGLE	AMPLI-TUDE	POINT	ANGLE	AMPLI-TUDE	POINT	ANGLE	AMPLI-TUDE
0	0°	0	0	0°	0	0	0°	0
1	155°	0.2	1	120°	0.10	1	215°	0.1
2	158°	0.4	2	124°	0.20	2	216°	0.2
3	162°	0.6	3	126°	0.24	3	218°	0.24
4	169°	0.8	4	130°	0.28	4	222°	0.28
5	180°	1.0	5	134°	0.30	5	226°	0.30
6	191°	0.8	6	138°	0.28	6	230°	0.28
7	198°	0.6	7	142°	0.24	7	234°	0.24
8	202°	0.4	8	144°	0.20	8	236°	0.20
9	205°	0.2	9	145°	0.10	9	240°	0.10
10	0°	0	10	0°	0	10	0°	0

12-18. The data listed in Table P provide a vertical plane radiation pattern for a $\frac{3}{4}$-wave antenna. The 0- to-180° line represents the axis of the antenna. Make a polar plot, using an irregular (French) curve wherever possible. (It may be necessary to round off the tips of the lobes by careful freehand drawing.) Do not number points or put plotting symbols around them. (You may put $0°$ at the left side of the graph if you wish.)

TABLE P

Point no.	0	1	2	3	4	5	6	7	8	9	10	11	12
Angle (°)	0	15	16	19	23	27	30	33	37	41	44	45	0
Amplitude	0	0.2	0.4	0.6	0.8	0.96	1.0	0.96	0.8	0.6	0.4	0.2	0

Point no.	0	1	2	3	4			0	1	2	3	4
Angle	0	60	64	70	0			0	75	81	86	0
Amplitude	0	0.18	0.3	0.18	0			0	0.18	0.3	0.18	0

Point no.	0	1	2	3	4			0	1	2	3	4
Angle	0	92	98	104	0			0	250	256	262	0
Amplitude	0	0.18	0.3	0.18	0			0	0.18	0.3	0.18	0

Point no.	0	1	2	3	4	5	6	7	8	9	10	11	12
Angle	0	135	136	139	143	147	150	153	157	161	164	165	0
Amplitude	0	0.2	0.4	0.6	0.8	0.96	1.0	0.96	0.8	0.6	0.4	0.2	0

The following graphical plots are of such a nature that each one yields a straight-line pattern if plotted on a certain type of graph paper. If the straight-line pattern can be achieved, then the student can write the equation for the line (and the data), using one of the equations listed in Sec. 12-12. Some experimentation of a trial-and-error nature will be required, in most cases, before the student will be able to draw the graph on the correct paper. Principles of graphical presentation should be followed, and the final graph should have a title and good line work and lettering, and the equation (if required by the instructor) should be shown prominently on the graph.

12-19. The following data (Table Q) provide a single-family transfer characteristic of a transistor, which is part of an integrated circuit on a silicon wafer. (The intercept quantity will be negative.)

TABLE Q

COLLECTOR CHARACTERISTIC (I_c IN MILLIAMPERES)	GATE VOLTAGE (V_G IN VOLTS)
−1	0.2
10	0.6
21	1.0
32	1.4
42	1.8

12-20. The following data (Table R) are from test records on 2409 integrated circuits. There are two curves, one for 90 and one for 95 percent upper-confidence limits. Units are hours and failures per hour.

TABLE R

90 PERCENT CURVE		95 PERCENT CURVE	
OPERATING TIME (T IN HOURS)	FAILURE RATE (F IN FAILS/H)	OPERATING TIME (T IN HOURS)	FAILURE RATE (F IN FAILS/H)
40,400	10^{-4} (1 in 10,000)	50,400	10^{-4}
10^5	4.4×10^{-4}	10^5	5.2×10^{-4}
10^6	4.2×10^{-5}	10^6	5.2×10^{-5}
4.2×10^6	10^{-6}	5.2×10^6	10^{-6}
10^7	4.0×10^{-6}	10^7	5×10^{-6}

12-21. The following data (Table S) show the optimum frequency of different skin thickness of brass and iron. Plot frequency in megahertz. (The m quantities will be negative.)

TABLE S

	SKIN THICKNESS IN CENTIMETERS	
FREQUENCY (Hz)	BRASS	IRON
100 thousand	0.007	0.06
200 thousand	0.0042	0.042
400 thousand	0.0026	0.03
600 thousand	0.0019	0.024
1 million	0.0013	0.019

12-22. The following data (Table T) show the relationship of power loss to the armature voltage of a $\frac{1}{3}$-hp electric motor.

TABLE T

LOSS (WATTS)	ARMATURE VOLTAGE (VOLTS)
0.080	1.2
0.122	1.5
0.213	2
0.340	2.5
0.480	3
0.842	4

12-23. Table U provides data that show the resistance in ground connections according to how deep the grounding rod is placed in the soil.

TABLE U

R (Ω)	D (FEET)
90	2
68	3
51	4
33	7
24	10
17	15
11.5	25

12-24. The United States semiconductor industry has maintained a constant increase in productivity (Table V). Use 1950 for the y intercept.

TABLE V

YEAR	OUTPUT PER WORKER-HOUR (DOLLARS)
1950	81
1955	93
1960	106
1965	122
1970	140
1975	162
1980	189
1985	212

12-25. Failure to align the length standard when making linear measurements with a laser causes a cosine error which plots as follows (Table W):

TABLE W

MISALIGNMENT ANGLE X (MINUTES)	COSINE ERROR e_c (PPM)
2	2.0
3	4.5
4	8.0
6	18.0
10	51.0

12-26. The manufacturing tolerances of thin-film resistors, according to the present state of the art, are (Table X):

TABLE X

RESISTANCE VALUE	MANUFACTURING TOLERANCE (PERCENT)
200 Ω	±0.29%
600 Ω	0.48
1 kΩ	0.78
1.4 kΩ	1.33

12-27. The luminous intensity of an LED varies with the air temperature (Table Y). Use $0°C$ for the y intercept.

TABLE Y

LUMINOUS INTENSITY (LUMENS)	AIR TEMPERATURE (°C)
1.9	−40
1.6	−20
1.3	0
1.05	20
0.90	40
0.75	60
0.62	80

Appendix A

Glossary of Electronics and Electrical Terms

Actuator: A device that does the actual changing, if required, of a parameter (e.g., voltage) that has been sensed by the sensor and is being adjusted by the controller.

Address: An identifier or label of a discrete location in a computer's memory. An address is composed of alphanumeric or numeric characters.

Alternating Current (AC): Current that flows in two directions.

ALU (Arithmetic Logic Unit): The heart of the microprocessor that performs the arithmetic, logical, and other related functions.

Ampere: The practical unit of current. One ampere will flow through a resistance of one ohm when a difference of potential of one volt is applied across its terminals.

Amplification: The process of increasing the strength (current, power, or voltage) of a signal.

Amplifier: A device used to increase the signal voltage, current, or power, generally composed of a transistor or vacuum tube and an associated circuit called a *stage.* It may contain several stages in order to obtain a desired gain.

Amplitude: The maximum instantaneous value of an alternating voltage or current, measured in either the positive or negative direction.

Analog-to-Digital Conversion: The process of converting varying analog signals to discrete digital signals. The acronym for Analog to digital conversion is A/D.

Anode: An electrode at which negative ions are discharged or from which the forward current (diode rectifier) flows.

Appliqués: Symbols and other graphical shapes printed on a sheet, which can be cut out and affixed to a drawing by moderate pressure.

ASCII: An acronym for American Standard Code for Information Interchange. It is the 8-bit (7, information; 1, parity) standard code for interchanging information between communication devices or circuits, where each letter, number, and control character has a discrete value.

Attenuation: The reduction in the strength of a signal.

Average: The average value of a quantity, such as average current.

Base: One of (usually) three regions of a transistor. Also one of the terminals of a transistor. In some transistors the base acts much like the grid of an electron tube.

Baud: The speed at which discrete information is transmitted (bits per second in binary or half-dots per second in Morse Code).

Bias:
Vacuum tube: the difference in potential between the control grid and the cathode.
Transistor: the difference in potential between the base and the collector.
Magnetic amplifier: the level of flux density in the magnetic amplifier core under no-signal condition.

Binary Digit (bit): A numeral in binary arithmetic. Can have a value of one or, a bit may be represented as zero. (Also true or false, and on or off.)

Bipolar: One of two fundamental processes for the fabrication of integrated circuits. It involves the production of silicon layers having different electrical characteristics.

Bistable: A circuit or circuit element having the ability to assume either of two states.

Bug: An error in a computer program.

Bus: In a computer, a group (usually a multiple of 8) of parallel conductors (cables or PC-board circuit paths), each carrying one bit of the binary data.

Busbar: A primary power-distribution point connected to the main power source.

Byte: In computers, a series of 8 bits (see "Binary Digit") of binary data.

CAD: Acronym for computer-aided design. The term used to be called *computer-aided drafting.*

Capacitor: A device consisting of two conducting surfaces separated by an insulating material or dielectric such as air, paper, or mica. A capacitor stores electric energy, blocks the flow of alternating current, and permits the flow of alternating current to a degree depending on the capacitance and frequency.

Cathode: An electrode through which a primary stream of electrons enters the inter-electrode space or to which the forward current flows (semiconductor).

Choke Coil: A coil of low ohmic resistance and high impedance to alternating current.

Circuit Breaker: An electromagnetic or thermal device which opens a circuit automatically when the current in the circuit exceeds a predetermined amount. A circuit breaker can be reset.

Circular Mil: A unit of area equal to $\pi/4$ of a square mil, or 0.7854 sq mil. This is a unit of measure for wire sizes. (See Appendix D.)

Clock Pulse: A signal provided by the clock in a microprocessor to provide synchronization of other control pulses.

Clock Rate: How fast a clock operates, that is, pulses per second or frequency (e.g., 5 MHz).

Coaxial Cable: A transmission line consisting of two conductors concentric with and insulated from each other.

Cold Cathode: A cathode without a heater such as is found in most fluorescent lamps and some vacuum tubes.

Collector: That region of a transistor that collects electrons, or the terminal that corresponds to the anode (plate) of the electron tube in the normal mode of operation.

Commutator: The copper segments on the armature of a dc motor or generator. It is cylindrical in shape and is used to pass power into or from the brushes. It is a switching device.

Conductance: The ability of a material to conduct or carry an electric current. It is the reciprocal of the resistance of the material and is expressed in siemens (formerly called *mhos*).

Conductor: Any material suitable for carrying electric current. A conductor can be insulated (e.g., 120 V ac house cord) or noninsulated (e.g., some PC-board paths or large power conductors).

CPU: An acronym for the central processing unit which, essentially, is the heart of the computer. The CPU contains the ALU, registers, input-output circuits, and control circuits.

Cryogen (Cryogenic): A device that becomes a superconductor (has practically no resistance) at extremely cold temperatures. A circuit having such devices.

Current: The rate of transfer of electricity. An amount of electricity. The basic unit is the ampere.

Decibel: A unit (abbreviated dB) that expresses the logarithmic relationship between two signal power levels. Most often used to express power gain or attenuation.

Deflecting Plate: That part of a certain type of electron tube which deflects the electron beam within the tube itself.

Detection: The process of separating the modulation component from the received signal.

Dielectric: An insulator; a term that refers to the insulating material between the plates of a capacitor.

Diffusion: A high-temperature process involving the movement of impurities into a silicon slice to change its electrical properties, used in fabrication of transistors, diodes, and ICs.

Digital-to-Analog Conversion: The process of converting discrete digital circuits to varying analog signals. The acronym for digital-to-analog conversion is D/A.

Diode:
Vacuum tube: a two-element tube that contains a cathode and plate.
Semiconductor: a material of either germanium or silicon that is manufactured to allow current to flow in only one direction. Diodes are used as rectifiers and detectors.
Direct Current (DC): Current that flows in only one direction.
Drain: One end of a channel in a field-effect transistor (FET). It compares with the collector of a bipolar transistor.
Duplex: In communication, data being transmitted and received simultaneously.
Electric: A description of any circuit or system that uses electrical generation, transmission, and distribution equipment or devices.
Electrode: A terminal used to emit, collect, or control electrons and ions; a terminal at which electric current passes from one medium into another.
Electron: A negatively charged particle of matter.
Electron Emission: The liberation of electrons from a body into space under the influence of heat, light, impact, chemical disintegration, or potential difference.
Electronic: A description of any circuit or system that uses solid-state or vacuum-tube devices.
Emitter: That part or element of a transistor that emits electrons; it corresponds to the cathode of an electron tube, in the most common form of operation.
Energy: The capacity to do work or the work being done by a system. There are many types of energy (e.g., potential, electrical, magnetic, kinetic). It is measured in ergs, joules, kilowatts, and many or equivalent terms.
Epitaxial: A thin-film type of deposition for producing certain devices in microcircuits. It involves a realignment of molecules and hence has a deeper significance than simply thin-film manufacture.
Eyelet: Used on PC boards to make reliable connections from one side of the board to the other side.
Farad: The unit of capacitance.
Feedback: A transfer of energy from the output circuit of a device back to its input.
Filament: An electrically heated wire that emits electrons or heats a cathode, which then emits electrons.
Filter: A combination of circuit elements designed to pass a definite range of frequencies, attenuating all others.
Frequency: The number of complete cycles per second existing in any form of wave motion, such as the number of cycles per second of an alternating current.
Fuse: A protective device inserted in series with a circuit. It contains a metal that will melt or break when current is increased beyond a specific value for a definite time period.
Gain: The ratio of the output power, voltage, or current to the input power, voltage, or current, respectively.
Gate: A device or element that has one output channel and one or more input channels whose state(s) determines (determine) the state of the output. The gate electrode of an FET which controls the current flow in the channel.
Grid: A wire, usually in the form of a spiral, that controls the electron flow in a vacuum tube.
Ground: A conductive connection with the earth to establish ground potential. Also, a common return to a point of zero potential. The chassis of a receiver or a transmitter is sometimes the common return and is, therefore, the "ground" of the unit.
Gun: The group of electrodes within a cathode ray tube (CRT) that emits the beam of electrons.
Henry: The basic unit of inductance. The inductance of a circuit is one henry when a current variation of one ampere per second induces one volt. (The plural is henrys.)
Hole: In semiconductors, the space in an atom left vacant by a departed electron. Holes

flow in a direction opposite to that of electrons, are considered to be current carriers, and bear a positive charge.

Impedance: The total opposition offered to the flow of an alternating current. It may consist of any combination of resistance, inductive reactance, and capacitive reactance.

Inductance: The property of a circuit or two neighboring circuits which determines how much electromotive force will be induced in one circuit by a change of current in either circuit.

Inductor: A circuit element designed so that its inductance is its most important electrical property; a coil.

Integrated Circuit (IC): A circuit in which devices of several different types such as resistors, capacitors, and transistors are made from a single piece of material such as a silicon chip and are then connected to form a circuit.

Inverters: Devices used to change direct current to alternating current.

LED: A diode that emits light when current is passed through it.

Light Pen: A photosensitive penlike device, used primarily in CAD, that can cause the computer to change or modify the display on a CRT.

Logic: The arrangement of circuitry designed to accomplish certain objectives such as the addition of two signals. Used largely in computer circuits, but also used in other equipment such as automated machine tools and electric controls.

Magnetron: A vacuum-tube oscillator containing two electrodes, in which the flow of electrons from cathode to anode is controlled by an externally applied magnetic field.

Micron: One millionth of a meter, now called a *micrometer* in the International System of Units (SI).

Modem: An electronic device that performs the functions of modulation and demodulation required by communications.

Modulation: The process of varying the amplitude (amplitude modulation), the frequency (frequency modulation), or the phase (phase modulation) of a carrier wave in accordance with other signals in order to convey intelligence. The modulating signal may be an audio-frequency signal, video signal (as in television), electric pulses or tones to operate relays, or similar.

Mylar: A trademark used for a type of polyester film or sheet that is used for magnetic tape and capacitor dielectrics and as a drawing medium.

Nibble: In computers, a series of 4 bits of binary data (half a byte).

Oscillator: A circuit that is designed to generate an audio or radio frequency; a mode of amplification. Also the main device in such a circuit.

Oscilloscope: An instrument for showing, visually, graphical representations of the waveforms encountered in electric circuits.

Overload: The condition where the electrical or electronic circuit is drawing more current than normal. Depending on the size of the overload, the circuit may be interrupted by a protective device such as a fuse or a circuit breaker.

Peak: The maximum instantaneous value of a quantity, such as peak power.

Permalloy: An alloy of nickel and iron with an abnormally high magnetic permeability.

Photoconductivity: Property of a device or material to experience a change in conductivity on exposure to light.

Pinboard: A board given to drafters that contains equipment plan views pinned in a particular arrangement. From this pinboard, a drafter can make a layout.

Pixel: The smallest dot being displayed on a CRT screen. A greater number of pixels per unit area allows more resolution.

Plate: The principal anode (electrode) in an electron tube to which the electron stream is attracted. Also, one of the conductive electrodes in a capacitor or battery.

Potential: The degree of electrification as referred to some standard such as the earth. The amount of work required to bring a unit quantity of electricity from infinity to the point in question.

Potentiometer: A variable-voltage divider; a resistor which has a variable contact arm so that any portion of the potential applied between its ends may be selected.

Power: The rate of doing work or the rate of expending energy. The unit of electric power is the watt.

Printed Circuit: A circuit in which the wires have been replaced by conductive strips on an insulating board, abbreviated with uppercase letters, PC.

Raceway: Any channel for enclosing conductors which is designed expressly and used solely for this purpose.

RAM: Acronym for random-access memory, which is memory that can be written to (data inputted) or read from (data outputted).

Relay: An electromechanical or electronic switching device that can be used as a remote control, usually to open or close a circuit.

Resistance: The opposition that a device or material offers to the flow of current. It determines the rate at which electric energy is converted into heat or radiant energy (the basic unit is ohms).

Resistor: A circuit element whose chief characteristic is resistance; used to oppose the flow of current.

Resonance: The condition existing in a circuit in which the inductive and capacitive reactances cancel each other.

ROM: Read-only memory which permits only reading from memory as opposed to RAM, which is both read and write memory.

Saturation: The condition existing in any circuit when an increase in the driving signal produces no further change in the resultant effect.

Semiconductor: An element, such as germanium or silicon, from which transistors or diodes are made; the device itself. The resistivity of the element is in the range between those of metals and insulators.

Simplex: In communications, data being either transmitted or received but not simultaneously.

Sensor: A device that measures a change in a parameter (e.g., voltage) and sends this information to some type of recorder, controller (e.g., computer, programmable controller), or both.

Solenoid: An electromagnetic coil that contains a movable plunger.

Synchronous: Happening at the same time; having the same period and phase.

Tachometer: An instrument for indicating revolutions per minute.

Terminal: A combination of an output CRT (display) and an input keyboard.

Thick-Film Circuit: A film-type circuit. The range of thickness of deposited patterns is from 0.01 to 0.05 mm.

Thin-Film Circuit: A circuit made by depositing material on a substrate, such as glass or quartz, to form patterns that make devices, such as resistors and capacitors, and their connections. The thickness of the film forming these devices is only a few micrometers (0.001 mm).

Thyristor: A bistable semiconductor device having three or more junctions that can be switched from OFF to ON or vice versa. Silicon-controlled rectifiers and triacs are in this class.

Transducer: A device that converts an input into a different type of output. Examples are microphones, speakers, lamps, vibrators, strain gauges, pneumatic to electric (p/e), and so on.

Transformer: A device composed of two or more coils linked by magnetic lines of force; used to transfer energy from one circuit to another (i.e., mutual coupling between circuits).

Triode: A three-electrode vacuum tube containing a cathode, control grid, and plate. Also a three-region semiconductor.

Volt: The unit of voltage, potential, or electromotive force (emf). One volt will send a current of one ampere through a resistance of one ohm.

Voltage: Used interchangeably with *potential.* (See "Potential.")

Watt: The unit of electric power. In a direct current one watt is equal to volts multiplied by amperes. In an alternating current the true power in watts is equal to effective volts multiplied by effective amperes.

Word: In computers, a series of one to several bytes. Word length is expressed in bits and bytes, such as a 2-byte word = a 4-nibble word = a 16-bit word. Word length is not standard.

Electrical Part (Device) Reference Designations

FROM ANSI Y32.2-1975

Alarm, audible	LS	Connector, plug, affixed to	
Amplifier	AR	end of cable, wire	P
Amplifier, rotating	G	Contact, electrical	E*
Annunciator	DS	Contactor, electrically	
Antenna	E*	operated	K
Arrester, lightning	E*	Contactor, mechanically or	
Assembly	A	thermally operated	S
Attenuator, fixed	AT*	Coupler, directional	DC
Audible signaling device	LS	Crystal detector	CR
Autotransformer	T	Crystal diode	CR
Battery	BT	Crystal, piezoelectric	Y
Bell, electric	LS*	Cutout, fuse	F
Blower	B	Cutout, thermal	S
Board, terminal	TB	Detector, crystal	CR
Breaker, circuit	CB*	Device, indicating	DS
Buzzer	LS*	Dipole antenna	E*
Cable	W	Disconnecting device	S
Capacitor	C*	Earphone	HT*
Cell, aluminum or electrolytic	E	Electron tube	V
Cell, light-sensitive,		Exciter	G
photoemissive	V	Fan	B
Choke	L	Filter	FL*
Circuit breaker	CB*	Fuse	F*
Clock	M*	Fuse holder, lamp holder, or	
Coil, hybrid	HY	socket	X*
Coil, induction, relay,		Generator	G*
tuning, operating	L	Handset	HS*
Coil, repeating	T	Head, erasing, recording,	
Computer	A	reproducing	PU*
Connector, receptacle,		Heater	HR*
affixed to wall, chassis,		Horn, howler	LS*
panel	J	Indicator, visual	DS*

* Those parts marked with an asterisk (*) are also approved in the Federal Item Identification Guide Cataloging Handbook H6-1.

Inductor	L	Radio transmitter	TR*
Instrument	M	Receiver, telephone	HT
Insulator	E*	Receptacle (fixed connector)	J
Interlock, mechanical	MP	Regulator, voltage (except	
Interlock, safety, electrical	S	electron tube)	VR
Jack (see connector,		Relay, electrically operated	
receptacle, electrical)	J*	contactor or switch	K*
Junction, coaxial or		Repeater	AR
waveguide (tee or wye)	CP	Resistor	R*
Junction, hybrid	HY	Rheostat	R*
Key, switch	S	Selenium cell	CR
Lamp, pilot or illuminating	DS	Shunt, relay	R
Lamp, signal	DS	Solenoid	L*
Line, delay	DL*	Speaker	LS
Loop antenna	E*	Speed regulator	S
Magnet, permanent	E*	Subassembly	A
Meter	M	Switch	S
Microphone	MK*	Terminal board	TB*
Mode transducer	MT	Terminal strip	TB*
Modulator	A	Test point	TP
Motor	B*	Thermistor	RT
Motor-generator	MG	Thermocouple	TC*
Oscillator (excluding electric		Thermostat	S
tube used in oscillator)	Y	Timer	M
Oscilloscope	M*	Transducer	MT
Pad	AT	Transformer	T*
Pair	W	Transistor	Q*
Path, guided, transmission	W	Transmission path	W
Phototube	V	Tube, electron	V*
Pickup, erasing, recording,		Varistor, symmetrical	RV
or reproducing head	PU	Voltage regulator (except an	
Plug	P	electron tube)	VR
Potentiometer	R	Waveguide	W
Power supply	PS*	Winding	L*
Radio receiver	RE*	Wire	W*

The preceding list is not the complete list of devices shown in ANSI Y32.2-1975. This listing contains more commonly used parts. This listing or ANSI Y32.2-1975 does not contain device function designations for power switchgear, industrial control, and industrial equipment. For these function designations, please consult:

1. American National Standard Manual and Station Control, Supervisory and Associated Telemetering Equipment, C37.2-1970.

2. NEMA Standard, Industrial Controls ICS-1970 (R1975)

3. Joint Industrial Council Electrical Standards for Mass Production Equipment, EMP-1-1967 and General Purpose Machine Tools, EGP-1-1976

4. Military Standard, Designations for Electric Power Switch Devices and Industrial Control Devices, MIL STD 27 (see Table 9-1).

Control-Device Designations[1]

Brake relay	BR	Magnetic clutch	MC
Control relay	CR	Manual	MN
Control relay manual	CRH	Overload relay	OL
Control relay master	CRM	Pushbutton	PB
Down	D	Reverse	R
Disconnect switch	DISC	Rheostat	RH
Electron tube	ET	Switch	S
Flow switch	FLS	Solenoid	SOL
Float switch	FS	Selector switch	SS
Instantaneous overload	IOL	Transformer	T
Limit switch	LS	Time delay relay	TR
Motor starter	M	Reactor	X
Magnetic brake	MB		

Abbreviations for Drawings and Technical Publications

Primarily from MIL STD 12D and other sources where Mil Std 12D does not list an abbreviation.

Adapter	ADPTR	Anode (electron devices)	A
Air circuit breaker	ACB		
Alternating current	AC	Arrester	ARSR
Alternating current volts	VAC	Attenuation, attenuator	ATTEN
Aluminum	AL	Audio frequency	AF
American Society of Mechanical Engineers	ASME	Auto frequency control	AFC
		Automatic gain control	AGC
Ammeter	AMM		
Ampere	A	Base (electron device)	B
Amplifier	AMPL	Battery	BTRY
Antenna	ANT	Beat-frequency oscillator	BFO
Armature	ARM		
Anode	AD	Bottom	BOT

[1] The Joint Industrial Council (JIC), "Electrical Standards for Industrial Equipment." See Table 9-1 for expanded listing.

Cabinet	CAB	Federal Power	
Capacitor	CAP	Commission	FPC
Cathode	CATH	Field	FLD
Cathode (electron		Field reversing	FFR
device)	K	Flat head	FLH
Cathode-ray tube	CRT	Fluorescent	FLUOR
Circuit	CKT	Fuse	FU
Coaxial	COAX	Gage	GA
Collector (electron		Gate (electron device)	G
device)	C	Grid	G
Compress	CPRS	Grommet	GROM
Condenser	COND	Guided missile	GM
Conductor	CNDCT	Heater	HTR
Conduit	CND	Heat-treat	HT TR
Counterclockwise	CCW	Heptode	HEPT
Current relay	CR	High frequency	HF
Current transformer	CT	High-frequency	
Cycles per second	Hz	oscillator	HFO
Decibel	DB	High voltage	HV
Diameter	DIA	Horizon, horizontal	HORIZ
Diode	DIO	Ignition	IGN
Direct current	DC	Indicator	IND
Double-pole, double-		Induction	IND
throw	DPDT	Induction-	
Double-pole, single-		capacitance	LC
throw	DPST	Institute of Electrical	
Drain (electron		and Electronic	
device)	D	Engineers	IEEE
Drawing	DWG	Instrument	INSTR
Dynamometer	DYNMT	Integrated circuit	IC
Dynamotor	DYNM	Intermediate	
Electric	ELEC	frequency	IF
Electrical	ELEC	Junction box	JB
Electric horsepower	EHP	Kilocycles per second	kHz
Electrolytic	ELCTLT	Kilovolt	kV
Electronic	ELEK	Kilovolt-ampere	kVA
Electronic Industries		Kilowatt	kW
Association	EIA	Kilowatt-hour	kWh
Emitter (electron		Knockout	KO
device)	E	Lighting	LTG
Engineer	ENGR	Lightning arrester	LA
Engineering	ENGRG	Low frequency	LF
Escutcheon	ESC	Low voltage	LV
Exciter	EXCTR	Magnetic amplifier	MAGAMPL
Federal Communica-		Magnetic modulator	MAGMOD
tions Commission	FCC	Manual	MNL

Master switch	MSW	capacitance	RC
Medium frequency	MF	Resistance-	
Mega (10⁶)	M	capacitance-	
Megacycles per		coupled	RC CPLD
second	MHz	Resistor	RES
Meter	MTR	Rotor	RTR
Metering	MTRG	Roundhead	RDH
Missile	MSL	Saturable reactor	SR
Modify	MOD	Schedule	SCHED
Modulator	MOD	Screw	SCR
Modulator		Secondary	SEC
demodulator	MODEM	Selector	SEL
Motor	MOT	Selenium	Se
Mounting	MTG	Semiconductor	
Multiplex	MUX	controlled rectifier	SCR
Multivibrator	MV	Semiconductor	
National Aeronautics		controlled switch	SCS
and Space		Series relay	SRLY
Administration	NASA	Servomechanism	SVO
National Electrical		Shield	SHLD
Code	NEC	Signal	SIG
Not to scale	NTS	Single-pole, double-	
Oil circuit breaker	OCB	throw	SPDT
Oscillator	OSC	Single-pole, single-	
Oscilloscope	SCOPE	throw	SPST
Overload	OVLD	Slow operate (relay)	SO
Pentode	PENT	Slow release (relay)	SR
Phase	PH	Solenoid	SOL
Piezoelectric-crystal		Source (electron	
unit	CU	device)	S
Polarity	PLRT	Speaker	SPKR
Potential transformer	PT	Specification	SPEC
Potentiometer	POT	Standoff	STOF
Power supply	PWR SPLY	Stator	STTR
Primary	PRI	Suppress	SUPPR
Quick-opening device	QOD	Surge	SRG
Radar	RDR	Switch	SW
Radio	RAD	Switchboard	SWBD
Radio frequency	RF	Switchgear	SWGR
Reactive volt-amp	VAR	Synchronous	SYN
Receiver	RCVR	Tachometer	TACH
Receptacle	RCPT	Technical circular	TC
Reference	REF	Technical manual	TM
Reference line	REFL	Telemeter	TLM
Resistance	RES	Terminal	TERM
Resistance		Test switch	TSW

Thermocouple	TC	Very high frequency	VHF
Three-conductor	3/C	Very low frequency	VLF
Three-phase	3PH	Video	VID
Three-pole	3P	Video frequency	VIDF
Time delay	TD	Video tape recorder	VTR
Transceiver	XCVR	Volt	V
Transformer	XFMR	Volt alternating	
Transistor	XSTR	current	VAC
Transmitter	XMTR	Volt direct current	VDC
Triode	TRI	Voltage regulator	VR
Tuning	TUN	Voltage relay	VRLY
Twisted	TW	Voltmeter	VM
Ultrahigh frequency	UHF	Volume	VOL
Unfused	UNF	Watt	W
Vacuum tube	VT	Watt-hour	Wh
Vacuum tube		Watt-hour demand	
voltmeter	VTVM	meter	WHDM
Var-hour meter	VARHM	Watt-hour meter	WHM
Variable-frequency		Wattmeter	WM
oscillator	VFO	Wire-wound	WW

Note: These symbols are authorized for use in drawings for the Military and in specifications; however, there are abbreviations listed here that are not formed in Mil Std 12D. Therefore, the authors use their preferred abbreviation. A similar list has been compiled by the American National Standards Institute. The preceding list is not the complete Military Standard; actually, there are approximately 10,000 abbreviations in the standard. If in doubt about abbreviation, spell out the word.

The Frequency Spectrum

			WAVELENGTH	
FREQUENCY	DESIGNATION	ABBREVIATION	METERS	CENTIMETERS
Below 3 kHz	Extremely low frequency	elf	30,000	
3–30 kHz	Very low frequency	vlf	30,000–10,000	
30–300 kHz	Low frequency	lf	10,000–1,000	
300–3000 kHz	Medium frequency	mf	1,000–100	
3–30 MHz	High frequency	hf	100–10	
30–300 MHz	Very high frequency	vhf	10–1	
300–3000 MHz	Ultrahigh frequency	uhf	1–0.1	
3000 MHz–30	Superhigh frequency	shf		10–1
30–300 GHz	Extremely high frequency	ehf		1–0.1
300–3000 GHz	As yet unnamed			0.1–0.01

Note: The current IEEE and international standards utilize the term *hertz* for cycles per second. Thus KC (old form) becomes kHz, MC (old form) becomes MHz, etc.; kHz represents kilocycles; MHz, megacycles; and GHz, gigacycles (1 billion cps).

Appendix B

Symbols for Electrical and Electronics Devices

This list is abridged from ANS Y32.2, "Graphical Symbols for Electrical and Electronics Diagrams," which is now ANSI/IEEE Y32E, "Electrical and Electronics Graphics Symbols and Reference Designations."

Certain specialized build-up applications of basic symbols are omitted. Where the American National Standard shows both single-line and complete symbol equivalents, this chart shows only the single-line symbol. Consult the basic standard for complete symbols in these cases.

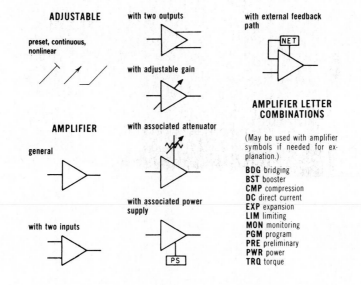

ADJUSTABLE

preset, continuous, nonlinear

AMPLIFIER

general

with two inputs

with two outputs

with adjustable gain

with associated attenuator

with associated power supply

with external feedback path

AMPLIFIER LETTER COMBINATIONS

(May be used with amplifier symbols if needed for explanation.)

BDG bridging
BST booster
CMP compression
DC direct current
EXP expansion
LIM limiting
MON monitoring
PGM program
PRE preliminary
PWR power
TRQ torque

ANTENNA

general

 OR

dipole

loop

OR

loop antenna (alternate symbol)

counterpoise, antenna

ARRESTER, LIGHTNING

general

carbon block

electrolytic or aluminum cell

horn gap

protective gap

sphere gap

valve or film element

multigap

ATTENUATOR, FIXED

See also **PAD** (same symbols as variable attenuator without adjustment arrow.)

ATTENUATOR VARIABLE

general

balanced

unbalanced

AUDIBLE SIGNALING DEVICE

bell

buzzer

loudspeaker

LOUDSPEAKER LETTER COMBINATIONS

* **HN** horn, electrical
* **HW** howler
* **LS** loudspeaker
* **SN** siren
† **EM** electromagnetic with moving coil
† **EMN** electromagnetic, moving coil and neutralized winding
† **MG** magnetic armature
† **PM** permanent magnet

(Asterisk (*) and dagger (†) are not part of symbol.)

sounder, telegraph

BATTERY

one cell

multicell

multicell with taps

multicell with adjustable tap

CAPACITOR

general

 OR —||— IEC

polarized

 OR —||— IEC

adjustable or variable

adjustable or variable with mechanical linkage

continuously adjustable or variable differential

phase shifter

split stator

feedthrough

CELL, PHOTOSENSITIVE

asymmetrical photoconductive transducer

symmetrical photoconductive transducer

photovoltaic transducer

CIRCUIT BREAKER

general

CIRCUIT ELEMENT

general

LETTER COMBINATIONS FOR CIRCUIT ELEMENTS

(* Asterisk not part of symbol.)
CB circuit breaker
DIAL telephone dial
EQ equalizer
FAX facsimile set
FL filter
FL-BE filter, band elimination
FL-BP filter, band pass
FL-HP filter, high pass
FL-LP filter, low-pass
NET network
PS power supply
RU reproducing unit
RG recording unit
TEL telephone station
TPR teleprinter
TTY teletypewriter

ADDITIONAL LETTER COMBINATIONS

(Specific graphical symbols preferred.)

AR amplifier
AT attenuator
C capacitor
HS handset
I indicating lamp
L inductor
LS loudspeaker
J jack
MIC microphone
OSC oscillator
PAD pad
P plug
HT receiver, headset
K relay
R resistor
S switch
T transformer
WR wall receptacle

GROUND

earth ground

chassis connection

common connections

OR

(Identifying marks to denote points tied together shall replace (*) asterisks.)

CLUTCH; BRAKE

clutch disengaged when operating means deenergized

OR

clutch engaged when operating means deenergized

OR

brake applied when operating means energized

OR

brake released when operating means energized

OR

COIL, OPERATING (RELAY)

(Replace asterisk (*) with device designation.)

 OR OR

dot shows inner end of winding

OR

CONNECTION, MECHANICAL (INTERLOCK)

with fulcrum

CONNECTOR

female contact

male contact

separable connectors (engaged)

OR

separable connectors (alternate symbol)

coaxial connector with outside conductor carried through

two-conductor switchboard jack

two-conductor switchboard plug

female contact
(convenience outlets and mating connectors)

male contact (convenience outlets and mating connectors)

two-conductor nonpolarized connector with female contacts

two-conductor polarized connector with male contacts

WAVEGUIDE FLANGES

mated (symmetrical)

mated (asymmetrical)

mated (rectangular waveguide)

CONTACT, ELECTRICAL

fixed contact for jack, key or relay

 OR OR

fixed contact for switch

o OR

fixed contact for momentary switch

sleeve

OR OR

moving contact, adjustable

OR

moving contact, locking

moving contact, nonlocking

segment, bridging contact

OR

vibrator reed

vibrator split reed

rotating contact

closed contact, break

OR OR

open contact, make

 OR OR

transfer

OR OR

make-before-break

open contact with time-closing or time-delay-closing

TC OR TDC

closed contact with time-opening or time-delay-opening

TO OR TDO

time-sequential-closing

OR

CORE

air core

NO SYMBOL

magnetic core of inductor or transformer

core of magnet

COUNTER, ELECTROMECHANICAL

COUPLER, DIRECTIONAL

general

E-plane aperture coupling, 30-db loss

(E) 30DB

loop coupling, 30-db loss

30DB

probe coupling, 30-db loss

30DB

resistance coupling, 30-db loss

30DB

COUPLING

(by aperture of less than waveguide size)

(Replace asterisk (*) by E, H or HE depending upon type of coupling to guided transmission path.)

DELAY FUNCTION

general

tapped delay

(Replace asterisk (*) with value of delay.)

DIRECTION OF FLOW

one way

OR

both ways

OR

DISCONTINUITY

equivalent series element

capacitive reactance

inductive reactance

inductance-capacitance circuit, infinite reactance at resonance

inductance-capacitance circuit, zero reactance at resonance

resistance

equivalent shunt element

capacitive susceptance

conductance

inductive susceptance

inductance-capacitance circuit with infinite susceptance at resonance

inductance-capacitance circuit with zero susceptance at resonance

ELECTRON TUBE

directly heated cathode, heater

indirectly heated cathode

cold cathode
(including ionically heated cathode)

photocathode

pool cathode

ionically heated cathode with supplementary heating

grid

deflecting electrode

ignitor

exciter

anode or plate

target or x-ray anode

dynode

composite anode-photocathode

composite anode-cold cathode

composite anode-ionically heated cathode with supplementary heating

shield, within envelope and connected to a terminal

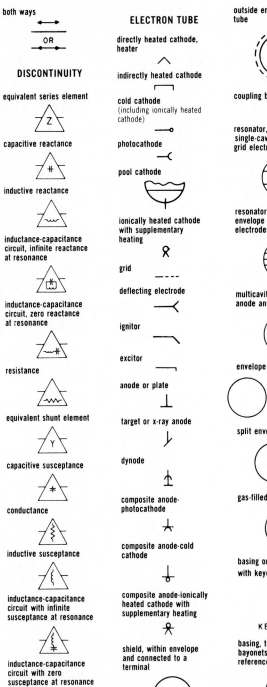

outside envelope of x-ray tube

coupling by loop

resonator, cavity type—single-cavity envelope with grid electrodes

resonator—double cavity envelope with grid electrodes

multicavity magnetron anode and envelope

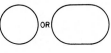

envelope

OR

split envelope

gas-filled envelope

basing orientation, tubes with keyed bases

KEY →

basing, tubes with bayonets, bosses or other reference points

base terminals

SMALL PIN

LARGE PIN

envelope terminals

RIGID TERMINAL

FLEXIBLE LEAD

triode with directly heated cathode and envelope connection to base terminal

pentode

twin triode equipotential cathode

cold-cathode voltage regulator

vacuum phototube

multiplier phototube

cathode-ray tube, electrostatic deflection

cathode-ray tube, magnetic deflection

mercury-pool tube with ignitor and control grid

mercury-pool tube with excitor, control grid and holding anode

single-anode pool-type vapor rectifier with ignitor

six-anode metal-tank pool-type rectifier with excitor

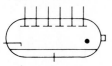

resonant magnetron with coaxial output

resonant magnetron with permanent magnet

transit-time magnetron

tunable magnetron

reflex klystron, integral cavity

double-cavity klystron, integral cavity

transmit-receive (t-r) tube

x-ray tube with directly heated cathode and focusing grid

x-ray tube with control grid

x-ray tube with grounded shield

double-focus x-ray tube with rotating anode

x-ray tube with multiple accelerating electrode

FUSE

general

OR

OR

isolating fuse switch

IEC

OR

high-voltage fuse, oil

OR

GOVERNOR

HALL GENERATOR

HANDSET

HYBRID

general

HYB

hybrid junction

circular hybrid

(Replace asterisk (*) with E, H or HE to denote transverse field.)

INDUCTOR

general

OR

magnetic-core inductor

tapped inductor

adjustable inductor

continuously adjustable inductor

shunt inductor

KEY, TELEGRAPH

LAMP

ballast tube

fluorescent lamp, two-terminal

fluorescent lamp, four terminal

cold-cathode glow lamp, a-c type

cold-cathode glow lamp, d-c type

incandescent lamp

MACHINE, ROTATING

generator

GEN

motor

MOT

1-phase

3-phase wye grounded

3-phase wye ungrounded

3-phase delta

MAGNET, PERMANENT

METER

METER LETTER COMBINATIONS

(Replace asterisk (*) with proper letter combination.)

A ammeter
AH ampere-hour
CMA contact-making or breaking ammeter
CMC contact-making or breaking clock
CMV contact-making or breaking voltmeter
CRO cathode-ray oscilloscope
DB decibel meter
DBM decibels referred to one milliwatt
DM demand meter
DTR demand-totalizing relay
F frequency meter
G galvanometer
GD ground detector
I indicating
μA or **UA** microammeter
MA milliammeter
NM noise meter
OHM ohmmeter
OP oil pressure
OSCG oscillograph, string
PH phasemeter

PI position indicator
PF power factor
RD recording demand meter
REC recording
RF reactive factor
SY synchroscope
T temperature
THC thermal converter
TLM telemeter
TT total time
V voltmeter
VA volt-ammeter
VAR varmeter
VARH varhour meter
VI volume indicating
VU standard volume indicating
W wattmeter
WH watthour meter

MICROPHONE

MODE SUPPRESSION

MODE TRANSDUCER

MOTION, MECHANICAL

translation, one direction

translation, both directions

rotation, one direction

rotation, both directions

NETWORK

NET

OSCILLATOR

OPTICAL COUPLER

PATH, TRANSMISSION

general

wire

two conductors

air or space path

dielectric path other than air

DIEL

crossing of conductors not connected

junction

junction of connected paths, conductors or wires

OR

OR ONLY IF REQUIRED BY SPACE LIMITATION

shielded single-conductor cable

coaxial cable

two-conductor cable

shielded two-conductor cable with shield grounded

twisted conductors

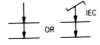

OR

IEC

grouping of leads

OR

alternate or conditional wiring

associated or future wiring

associated or future equipment
(amplifier shown)

circular waveguide

rectangular waveguide

PHASE SHIFTER

general

adjustable

PICKUP HEAD

general

recording

playback

erasing

writing, reading and erasing

stereo

PIEZOELECTRIC CRYSTAL

POLARITY

positive

$+$

negative

$-$

RECEIVER, TELEPHONE

general

headset

RECTIFIER

(Represents any method of rectification such as electron tube, solid-state device, electrochemical device, etc.)

general

controlled

bridge type

RELAY

alternating current or ringing

magnetically polarized

slow-operate

slow-release

RELAY LETTER COMBINATIONS

(Not required with specific symbol.)

AC alternating current

D differential
DB double biased
DP dashpot
EP electrically polarized
FO fast operate
FR fast release
MG marginal
NB no bias
NR nonreactive
P magnetically polarized
SA slow operate and slow release
SO slow operate
SR slow release
SW sandwich wound

RESISTOR

general

–W— OR

tapped resistor

tapped resistor with adjustable contact

adjustable or continuously adjustable

instrument or relay shunt

nonlinear resistor

symmetrical varistor

(Replace asterisks (*) with identification of symbol.)

RESONATOR, TUNED CAVITY

ROTARY JOINT

general (Replace asterisk (*) with transmission-path recognition symbol.)

coaxial in rectangular waveguide

circular in rectangular waveguide

SEMICONDUCTOR DEVICES

semiconductor region with one ohmic connection

semiconductor region with plurality of ohmic connections

OR   OR

rectifying junction, P on N region

OR

rectifying junction, N on P region

OR

emitter, P on N region

plurality of P emitters on N region

emitter, N on P region

plurality of N emitters on P region

collector

plurality of collectors

transition between regions of dissimilar conductivity

intrinsic region between regions of dissimilar conductivity

intrinsic region between regions of similar conductivity

intrinsic region between collector and region of dissimilar conductivity

intrinsic region between collector and region of similar conductivity

---

photosensitive

temperature dependent

$t°$

capacitive device

tunneling device

unidirectional

OR

PNP transistor (actual device and construction of symbol)

PNINIP device (ac device and construction of symbol)

semiconductor diode (also: rectifier)

OR

OR

capacitive diode (also: Varicap, varactor, reactance diode, parametric diode)

OR

breakdown diode, unidirectional (also: backward diode, avalanche diode, voltage regulator diode, zener diode, voltage reference diode)

OR

breakdown diode, bidirectional and backward diode (also: bipolar voltage limiter)

OR

tunnel diode (also esaki diode)

 OR

temperature dependent diode

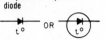 OR

photodiode (also: solar cell)

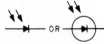 OR

silicon controlled rectifier

PNP transistor (also: junction, point-contact, mesa, epitaxial, planar surface-barrier)

PNP transistor with one electrode connected to envelope

NPN transistor (see other names under PNP transistor)

unijunction transistor, N-type base (also: double-base diode, filamentary transistor)

unijunction transistor, P-type base (see other names above)

field-effect transistor N-type base

 OR

field-effect transistor, P-type base

 OR

semiconductor triode, PNPN switch (also: controlled rectifier)

semiconductor triode, NPNP switch (also: controlled rectifier)

NPN transistor with transverse-biased base

 OR

depletion-type MOSFET, N-channel

depletion-type MOSFET, P-channel

enhancement-type MOSFET, N-channel

enhancement-type MOSFET, P-channel

SHIELD

SQUIB

explosive

igniter

sensing link

SWITCH

single-throw

double-throw

double-pole, double-throw with terminals shown

with horn gap

knife switch

push button, circuit closing (make)

push button, circuit opening (break)

nonlocking; momentary or spring return—circuit closing (make)

 OR

nonlocking; momentary or spring return—circuit opening (break)

OR

nonlocking; momentary or spring return—transfer

OR

locking—circuit closing (make)

OR

locking—circuit opening (break)

OR

locking—transfer, three-position

OFF

selector switch

OR

selector, shorting during contact transfer

wafer (example shown: 3-pole, 3-circuit with 2 nonshorting and 1 shorting moving contacts)

safety interlock—circuit opening

safety interlock—circuit closing

SWITCHING FUNCTION

conducting, closed contact (break)

nonconducting, open contact (make)

transfer

 OR

SYNCHRO

general

SYNCHRO LETTER COMBINATIONS

CDX control-differential transmitter
CT control transformer
CX control transmitter
TDR torque-differential receiver
TDX torque-differential transmitter
TR torque receiver
TX torque transmitter
RS resolver
B outer winding rotable in bearings

TERMINATION

cable

open circuit

short circuit

movable short

terminating series
capacitor, path open

terminating series
capacitor, path shorted

terminating series
inductor, path open

terminating series
inductor, path shorted

terminating resistor

series resistor, path open

series resistor, path
shorted

THERMAL ELEMENT

actuating device

OR

thermal cutout

OR

**thermal relay with
normally open contact**

**thermostat (closing on
rising temp.)**

**thermostat with contact
motion clarified**

thermostat with integral
heater and transfer
contacts

THERMISTOR

general OR

with integral heater

THERMOCOUPLE

general

with integral heater
internally connected

heater

with integral insulated
heater

heater

semiconductor
thermocouple, temperature
measuring

semiconductor
thermocouple, current
measuring

TRANSFORMER

general OR

OR

one winding with
adjustable inductance

each winding with
adjustable inductance

adjustable mutual inductor

adjustable transformer

current transformer with
polarity marking

 OR

bushing type current
transformer

OR

potential transformer

OR

TRANSFORMER
CONNECTION
WINDING

3-phase 3-wire Delta or
mesh

3-phase 3-wire Delta
grounded

3-phase open Delta
grounded at common
point

3-phase wye or star
ungrounded

TERMINAL BOARD
OR STRIP

VIBRATOR

shunt drive

separate drive

VISUAL SIGNALING
DEVICE

annunciator, general

annunciator drop or
signal, shutter type

annunciator drop or
signal, ball type

manually restored drop

electrically restored drop

switchboard-type lamp

indicating lamp

⬭ OR ⬭ OR ✳

jeweled signal light

INDICATING LIGHT
LETTER
COMBINATIONS

(Replace asterisk (*) with
proper letter combination.)

A amber
B blue
C clear
G green
NE neon
O orange
OP opalescent
P purple
R red
W white
Y yellow

Symbols for Electrical and Electronics Devices for JIC-Oriented Drawings*

SWITCHES				
DISCONNECT	CIRCUIT INTERRUPTER	CIRCUIT BREAKER	LIMIT	

Table contents (symbols): DISCONNECT (DISC), CIRCUIT INTERRUPTER (CI), CIRCUIT BREAKER (CB), LIMIT — NORMALLY OPEN (LS), NORMALLY CLOSED (LS), NEUTRAL POSITION (LS / NP), ACTUATED (LS), HELD CLOSED (LS), HELD OPEN (LS), NP

LIMIT (CONTINUED): MAINTAINED POSITION (LS), PROXIMITY SWITCH CLOSED (PRS), PROXIMITY SWITCH OPEN (PRS) | LIQUID LEVEL — NORMALLY OPEN (FS), NORMALLY CLOSED (FS) | VACUUM & PRESSURE — NORMALLY OPEN (PS), NORMALLY CLOSED (PS) | TEMPERATURE — NORMALLY OPEN (TAS), NORMALLY CLOSED (TAS)

FLOW (AIR, WATER ETC.) — NORMALLY OPEN (FLS), NORMALLY CLOSED (FLS) | FOOT — NORMALLY OPEN (FTS), NORMALLY CLOSED (FTS) | TOGGLE (TGS) | CABLE OPERATED (EMERG.) SWITCH (COS) | PLUGGING (PLS, F / R) | NON-PLUG (PLS, F / R)

PLUGGING W/LOCK-OUT COIL (PLS, F / LO) | SELECTOR — 2-POSITION (SS), 3-POSITION (SS) | ROTARY SELECTOR — †NON-BRIDGING CONTACTS (RSS), †BRIDGING CONTACTS (RSS), OR | †TOTAL CONTACTS TO SUIT NEEDS

THERMOCOUPLE SWITCH (TCS) OFF / 1 / 2 | PUSHBUTTONS — SINGLE CIRCUIT (NORMALLY OPEN PB, NORMALLY CLOSED PB), DOUBLE CIRCUIT (PB, MUSHROOM HEAD PB), MAINTAINED CONTACT (PB, PB) | CONNECTIONS, ETC. — CONDUCTORS (NOT CONNECTED, CONNECTED)

* *Source:* Joint Industrial Council (JIC), ''Electrical Standards for Mass Production Equipment No. EMP-1–67 and General Purpose Machine Tools EPG-1–67.''

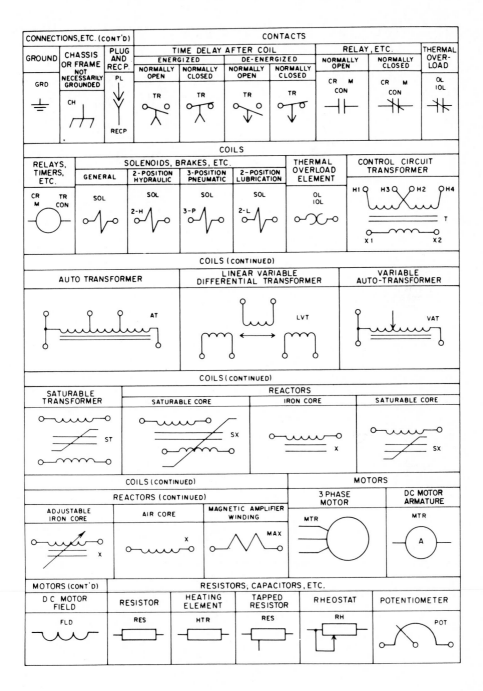

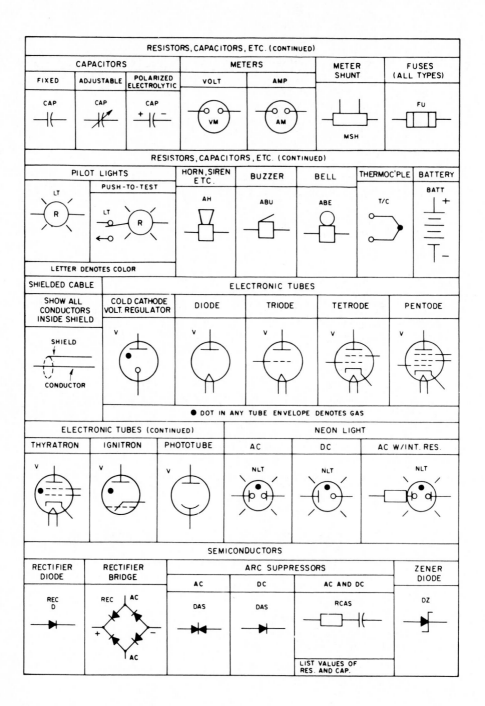

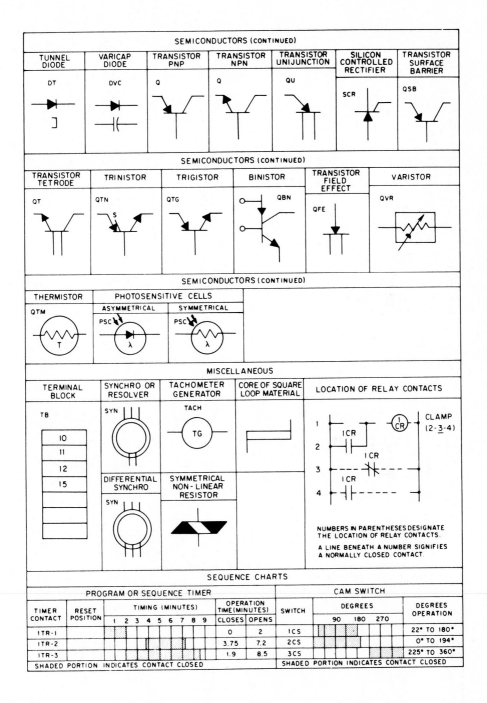

		SEMICONDUCTORS (CONTINUED)				
TUNNEL DIODE	VARICAP DIODE	TRANSISTOR PNP	TRANSISTOR NPN	TRANSISTOR UNIJUNCTION	SILICON CONTROLLED RECTIFIER	TRANSISTOR SURFACE BARRIER
DT	DVC	Q	Q	QU	SCR	QSB

	SEMICONDUCTORS (CONTINUED)				
TRANSISTOR TETRODE	TRINISTOR	TRIGISTOR	BINISTOR	TRANSISTOR FIELD EFFECT	VARISTOR
QT	QTN	QTG	QBN	QFE	QVR

	SEMICONDUCTORS (CONTINUED)	
THERMISTOR	PHOTOSENSITIVE CELLS	
	ASYMMETRICAL	SYMMETRICAL
QTM	PSC	PSC

MISCELLANEOUS

TERMINAL BLOCK	SYNCHRO OR RESOLVER	TACHOMETER GENERATOR	CORE OF SQUARE LOOP MATERIAL	LOCATION OF RELAY CONTACTS
TB 10 11 12 15	SYN	TACH TG		1 ICR CLAMP (2-3-4) 2 ICR 3 ICR 4 ICR
	DIFFERENTIAL SYNCHRO SYN	SYMMETRICAL NON-LINEAR RESISTOR		NUMBERS IN PARENTHESES DESIGNATE THE LOCATION OF RELAY CONTACTS. A LINE BENEATH A NUMBER SIGNIFIES A NORMALLY CLOSED CONTACT.

SEQUENCE CHARTS

		PROGRAM OR SEQUENCE TIMER						CAM SWITCH			
TIMER CONTACT	RESET POSITION	TIMING (MINUTES)			OPERATION TIME (MINUTES)		SWITCH	DEGREES			DEGREES OPERATION
		1 2 3 4 5 6 7 8 9			CLOSES	OPENS		90	180	270	
1TR-1					0	2	1CS				22° TO 180°
1TR-2					3.75	7.2	2CS				0° TO 194°
1TR-3					1.9	8.5	3CS				225° TO 360°
SHADED PORTION INDICATES CONTACT CLOSED							SHADED PORTION INDICATES CONTACT CLOSED				

The Relationship of Basic Logic Symbology* between Various Standards

FUNCTION (TWO INPUTS SHOWN WHERE APPLICABLE)	TRUTH TABLE	ANSI Y32.14-1973 IEEE STD. 91-1973		NEMA ICS 1-102 ICS-1-103 IS5 PART 11B		ANSI Y32.14-1962 IEEE 91-1962 MIL STD. 806C (NAVY)		MIL STD.[2] 806B
		RECTANGULAR SHAPE SYMBOLS	DISTINCTIVE SHAPE SYMBOLS	RECTANGULAR SHAPE SYMBOLS	DISTINCTIVE SHAPE SYMBOLS	UNIFORM SHAPE	DISTINCT SHAPE	
AND	A B Y / 0 0 0 / 0 1 0 / 1 0 0 / 1 1 1	A, B & Y	(distinctive shape)	A	(distinctive shape)	A	(distinctive shape)	(distinctive shape)
OR	A B Y / 0 0 0 / 0 1 1 / 1 0 1 / 1 1 1	≥1	(distinctive shape)	OR	(distinctive shape)	OR	(distinctive shape)	(distinctive shape)
EXCLUSIVE OR	A B Y / 0 0 0 / 0 1 1 / 1 0 1 / 1 1 0	=	(distinctive shape)	OE	OE	OE	(distinctive shape)	(distinctive shape)
AND INVERT (NEGATED OUTPUTS)	A B Y / 0 0 1 / 0 1 1 / 1 0 1 / 1 1 0	&	(distinctive shape)	A / NAND	(distinctive shape)	A	(distinctive shape)	
OR INVERT (NEGATED INPUTS)	A B Y / 0 0 1 / 0 1 1 / 1 0 1 / 1 1 0	≥1	(distinctive shape)	OR	(distinctive shape)	OR	(distinctive shape)	

* All logic symbols or combinations are not shown here; the respective standards should be consulted for additional symbology and its application.
† Superseded standards.

504

AND INVERT (NEGATED INPUTS)	OR INVERT (NEGATED OUTPUTS)	NEGATOR (NOT)	ELECTRIC INVERTOR	AMPLIFIER	OSCILLATOR
A	OR	N	N	AR	OSC
A / NOR / OR		A / NOR / OR		AR	OSC
&	≥1	1	1	△	G

AND INVERT (NEGATED INPUTS)

A	B	Y
0	0	1
0	1	0
1	0	0
1	1	0

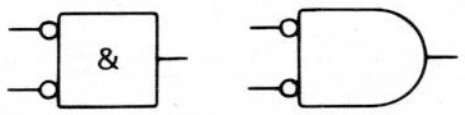

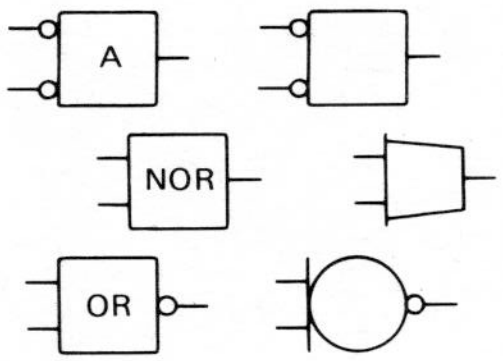

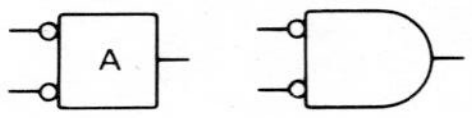

OR INVERT (NEGATED OUTPUTS)

A	B	Y
0	0	1
0	1	0
1	0	0
1	1	0

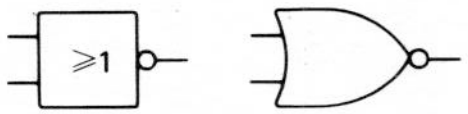

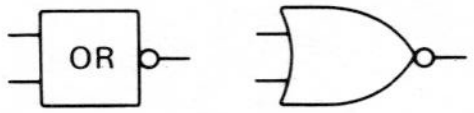

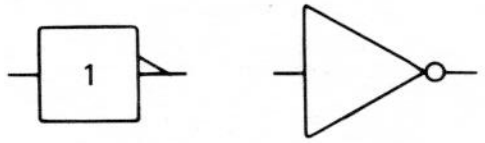

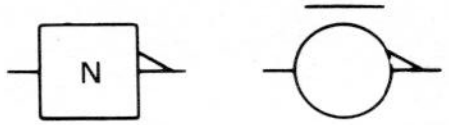

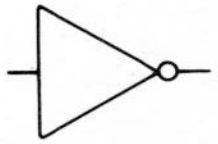

NEGATOR (NOT)

A	Y
0	1
1	0

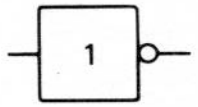

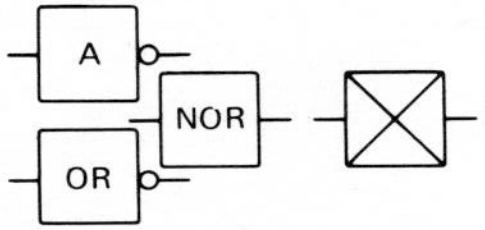

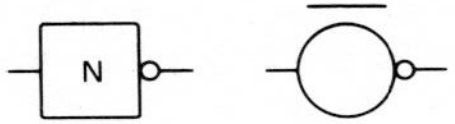

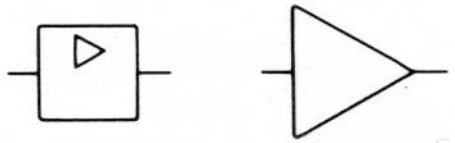

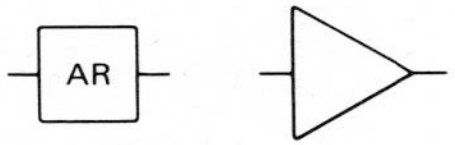

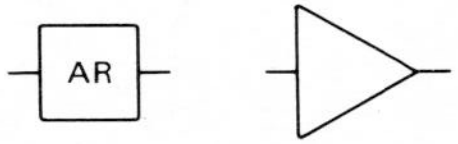

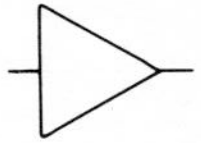

AMPLIFIER

A	Y
0	0
1	1

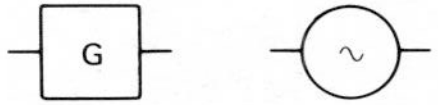

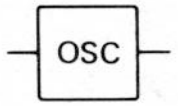

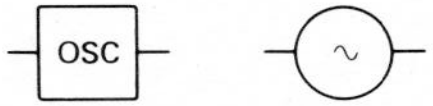

SINGLE SHOT	1Ω / ⊓	SS	SS	SS / SS 1 0
SCHMITT TRIGGER	⊓	ST	ST	ST / ST 1 0
FLIP-FLOP LATCH	S R Z / S - - R	S FL 1 C 0	S FF 1 C 0	FL 1 0 / S FL 1 C 0	S FF 1 C 0
FLIP-FLOP COMPLEMENTARY	S T C / S T - - C	S T FF 1 C 0	S T FF 1 C 0	FF 1 0 / S T FF 1 C 0	S T FF 1 C 0
TIME DELAY (ON/OFF- SAME TIME)	(t) / (t)	TD (t)	TD (t)	(t) / TD (t)	(t)
TIME DELAY (ON/OFF- SAME TIME) ADJUSTABLE	/	TD t_0-t_1	TD t_0-t_1	t_0-t_1 / TD t_0-t_1	

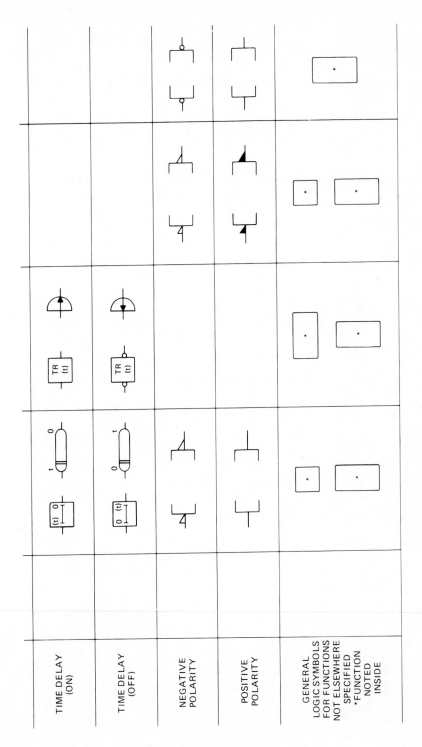

TIME DELAY (ON)				
TIME DELAY (OFF)				
NEGATIVE POLARITY				
POSITIVE POLARITY				
GENERAL LOGIC SYMBOLS FOR FUNCTIONS NOT ELSEWHERE SPECIFIED *FUNCTION NOTED INSIDE				

Electrical Symbols for Architectural Drawings*

1-0 Lighting Outlets

	CEILING	WALL	
1-1	○	─○	Surface or pendant incandescent mercury vapor or similar lamp fixture
1-2	(R)	─(R)	Recessed incandescent mercury vapor or similar lamp fixture
1-3	▭		Surface or pendant individual fluorescent fixture
1-4	▭R		Recessed individual fluorescent fixture
1-5	▭		Surface or pendant continuous-row fluorescent fixture
1-6	▭R		Recessed continuous-row fluorescent fixture†
1-7	├──┼──┼──┤		Bare-lamp fluorescent strip‡
1-8	(X)	─(X)	Surface or pendant exit light
1-9	(RX)	─(RX)	Recessed exit light
1-10	(B)	─(B)	Blanked outlet
1-11	(J)	─(J)	Junction box
1-12	(L)	─(L)	Outlet controlled by low-voltage switching when relay is installed in outlet box

* These symbols are taken from ANS Y32.9–1972, "Graphical Electrical Wiring Symbols for Architectural and Electrical Layout Drawings," published by the American National Standards Institute and sponsored by the Institute of Electrical and Electronics Engineers and the American Society of Mechanical Engineers.

† In the case of combination continuous-row fluorescent and incandescent spotlights, use combinations of the above standard symbols.

‡ In the case of continuous-row bare-lamp fluorescent strip above an area-wide diffusing means, show each fixture run, using the standard symbol; indicate area of diffusing means and type by light shading and/or drawing notation.

2-0 Receptacle Outlets

American National Standard C1–1971, National Electric Code (NFPA 70–1978) requires that grounded receptacles be used in most installations. Therefore, when a majority of the receptacles are to be of the grounded type, the ungrounded receptacles should be identified by the notation UNG at the outlet location, and the types of receptacles required noted in the drawing list of symbols and in the specifications.

Where weatherproof, explosionproof, or other specific types of device are to be required, use the type of uppercase subscript letters referred to under Sec. I3.2.1.2 of this standard. For example, weatherproof single or duplex receptacles would have the uppercase subscript letters noted alongside the symbol (WP, UNGWP).

	GROUNDED	UNGROUNDED	
2-1		UNG	Single receptacle outlet
2-2		UNG	Duplex receptacle outlet
2-3		UNG	Triplex receptacle outlet
2-4		UNG	Quadruplex receptacle outlet
2-5		UNG	Duplex receptacle outlet, split-wired
2-6		UNG	Triplex receptacle outlet, split-wired
2-7	*	* UNG	Single special-purpose receptacle outlet*
2-8	*	* UNG	Duplex special-purpose receptacle outlet*
2-9	R	UNG R	Range outlet

* Use numeral or letter either within the symbol or as a subscript alongside the symbol keyed to explanation in the drawing list of symbols to indicate type of receptacle or usage.

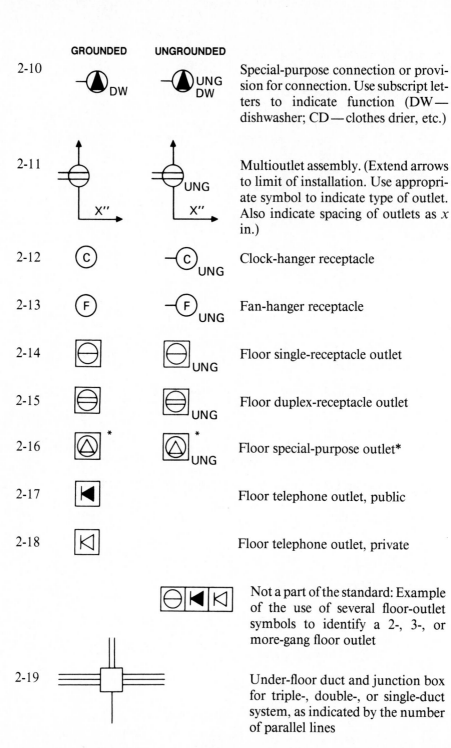

	GROUNDED	UNGROUNDED	
2-10	DW	UNG DW	Special-purpose connection or provision for connection. Use subscript letters to indicate function (DW—dishwasher; CD—clothes drier, etc.)
2-11	X″	UNG X″	Multioutlet assembly. (Extend arrows to limit of installation. Use appropriate symbol to indicate type of outlet. Also indicate spacing of outlets as x in.)
2-12	C	C UNG	Clock-hanger receptacle
2-13	F	F UNG	Fan-hanger receptacle
2-14		UNG	Floor single-receptacle outlet
2-15		UNG	Floor duplex-receptacle outlet
2-16	*	* UNG	Floor special-purpose outlet*
2-17			Floor telephone outlet, public
2-18			Floor telephone outlet, private
			Not a part of the standard: Example of the use of several floor-outlet symbols to identify a 2-, 3-, or more-gang floor outlet
2-19			Under-floor duct and junction box for triple-, double-, or single-duct system, as indicated by the number of parallel lines

* Use numeral or letter either within the symbol or as a subscript alongside the symbol keyed to explanation in the drawing list of symbols to indicate type of receptacle or usage.

GROUNDED   UNGROUNDED

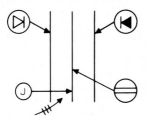

Not a part of the standard: Example of use of various symbols to identify location of different types of outlet or connection for underfloor duct or cellular floor systems

2-20      Cellular floor header duct

3-0 Switch Outlets

3-1	S	Single-pole switch	3-9	$\ominus$S	Switch and single receptacle
3-2	S_2	Double-pole switch	3-10	$\ominus$S	Switch and double receptacle
3-3	S_3	Three-way switch	3-11	S_D	Door switch
3-4	S_4	Four-way switch	3-12	S_T	Time switch
3-5	S_K	Key-operated switch	3-13	S_{CB}	Circuit-breaker switch
3-6	S_P	Switch and pilot lamp	3-14	S_{MC}	Momentary contact switch for pushbutton for other than signaling system
3-7	S_L	Switch for low-voltage switching system			
3-8	S_{LM}	Master switch for low-voltage switching system	3-15	Ⓢ	Ceiling pull switch

Signaling-System Outlets

4-0 Institutional, Commercial, and Industrial Occupancies

	Basic symbol	**Examples of individual item identification (not a part of the standard)**	

4-1

I. Nurse-call-system devices (any type)

Nurses' annunciator (can add a number after it as ⊢①
24 to indicate number of lamps)

Call station, single cord, pilot light

Call station, double cord, microphone-speaker

Corridor dome light, one lamp

Transformer

Any other item on same system: use numbers as required

4-2

II. Paging-system devices (any type)

Keyboard

Flush annunciator

Two-face annunciator

Any other item on same system: use numbers as required

Basic symbol	Examples of individual item identification (not a part of the standard)	

4-3

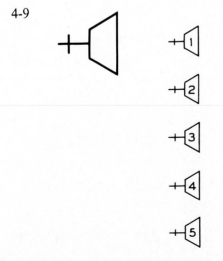

III. Fire-alarm-system devices (any type) including smoke and sprinkler alarm devices

	1	Control panel
	2	Station
	3	10-in. gong
	4	Presignal chime
	5	Any other item on same system: use numbers as required

4-9

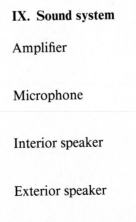

IX. Sound system

	1	Amplifier
	2	Microphone
	3	Interior speaker
	4	Exterior speaker
	5	Any other item on same system: use numbers as required

	Basic symbol	Examples of individual item identification (not a part of the standard)

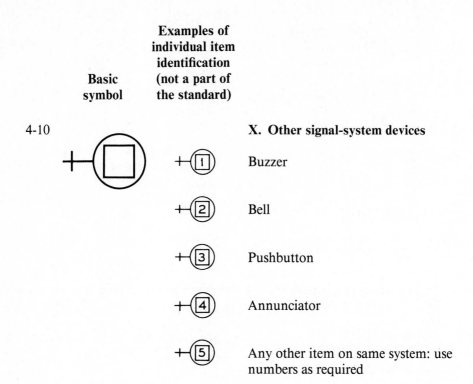

4-10

X. Other signal-system devices

Buzzer

Bell

Pushbutton

Annunciator

Any other item on same system: use numbers as required

5-0 Residential Occupancies

Signaling-system symbols for use in identifying standardized residential-type signal-system items on residential drawings where a descriptive symbol list is not included on the drawing. When other signal-system items are to be identified, use the above basic symbols for such items together with a descriptive symbol list.

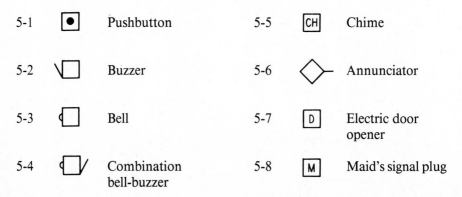

5-1	Pushbutton	5-5	Chime
5-2	Buzzer	5-6	Annunciator
5-3	Bell	5-7	Electric door opener
5-4	Combination bell-buzzer	5-8	Maid's signal plug

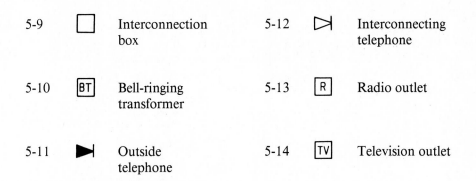

5-9 □ Interconnection box

5-10 [BT] Bell-ringing transformer

5-11 ▶◀ Outside telephone

5-12 ▷ Interconnecting telephone

5-13 [R] Radio outlet

5-14 [TV] Television outlet

6-0 Panelboards, Switchboards, and Related Equipment

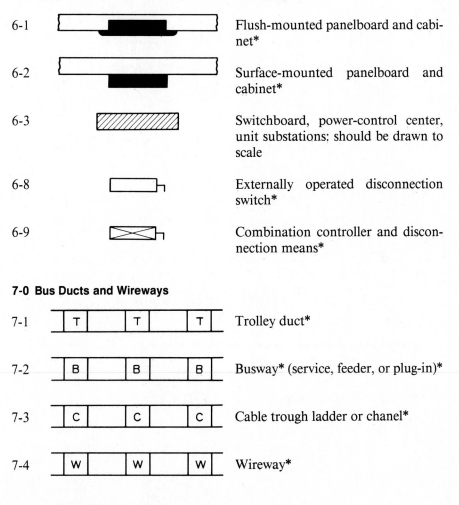

6-1 Flush-mounted panelboard and cabinet*

6-2 Surface-mounted panelboard and cabinet*

6-3 Switchboard, power-control center, unit substations: should be drawn to scale

6-8 Externally operated disconnection switch*

6-9 Combination controller and disconnection means*

7-0 Bus Ducts and Wireways

7-1 | T | T | T | Trolley duct*

7-2 | B | B | B | Busway* (service, feeder, or plug-in)*

7-3 | C | C | C | Cable trough ladder or chanel*

7-4 | W | W | W | Wireway*

* Identify by notation or schedule.

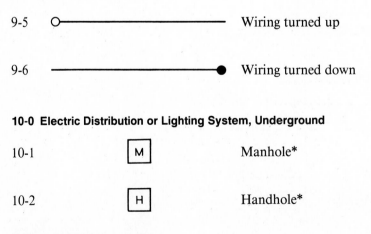

9-0 Circuiting

Wiring method identification by notation on drawing or in specifications

9-1 ——————————————— Wiring concealed in ceiling or wall

9-2 —— —— —— —— Wiring concealed in floor

9-3 — — — — — — — — — — — Wiring exposed

Note: Use heavy-weight line to identify service and feeders. Indicate empty conduit by notation CO (conduit only).

9-4 ———————————————⟶ Branch circuit home run to panelboard. Number of arrows indicates number of circuits. (A numeral at each arrow may be used to identify circuit number.)

Note: Any circuit without further identification indicates two-wire circuit. For a greater number of wires, indicate with cross lines, for example: —⫲— three wires; —⫲⫲— four wires, etc. Unless indicated otherwise, the wire size of the circuit is the minimum size required by the specification.

Identify different functions of wiring system, for example, signaling system, by notation or other means.

9-5 O——————————— Wiring turned up

9-6 ———————————● Wiring turned down

10-0 Electric Distribution or Lighting System, Underground

10-1 ☐M Manhole*

10-2 ☐H Handhole*

* Identify by notation or schedule.

Appendix C

Component Identification

A designer or drafter should have some knowledge of how to distinguish the ratings of a passive component (resistor, capacitor, and inductors). This knowledge is necessary to trace out existing circuits to develop schematics (if they are not available), perform layouts of component dimensions (if dimensions are not readily available), and find substitute components and for other aspects of circuit design and troubleshooting.

About 30 years ago it was relatively easy to identify passive components. These devices used to have some type of standardized identification means such as colored bands or dots. Today it is much more difficult to identify components. This is because there has been an increase in types of material used in each type of component (e.g., tantalum, ceramic, solid electrolyte, wet electrolyte capacitors, to mention a few), in the packaging of components (e.g., dip network, carbon film, inline network, square potentiometers resistors), in the precision and tolerances of these devices, and most of all the miniaturization of these devices makes it difficult at times to use a color code (e.g., there are capacitors no bigger than the head of a common pencil). For these reasons, many devices are not color coded but their values are stamped on them.

In the following pages there are examples of the standard color coding of these devices. The standard color coding was developed by the Electronics Industry Association (EIA) in standards RS-236B, RS-279, and RS-359. These standards are advisory and not mandatory; therefore, some manufacturers modify the meaning of the colors or, as mentioned before, don't use them. These color code standards are used, but with each type of device the color may represent different values, tolerances, and other parameters associated with the device. The carbon composition resistor will be discussed in great extent to give a complete understanding of color codes; then the other components are presented with a briefer explanation.

Resistor Identification

Color Code for Carbon Composition Resistors

The colored bands around the body of a carbon composition resistor indicate its value in ohms and, if there are four bands, its tolerance (see Table C-1). As the drawing in Fig. C-1 shows, the first band represents the first digit of the value and the second band, the second digit of the value. (The first band is near one edge and is often at that edge of the device.) The third band represents the number by which the two digits are multiplied. A fourth band of gold or silver

TABLE C-1 Color Code for Carbon Composition Resistors

COLOR	DIGIT		MULTIPLIER	TOLERANCE
	1st	2nd		
Black	0	0	1	—
Brown	1	1	10	—
Red	2	2	100	—
Orange	3	3	1,000	—
Yellow	4	4	10,000	—
Green	5	5	100,000	—
Blue	6	6	1,000,000	—
Violet	7	7	10,000,000	—
Gray	8	8	100,000,000	—
White	9	9	1,000,000,000	—
Gold	—	—	—	± 5%
Silver	—	—	—	±10%
No band	—	—	—	±20%

represents a tolerance of 5 or 10 percent, respectively. If no fourth band is shown, the tolerance is 20 percent of the indicated value of the resistance. Color codes always refer to resistance in ohms (Ω). Also, resistors are read from left to right, with the first digit appearing on the left.

The physical size of the carbon composition resistor is related to its voltage rating. Size increases as the wattage rating increases. The diameters of $-\frac{1}{2}$-, -1-, and 2-W carbon composition resistors are approximately $\frac{1}{8}$, $\frac{1}{4}$, and $\frac{5}{16}$ in., respectively!

The carbon film resistor color code (see Table C-2 and Fig. C-2) uses a five-color band identification having three bands for three digits, and one band for the multiplier, and one band for the tolerance. There is visibly wider spacing between the multiplier band and the tolerance band to identify direction of reading the values.

Again it should be emphasized that all resistors are not color coded; in fact, many types have a value stamped on them and some of the new small resistors do not have any resistance identification because of their size.

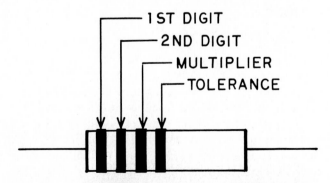

FIG. C-1 Color coding on a carbon composition resistor.

TABLE C-2   Color Code for Carbon Film Resistors

COLOR	DIGITS			MULTIPLIER	TOLERANCE
	1st	2nd	3rd		
Black	0	0	0	1	
Brown	1	1	1	10	±1%
Red	2	2	2	100	±2%
Orange	3	3	3	1,000	
Yellow	4	4	4	10,000	
Green	5	5	5	100,000	±5%
Blue	6	6	6	1,000,000	±0.25%
Violet	7	7	7	10,000,000	±0.10%
Gray	8	8	8		±0.05%
White	9	9	9	0.1	±5%
Gold	—	—	—	0.01	±10%
Silver	—	—	—		

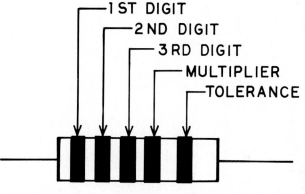

FIG. C-2   Color coding on a carbon film resistor.

Capacitor Identification

Capacitor identification is more difficult than resistor identification because there are more identifying color codes, types of capacitor, and physical shapes of capacitor. In fact, there are so many color codes and types that it would be better for individuals who work with capacitors infrequently to consult an identification reference.

Capacitors are named after some particular physical quantity, such as: a disk capacitor is shaped like a disk, and an electrolytic (solid or liquid) capacitor is named for the dielectric. The physical size of a capacitor increases with increases in capacitance, working voltage, or both. The information found in the color code or stamped identification can include some or all of these parameters (capacitance, tolerance, temperature coefficient, breakdown voltage, capacitance drift). Color codes always refer to capacitance in picofarads (pF).

The following are several types of capacitance identification.

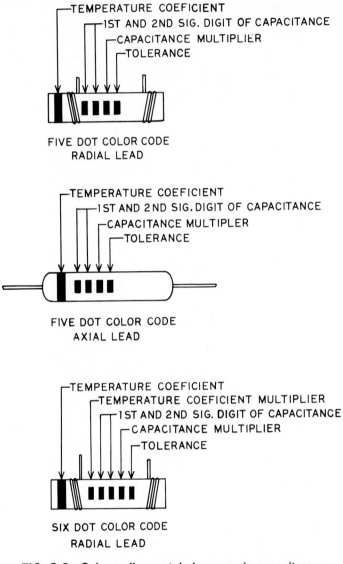

FIG. C-3 Color coding on tubular ceramic capacitors.

Color Codes for Tubular Ceramic Capacitors

Tubular ceramic capacitors (see Fig. C-3) use colored dots or colored bands in a five- or six-color identification. It is read from left to right with the temperature coefficient dot or band (the largest in the series) appearing on the left.

Ceramic Disk Capacitor Color Code

Utilizing the same color value as in the tubular ceramic capacitor, the ceramic disk capacitors (see Table C-3 and Fig. C-4) have a 3- or 5-dot identification.

TABLE C-3   Color Codes for Tubular and Disc Ceramic Capacitors

| BAND OR DOT COLOR | SIGNIF-ICANT DIGITS | | CAPACITANCE MULTIPLIER | TOLERANCE | | TEMPERATURE COEFFICIENT pp °C (5-DOT SYSTEM) | TEMPERATURE COEFFICIENT 6-DOT SYSTEM SIG. FIG. | TEMPERATURE COEFFICIENT MULTIPLIER |
	1st	2nd		≤10 pF	>10 pF			
Black	0	0	1	±2.0 pF	±20%	0	0.0	−1
Brown	1	1	10	±0.1 pF	±1%	−33		−10
Red	2	2	100		±2%	−75	1.0	−100
Orange	3	3	1000		±3%	−150	1.5	−1000
Yellow	4	4				−230	2.0	−10000
Green	5	5		±0.5 pF	±5%	−330	3.3	+1
Blue	6	6				−470	4.7	+10
Violet	7	7				−750	7.5	+100
Gray	8	8	0.01	±0.25 pF		+150 to −1500		+1000
White	9	9	0.1	±1.0 pF	±10%	+100 to −75		+10000
Silver	—	—						
Gold	—	—						

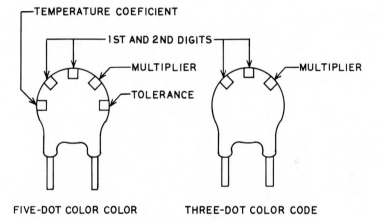

FIVE-DOT COLOR COLOR   THREE-DOT COLOR CODE

FIG. C-4   Color coding on a ceramic disk capacitor.

Color Code for Molded Paper Tubular Capacitors

Molded paper capacitors (see Table C-4 and Fig. C-5) are essentially obsolete; however, it might be necessary to identify them in some old equipment. Other than the standard color codes, there can be one or two bands used for coding the working voltage. If the capacitor has a working voltage of 1000 V dc or less, one color band is used. If the capacitor has a working voltage of more than 1000 V dc, two color bands are used. Working voltage is determined by taking the value of tolerance bands, such as brown-green = 15, and multiplying by 100, equaling a working voltage 1500 WVD.

Color Code for Mica Capacitors

Mica capacitors (see Fig. C-6) utilize a color dot arrangement to indicate capacitance. There are numerous color dot arrangements from 3 to 9. The 6- and

TABLE C-4   Color Code for Molded Paper Tubular Capacitors

COLOR	(CAPACITANCE IN PICOFARADS) SIGNIF-ICANT DIGITS 1st	2nd	MULTIPLIER	TOLERANCE	VOLTAGE
Black	0	0	1	±20%	—
Brown	1	1	10		100
Red	2	2	100		200
Orange	3	3	1000	±30%	300
Yellow	4	4	10000		400
Green	5	5			500
Blue	6	6			600
Violet	7	7			700
Gray	8	8			800
White	9	9			900
Gold	—	—			1000
Silver	—	—		±10%	—

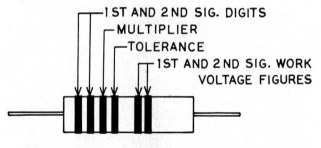

FIG. C-5   Color coding on a paper tubular capacitor.

9-dot arrangements are more prevelant. These arrangements are shown in Tables C-5 and C-6.

Marking of Capacitors

The ceramic disk capacitor shown in Fig. C-7 has three markings on it. Polarity is shown by a dot. The capacitance is marked picofarads (47) and the dc voltage is 35. Also, the positive lead wire is longer than the negative.

Figure C-8 shows an example of how one manufacturer identifies their miniature electrolytes. The manufacturer's name, original equipment manufacturer (OEM), number, type number, and the capacitance (in microfarads) and dc working voltage are stamped on the outside of the capacitor.

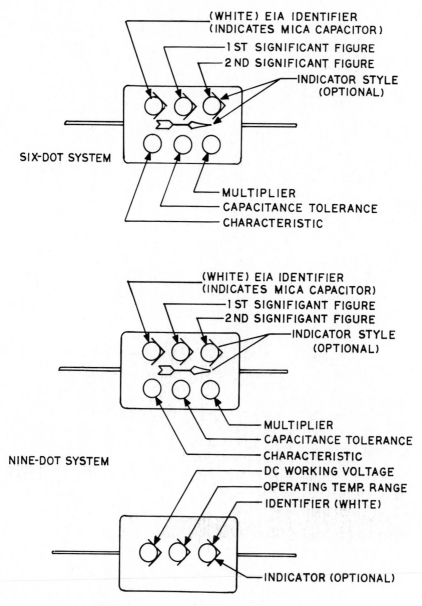

FIG. C-6 Color coding on mica capacitors.

Inductor Identification

In comparison to resistors and capacitors, inductors are not as prevalent in the electronics circuitry. Some inductors are color-coded, but most, primarily chokes, are not. Inductor sizes increase with an increase in value (microhenrys) and in resonant frequency (e.g., megahertz).

TABLE C-5   Color Code for Mica Capacitors

COLOR	CHARAC-TERISTIC	DIGITS 1st	DIGITS 2nd	MULTI-PLIER	TOLER-ANCE	DC WORKING VOLTAGE	OPERATING TEMPERATURE RANGE
Black		0	0	1	±20%	100	
Brown	B	1	1	10	±1%		−55°C to +85°C
Red	C	2	2	100	±2%	300	
Orange	D	3	3	1,000			−55°C to +125°C
Yellow	E	4	4	10,000		500	
Green	F	5	5		±5%		
Blue		6	6				
Violet		7	7				
Gray		8	8				
White		9	9				
Gold		—	—	0.1	±½%	1000	
Silver		—	—	0.01	±10%		

TABLE C-6   Mica Capacitor Characteristics

CHARAC-TERISTIC	TEMPERATURE COEFFICIENT OF CAPACITANCE (ppm/°C)	MAXIMUM CAPACITANCE DRIFT
B	Not specified	Not specified
C	±200	±(0.5% + 0.5 pF)
D	±100	±(0.3% + 0.1 pF)
E	−20 to +100	±(0.1% + 0.1 pF)
F	0 to +70	±(0.05% + 0.1 pF)

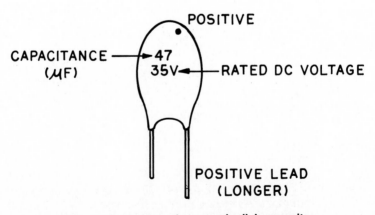

FIG. C-7   Marking of a ceramic disk capacitor.

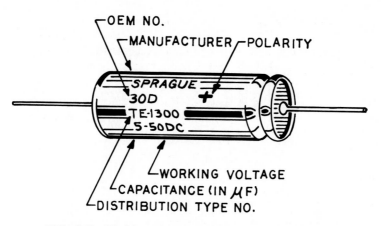

OEM NO.
MANUFACTURER POLARITY
SPRAGUE
30D
TE-1300
5-50DC
WORKING VOLTAGE
CAPACITANCE (IN μF)
DISTRIBUTION TYPE NO.

FIG. C-8 Marking of a tubular electrolytic capacitor.

Color Code for Inductors

Inductor color coding (see Table C-7 and Fig. C-9) is slightly different in that the color code uses a decimal point represented by a gold band. Inductor color code values are in microhenrys (μH).

Also, like capacitors and resistors, the values of many inductors are marked on the body of the inductor.

TABLE C-7 Color Code for Inductors (in Microhenrys)

COLOR	INDUCTANCE DIGITS 1st	2nd	MULTIPLIER	TOLERANCE
Black	0	0	1	
Brown	1	1	10	
Red	2	2	100	
Orange	3	3	1,000	
Yellow	4	4	10,000	
Green	5	5		
Blue	6	6		
Violet	7	7		
Gray	8	8		
White	9	9		
Gold	·	·		±5%
Silver	—	—	—	±10%
No band	—	—	—	±20%

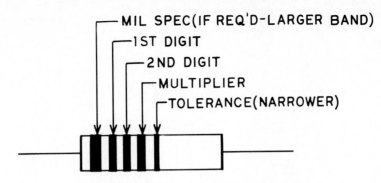

FIG. C-9 Color coding of an inductor.

Width of Copper-Foil Conductors for Printed Circuits*

WIDTH OF 0.00135-in.-THICK COPPER (INCHES)	WIDTH OF 0.0027-in.-THICK COPPER (INCHES)	CURRENT CAPABILITY (AMPERES)
$\frac{1}{64}$		1.5
$\frac{1}{32}$	$\frac{1}{64}$	2.5
$\frac{1}{16}$	$\frac{1}{32}$	3.5
	$\frac{1}{16}$	5.5
$\frac{1}{8}$		6.5
	$\frac{1}{8}$	8.0
	$\frac{3}{16}$	15.0

By permission of the Aerovox Corp.
* For a temperature rise of 40°C, the maximum recommended for lucite and nylon laminates.

Example: If the desired current is 3.5 A, a conductor 0.0027 in. thick and $\frac{1}{32}$ in. wide will carry it. Also, a 0.00135-in.-thick copper foil that is $\frac{1}{16}$ in. wide will carry 3.5 A with a rise of 40°C or less.

Color Code for Chassis Wiring

COLOR	ABBREV.	NUMERICAL CODE	CIRCUIT
Black	BK	0	Grounds, grounded elements and returns
Brown	BR	1	Heaters of filaments off ground
Red	R	2	Power supply B-plus
Orange	O	3	Screen grids
Yellow	Y	4	Cathodes, emitters
Green	GN	5	Control grids, base
Blue	BL	6	Plates (anodes), collectors
Violet (or purple)	V PR	7	Power supply, minus
Gray	GY	8	AC power lines
White	W	9	Miscellaneous, returns above or below ground, AVC, etc.

Source: Mil Std 122.

Circuit-Identification Color Code for Industrial Control Wiring

CIRCUIT	COLOR
Line, load, and control circuit at line voltage	Black
AC control circuit	Red
DC control circuit	Blue
Interlock panel control when energized from external force	Yellow
Equipment grounding conductor	Green
Grounded neutral conductor	White

Source: National Machine Tool Builders Association.

Transformer Color Codes

POWER TRANSFORMERS

1. Primary leads
   If tapped:
   - Common — Black
   - Tap — Black and yellow striped
   - Finish — Black and red striped
2. High-voltage plate winding — Red
   Center tap — Red and yellow striped
3. Rectifier filament winding — Yellow
   Center tap — Green and yellow striped
4. Filament winding No. 1 — Green
   Center tap — Green and yellow striped
5. Filament winding No. 2 — Brown
   Center tap — Brown and yellow striped
6. Filament winding No. 3 — Slate
   Center tap — Slate and yellow striped

AUDIO TRANSFORMERS

Blue — Plate (finish) lead of primary

Red — B + lead (no difference with center tap)

Brown — Plate (start) lead on center-tapped primaries.
(Blue may be used if polarity is not important.)

Green — Grid (finish) lead to secondary.

Black — Grid return (no difference with center tap.)

Yellow — Grid (start) lead on center-tapped secondaries.
(Green may be used if polarity is not important.)

IF TRANSFORMERS

Blue	Plate lead
Red	B + lead
Green	Grid (or diode) lead
Black	Grid (or diode) return

Appendix D

Approximate Radii for Aluminum Alloys for 90° Cold Bend*

ALLOY AND TEMPER	RADII FOR VARIOUS THICKNESSES IN TERMS OF THICKNESS t				
	$\frac{1}{64}$ in.	$\frac{1}{32}$ in.	$\frac{1}{16}$ in.	$\frac{1}{8}$ in.	$\frac{3}{16}$ in.
1100-0	0	0	0	0	0
1100-H12	0	0	0	0	0–1
1100-H14	0	0	0	0	0–1
1100-H16	0	0	0–1	$\frac{1}{2}$–$1\frac{1}{2}$	1–2
1100-H18	0–1	$\frac{1}{2}$–$1\frac{1}{2}$	1–2	$1\frac{1}{2}$–3	2–4
Alclad 2014-0	0	0–1	0–1	0–1	0–1
Alclad 2014-T3	1–2	$1\frac{1}{2}$–3	2–4	3–5	4–6
Alclad 2014-T4	1–2	$1\frac{1}{2}$–3	2–4	3–5	4–6
Alclad 2014-T6	2–4	3–5	3–5	4–6	5–7
2024-0	0	0–1	0–1	0–1	0–1
2024-T3	$1\frac{1}{2}$–3	2–4	3–5	4–6	4–6
2024-T36	2–4	3–5	4–6	5–7	5–7
2024-T4	$1\frac{1}{2}$–3	2–4	3–5	4–6	4–6
2024-T81	$3\frac{1}{2}$–5	$4\frac{1}{2}$–6	5–7	$6\frac{1}{2}$–8	7–9
2024-T86	4–$5\frac{1}{2}$	5–7	6–8	7–10	8–11
2219-T31			$\frac{1}{2}$–$1\frac{1}{2}$	1–$1\frac{1}{2}$	1–2
2219-T37			$\frac{1}{2}$–2	$1\frac{1}{2}$–3	2–$3\frac{1}{2}$
2219-T81			2–4	$2\frac{1}{2}$–$4\frac{1}{2}$	3–5
2219-T87			2–4	3–5	4–6
3003-0	0	0	0	0	0
3003-H12	0	0	0	0	0–1
3003-H14	0	0	0	0–1	0–1
3003-H16	0–1	0–1		1–2	$1\frac{1}{2}$–3
3003-H18	$\frac{1}{2}$–$1\frac{1}{2}$	1–2	$1\frac{1}{2}$–3	2–4	3–5

* *Alcoa Aluminum Handbook,* by permission of Aluminum Company of America.

Thickness of Wire and Metal-Sheet Gages (in Inches)

GAGE NUMBER	AMERICAN OR BROWN AND SHARPE GAGE 1*	UNITED STATES STANDARD GAGE 2†	GAGE NUMBER	AMERICAN OR BROWN AND SHARPE GAGE 1*	UNITED STATES STANDARD GAGE 2†
0	0.3249	0.3125	16	0.0508	0.0625
1	0.2893	0.2813	17	0.0453	0.0563
2	0.2576	0.2656	18	0.0403	0.0500
3	0.2294	0.2500	19	0.0359	0.0438
4	0.2043	0.2344	20	0.0320	0.0375
5	0.1819	0.2188	21	0.0285	0.0344
6	0.1620	0.2031	22	0.0253	0.0313
7	0.1443	0.1875	23	0.0226	0.0281
8	0.1285	0.1719	24	0.0201	0.0250
9	0.1144	0.1563	25	0.0179	0.0219
10	0.1019	0.1406	26	0.0159	0.0188
11	0.0907	0.1250	27	0.0142	0.0172
12	0.0808	0.1094	28	0.0126	0.0156
13	0.0720	0.0938	29	0.0113	0.0141
14	0.0641	0.0781	30	0.0100	0.0125
15	0.0571	0.0703	31	0.0089	0.0109

* For aluminum sheet, rod, and wire. Also for copper wire, and brass, alloy and nickel silver wire and sheet.
† For steel, nickel, and Monel metal sheets.

Metric Conversion Table

MILLIMETERS TO INCHES		INCHES TO MILLIMETERS		INCHES (DECIMALS) TO MILLIMETERS	
mm in.	mm in.	in. mm	in. mm	in. mm	in. mm
1 = 0.0394	17 = 0.6693	$\frac{1}{32}$ = 0.794	$\frac{17}{32}$ = 13.493	0.01 = 0.254	0.25 = 6.350
2 = 0.0787	18 = 0.7087	$\frac{1}{16}$ = 1.587	$\frac{9}{16}$ = 14.287	0.02 = 0.508	0.26 = 6.604
3 = 0.1181	19 = 0.7480	$\frac{3}{32}$ = 2.381	$\frac{19}{32}$ = 15.081	0.03 = 0.762	0.28 = 7.112
4 = 0.1575	20 = 0.7874	$\frac{1}{8}$ = 3.175	$\frac{5}{8}$ = 15.875	0.04 = 1.016	0.30 = 7.620
5 = 0.1969	21 = 0.8268	$\frac{5}{32}$ = 3.968	$\frac{21}{32}$ = 16.668	0.05 = 1.270	0.32 = 8.128
6 = 0.2362	22 = 0.8662	$\frac{3}{16}$ = 4.762	$\frac{11}{16}$ = 17.462	0.06 = 1.524	0.34 = 8.636
7 = 0.2756	23 = 0.9055	$\frac{7}{32}$ = 5.556	$\frac{23}{32}$ = 18.256	0.07 = 1.778	0.36 = 9.144
8 = 0.3150	24 = 0.9449	$\frac{1}{4}$ = 6.350	$\frac{3}{4}$ = 19.050	0.08 = 2.032	0.38 = 9.652
9 = 0.3543	25 = 0.9843	$\frac{9}{32}$ = 7.144	$\frac{25}{32}$ = 19.843	0.09 = 2.286	0.40 = 10.160
10 = 0.3937	26 = 1.0236	$\frac{5}{16}$ = 7.937	$\frac{13}{16}$ = 20.637	0.10 = 2.540	0.50 = 12.699
11 = 0.4331	27 = 1.0630	$\frac{11}{32}$ = 8.731	$\frac{27}{32}$ = 21.431	0.12 = 3.048	0.60 = 15.240
12 = 0.4724	28 = 1.1024	$\frac{3}{8}$ = 9.525	$\frac{7}{8}$ = 22.225	0.14 = 3.556	0.70 = 17.780
13 = 0.5118	29 = 1.1418	$\frac{13}{32}$ = 10.319	$\frac{29}{32}$ = 23.018	0.16 = 4.064	0.80 = 20.320
14 = 0.5512	30 = 1.1811	$\frac{7}{16}$ = 11.112	$\frac{15}{16}$ = 23.812	0.18 = 4.572	0.90 = 22.860
15 = 0.5906	50 = 1.9685	$\frac{15}{32}$ = 11.906	1 = 25.400	0.20 = 5.080	1.00 = 25.400
16 = 0.6299	100 = 3.9370	$\frac{1}{2}$ = 12.699	12 = 304.800	0.22 = 5.588	2.00 = 50.800

Minimum Radius of Conduit* (in Inches)

SIZE OF CONDUIT (INCHES)	NATIONAL MACHINE TOOL BUILDERS' ASSOCIATION	NEC—FOR CONDUCTORS WITHOUT LEAD SHEATH	NEC—FOR CONDUCTORS WITH LEAD SHEATH
$\frac{1}{2}$	4	4	6
$\frac{3}{4}$	$4\frac{1}{2}$	5	8
1	$5\frac{3}{4}$	6	11
$1\frac{1}{4}$	7	8	14
$1\frac{1}{2}$	$8\frac{1}{4}$	10	16
2	$9\frac{1}{2}$	12	21
$2\frac{1}{2}$	$10\frac{1}{2}$	15	25
3	13	18	31
$3\frac{1}{2}$	15	21	36
4	16	24	40
5	24	30	50
6	30	36	60

* The radius is that of the inner edge. Fittings shall be threaded unless structural difficulties prevent assembly. A run of conduit shall not contain more than the equivalent of four quarter bends (360°) total.

Decimal Equivalents of Fractions*

FRAC-TION	EQUIV-ALENT	FRAC-TION	EQUIV-ALENT	FRAC-TION	EQUIV-ALENT	FRAC-TION	EQUIV-ALENT
$\frac{1}{64}$	0.0156	$\frac{17}{64}$	0.2656	$\frac{33}{64}$	0.5156	$\frac{49}{64}$	0.7656
$\frac{1}{32}$	0.0312	$\frac{9}{32}$	0.2812	$\frac{17}{32}$	0.5312	$\frac{25}{32}$	0.7812
$\frac{3}{64}$	0.0468	$\frac{19}{64}$	0.2968	$\frac{35}{64}$	0.5468	$\frac{51}{64}$	0.7968
$\frac{1}{16}$	0.0625	$\frac{5}{16}$	0.3125	$\frac{9}{16}$	0.5625	$\frac{13}{16}$	0.8125
$\frac{5}{64}$	0.0781	$\frac{21}{64}$	0.3281	$\frac{37}{64}$	0.5781	$\frac{53}{64}$	0.8281
$\frac{3}{32}$	0.0937	$\frac{11}{32}$	0.3437	$\frac{19}{32}$	0.5937	$\frac{27}{32}$	0.8437
$\frac{7}{64}$	0.1093	$\frac{23}{64}$	0.3593	$\frac{39}{64}$	0.6093	$\frac{55}{64}$	0.8593
$\frac{1}{8}$	0.1250	$\frac{3}{8}$	0.3750	$\frac{5}{8}$	0.6250	$\frac{7}{8}$	0.8750
$\frac{9}{64}$	0.1406	$\frac{25}{64}$	0.3906	$\frac{41}{64}$	0.6406	$\frac{57}{64}$	0.8906
$\frac{5}{32}$	0.1562	$\frac{13}{32}$	0.4062	$\frac{21}{32}$	0.6562	$\frac{29}{32}$	0.9062
$\frac{11}{64}$	0.1718	$\frac{27}{64}$	0.4218	$\frac{43}{64}$	0.6718	$\frac{59}{64}$	0.9218
$\frac{3}{16}$	0.1875	$\frac{7}{16}$	0.4375	$\frac{11}{16}$	0.6875	$\frac{15}{16}$	0.9375
$\frac{13}{64}$	0.2031	$\frac{29}{64}$	0.4531	$\frac{45}{64}$	0.7031	$\frac{61}{64}$	0.9531
$\frac{7}{32}$	0.2187	$\frac{15}{32}$	0.4687	$\frac{23}{32}$	0.7187	$\frac{31}{32}$	0.9687
$\frac{15}{64}$	0.2343	$\frac{31}{64}$	0.4843	$\frac{46}{64}$	0.7343	$\frac{63}{64}$	0.9843
$\frac{1}{4}$	0.2500	$\frac{1}{2}$	0.5000	$\frac{3}{4}$	0.7500	1	1.000

* See page 530 for metric equivalents.

Small Drills — Metric*

METRIC DRILL DIAMETER	DIAMETER (in.)	METRIC DRILL DIAMETER	DIAMETER (in.)	METRIC DRILL DIAMETER	DIAMETER (in.)	METRIC DRILL DIAMETER	DIAMETER (in.)
0.200	0.0078	2.300	0.0905	5.70	0.2244	9.20	0.3622
0.250	0.0098	2.35	0.0925	5.75	0.2264	9.25	0.3642
0.300	0.0118	2.400	0.0945	5.80	0.2283	9.30	0.3661
0.350	0.0138	2.45	0.0964	5.90	0.2323	9.40	0.3701
0.400	0.0157	2.50	0.0984	6.00	0.2362	9.50	0.3740
0.450	0.0177	2.60	0.1024	6.10	0.2401	9.60	0.3779
0.500	0.0197	2.70	0.1063	6.20	0.2441	9.70	0.3819
0.550	0.0216	2.75	0.1083	6.25	0.2461	9.75	0.3838
0.600	0.0236	2.80	0.1102	6.30	0.2480	9.80	0.3858
0.650	0.0256	2.90	0.1142	6.40	0.2520	9.90	0.3898
0.700	0.0275	3.00	0.1181	6.50	0.2559	10.00	0.3937
0.750	0.0295	3.10	0.1220	6.60	0.2598	10.50	0.4134
0.800	0.0315	3.20	0.1260	6.70	0.2638	11.00	0.4331
0.850	0.0335	3.25	0.1279	6.75	0.2657	11.50	0.4527
0.900	0.0354	3.30	0.1299	6.80	0.2677	12.00	0.4724
0.950	0.0374	3.40	0.1338	6.90	0.2716	12.50	0.4921
1.000	0.0394	3.50	0.1378	7.00	0.2756	13.00	0.5118
1.050	0.0413	3.60	0.1417	7.10	0.2795	13.50	0.5315
1.100	0.0433	3.70	0.1457	7.20	0.2835	14.00	0.5512

mm	in	mm	in	mm	in	mm	in
1.150	0.0453	3.75	0.1476	7.25	0.2854	14.50	0.5709
1.200	0.0472	3.80	0.1496	7.30	0.2874	15.00	0.5905
1.250	0.0492	3.90	0.1535	7.40	0.2913	15.50	0.6102
1.300	0.0512	4.00	0.1575	7.50	0.2953	16.00	0.6299
1.350	0.0531	4.10	0.1614	7.60	0.2992	16.50	0.6496
1.400	0.0551	4.20	0.1653	7.70	0.3031	17.00	0.6693
1.450	0.0571	4.25	0.1673	7.75	0.3051	17.50	0.6890
1.500	0.0590	4.30	0.1693	7.80	0.3071	18.00	0.7087
1.550	0.0610	4.40	0.1732	7.90	0.3110	18.50	0.7283
1.600	0.0630	4.50	0.1772	8.00	0.3150	19.00	0.7480
1.650	0.0650	4.60	0.1811	8.10	0.3189	19.50	0.7677
1.700	0.0669	4.70	0.1850	8.20	0.3228	20.00	0.7874
1.750	0.0689	4.75	0.1870	8.25	0.3248	20.50	0.8071
1.800	0.0709	4.80	0.1890	8.30	0.3268	21.00	0.8268
1.850	0.0728	4.90	0.1929	8.40	0.3307	21.50	0.8464
1.900	0.0748	5.00	0.1968	8.50	0.3346	22.00	0.8661
1.950	0.0768	5.10	0.2008	8.60	0.3386	22.50	0.8858
2.000	0.0787	5.20	0.2047	8.70	0.3425	23.00	0.9055
2.050	0.0807	5.25	0.2067	8.75	0.3445	23.50	0.9252
2.100	0.0827	5.30	0.2087	8.80	0.3464	24.00	0.9449
2.150	0.0846	5.40	0.2126	8.90	0.3504	24.50	0.9646
2.200	0.0866	5.50	0.2165	9.00	0.3543	25.00	0.9842
2.250	0.0886	5.60	0.2205	9.10	0.3583	25.50	1.0039

* Drills beyond the range of this table increase in diameter by increments of 0.50 mm —26.00, 26.50, 27.00, etc.

Twist-Drill Sizes*

				NUMBER SIZES			
NO. SIZE	DECIMAL EQUIVALENT	METRIC EQUIVALENT	CLOSEST METRIC DRILL (mm)	NO. SIZE	DECIMAL EQUIVALENT	METRIC EQUIVALENT	CLOSEST METRIC DRILL (mm)
1	0.2280	5.791	5.80	41	0.0960	2.438	2.45
2	0.2210	5.613	5.60	42	0.0935	2.362	2.35
3	0.2130	5.410	5.40	43	0.0890	2.261	2.25
4	0.2090	5.309	5.30	44	0.0860	2.184	2.20
5	0.2055	5.220	5.20	45	0.0820	2.083	2.10
6	0.2040	5.182	5.20	46	0.0810	2.057	2.05
7	0.2010	5.105	5.10	47	0.0785	1.994	2.00
8	0.1990	5.055	5.10	48	0.0760	1.930	1.95
9	0.1960	4.978	5.00	49	0.0730	1.854	1.85
10	0.1935	4.915	4.90	50	0.0700	1.778	1.80
11	0.1910	4.851	4.90	51	0.0670	1.702	1.70
12	0.1890	4.801	4.80	52	0.0635	1.613	1.60
13	0.1850	4.699	4.70	53	0.0595	1.511	1.50
14	0.1820	4.623	4.60	54	0.0550	1.397	1.40
15	0.1800	4.572	4.60	55	0.0520	1.321	1.30
16	0.1770	4.496	4.50	56	0.0465	1.181	1.20
17	0.1730	4.394	4.40	57	0.0430	1.092	1.10
18	0.1695	4.305	4.30	58	0.0420	1.067	1.05
19	0.1660	4.216	4.20	59	0.0410	1.041	1.05
20	0.1610	4.089	4.10	60	0.0400	1.016	1.00
21	0.1590	4.039	4.00	61	0.0390	0.991	1.00
22	0.1570	3.988	4.00	62	0.0380	0.965	0.95
23	0.1540	3.912	3.90	63	0.0370	0.940	0.95
24	0.1520	3.861	3.90	64	0.0360	0.914	0.90
25	0.1495	3.797	3.80	65	0.0350	0.889	0.90
26	0.1470	3.734	3.75	66	0.0330	0.838	0.85
27	0.1440	3.658	3.70	67	0.0320	0.813	0.80
28	0.1405	3.569	3.60	68	0.0310	0.787	0.80
29	0.1360	3.454	3.50	69	0.0292	0.742	0.75
30	0.1285	3.264	3.25	70	0.0280	0.711	0.70
31	0.1200	3.048	3.00	71	0.0260	0.660	0.65
32	0.1160	2.946	2.90	72	0.0250	0.635	0.65
33	0.1130	2.870	2.90	73	0.0240	0.610	0.60
34	0.1110	2.819	2.80	74	0.0225	0.572	0.55
35	0.1100	2.794	2.80	75	0.0210	0.533	0.55
36	0.1065	2.705	2.70	76	0.0200	0.508	0.50
37	0.1040	2.642	2.60	77	0.0180	0.457	0.45
38	0.1015	2.578	2.60	78	0.0160	0.406	0.40
39	0.0995	2.527	2.50	79	0.0145	0.368	0.35
40	0.0980	2.489	2.50	80	0.0135	0.343	0.35

* Fraction-size drills range in size from one-sixteenth to 4 in. and over in diameter, by 64ths.

Twist-Drill Sizes *(Continued)*

LETTER SIZES

SIZE LETTER	DECIMAL EQUIVALENT	METRIC EQUIVALENT	CLOSEST METRIC DRILL (mm)
A	0.234	5.944	5.90
B	0.238	6.045	6.00
C	0.242	6.147	6.10
D	0.246	6.248	6.25
E	0.250	6.350	6.40
F	0.257	6.528	6.50
G	0.261	6.629	6.60
H	0.266	6.756	6.75
I	0.272	6.909	6.90
J	0.277	7.036	7.00
K	0.281	7.137	7.10
L	0.290	7.366	7.40
M	0.295	7.493	7.50
N	0.302	7.671	7.70
O	0.316	8.026	8.00
P	0.323	8.204	8.20
Q	0.332	8.433	8.40
R	0.339	8.611	8.60
S	0.348	8.839	8.80
T	0.358	9.093	9.10
U	0.368	9.347	9.30
V	0.377	9.576	9.60
W	0.386	9.804	9.80
X	0.397	10.084	10.00
Y	0.404	10.262	10.50
Z	0.413	10.491	10.50

Standard Unified Thread Series*

PRESENT UNIFIED THREAD NOMINAL SIZE — DIAMETER		COARSE (NC)(UNC)		FINE (NF)(UNF)		EXTRA-FINE (NEF)(UNEF)	
INCH	METRIC EQUIV.‡	THREADS PER INCH	TAP DRILL†	THREADS PER INCH	TAP DRILL†	THREADS PER INCH	TAP DRILL†
0.060	1.52	—		80	$\frac{3}{64}$	—	—
0.073	1.85	64	No. 53	72	No. 53	—	—
0.086	2.18	56	No. 50	64	No. 50	—	—
0.099	2.51	48	No. 47	56	No. 45	—	—
0.112	2.84	40	No. 43	48	No. 42	—	—
0.125	3.17	40	No. 38	44	No. 37	—	—
0.138	3.50	32	No. 36	40	No. 33	—	—
0.164	4.16	32	No. 29	36	No. 29	—	—
0.190	4.83	24	No. 25	32	No. 21	—	—
0.216	5.49	24	No. 16	28	No. 14	32	No. 13
0.250 $\frac{1}{4}$	6.35	20	No. 7	28	No. 3	32	No. 2
0.3125 $\frac{5}{16}$	7.94	18	F	24	I	32	K
0.375 $\frac{3}{8}$	9.52	16	$\frac{5}{16}$	24	Q	32	S
0.4375 $\frac{7}{16}$	11.11	14	U	20	$\frac{25}{64}$	28	Y
0.500 $\frac{1}{2}$	12.70	13	$\frac{27}{64}$	20	$\frac{29}{64}$	28	$\frac{15}{32}$
0.5625 $\frac{9}{16}$	14.29	12	$\frac{31}{64}$	18	$\frac{33}{64}$	24	$\frac{17}{32}$
0.625 $\frac{5}{8}$	15.87	11	$\frac{17}{32}$	18	$\frac{37}{64}$	24	$\frac{19}{32}$
0.6875 $\frac{11}{16}$	17.46	—		—		24	$\frac{41}{64}$
0.750 $\frac{3}{4}$	19.05	10	$\frac{21}{32}$	16	$\frac{11}{16}$	20	$\frac{45}{64}$
0.8125 $\frac{13}{16}$	20.64	—		—		20	$\frac{49}{64}$
0.875 $\frac{7}{8}$	22.22	9	$\frac{49}{64}$	14	$\frac{13}{16}$	20	$\frac{53}{64}$
0.9375 $\frac{15}{16}$	23.81	—		—		20	$\frac{57}{64}$
1.000 1	25.40	8	$\frac{7}{8}$	12	$\frac{59}{64}$	20	$\frac{61}{64}$

Decimal	Fraction	mm ‡	Threads	Tap Drill †	Threads	Tap Drill †	Threads	Tap Drill †
1.0625	1 1/16	26.99	—	—	—	—	18	1
1.125	1 1/8	28.57	**7**	63/64	**12**	1 3/64	18	1 5/64
1.1875	1 3/16	30.16	—	—	—	—	18	1 9/64
1.250	1 1/4	31.75	**7**	1 7/64	**12**	1 11/64	18	1 13/64
1.3125	1 5/16	33.34	—	—	—	—	18	1 17/64
1.375	1 3/8	34.92	**6**	1 13/64	**12**	1 19/64	18	1 5/16
1.4375	1 7/16	36.51	—	—	—	—	18	1 3/8
1.500	1 1/2	38.10	**6**	1 21/64	**12**	1 27/64	18	1 29/64
1.5625	1 9/16	39.69	—	—	—	—	18	1 1/2
1.625	1 5/8	41.27	—	—	—	—	18	1 9/16
1.6875	1 11/16	42.86	—	—	—	—	18	1 5/8
1.750	1 3/4	44.45	**5**	1 35/64	—	—	**16**	1 11/16
2.000	2	50.80	**4½**	1 25/32	—	—	**16**	1 15/16
2.250	2 1/4	57.15	**4½**	2 1/32	—	—	—	—
2.500	2 1/2	63.50	**4**	2 1/4	—	—	—	—
2.750	2 3/4	69.85	**4**	2 1/2	—	—	—	—
3.000	3	76.20	**4**	2 3/4	—	—	—	—
3.250	3 1/4	82.55	**4**	3	—	—	—	—
3.500	3 1/2	88.90	**4**	3 1/4	—	—	—	—
3.750	3 3/4	95.25	**4**	3 1/2	—	—	—	—
4.000	4	101.60	**4**	3 3/4	—	—	—	—

* Adapted from ANS B1.1-1960.

Bold type indicates Unified threads. To be designated UNC or UNF. Unified Standard — Classes 1A, 2A, 3A, 1B, 2B, and 3Ba.

For recommended hole-size limits before threading, see Tables 38 and 39, ANS B1.1–1960.

† Tap drill for a 75% thread (not Unified — American Standard). Bold-type sizes smaller than 1/4 in. are accepted for limited applications by the British, but the symbols NC or NF, as applicable, are retained.

‡ The values listed as metric equivalents of decimal inch values have been given to assist user in selecting the closest metric size to be found in the following table, Metric Screw Threads. Adherence to diameter preference is recommended, if feasible. For a metric thread use the tap-drill size recommended in the following table, Metric Screw Threads.

Metric Screw Threads

NOMINAL SIZE (mm)	SERIES WITH GRADED PITCHES*				THREAD-DIAMETER PREFERENCE†		
	COARSE	TAP DRILL‡	FINE	TAP DRILL‡	1	2	3¶
1.6	0.35	1.25	—	—	1.6	—	—
1.8	0.35	1.45	—	—	—	1.8	—
2	0.4	1.60	—	—	2	—	—
2.2	0.45	1.75	—	—	—	2.2	—
2.5	0.45	2.05	—	—	2.5	—	—
3	0.5	2.50	—	—	3	—	—
3.5	0.6	2.90	—	—	—	3.5	—
4	0.7	3.30	—	—	4	—	—
4.5	0.75	3.75	—	—	—	4.5	—
5	0.8	4.20	—	—	5	—	—
5.5	—	—	—	—	—	—	5.5
6	1	5.00	—	—	6	—	—
7	1	6.00	—	—	—	—	7
8	1.25	6.75	1	7.00	8	—	—
9	1.25	7.75	—	—	—	—	9
10	1.5	8.50	1.25	8.75	10	—	—
11	1.5	9.50	—	—	—	—	11
12	1.75	10.00	1.25	10.50	12	—	—
14	2	12.00	1.5	12.50	14	—	—
15	—	—	—	—	—	—	15
16	2	14.00	1.5	14.50	16	—	—
17	—	—	—	—	—	—	17
18	2.5	15.50	1.5	16.50	—	18	—
20	2.5	17.50	1.5	18.50	20	—	—
22	2.5	19.50	1.5	20.50	—	22	—
24	3	21.00	2	22.00	24	—	—
25	—	—	—	—	—	—	25
26	—	—	—	—	—	—	26
27	3	24.00	2	25.00	—	27	—
28	—	—	—	—	—	—	28
30	3.5	26.50	2	28.00	30	—	—
32	—	—	—	—	—	—	32
33	3.5	29.50	2	31.00	—	33	—
35	—	—	—	—	—	—	35
36	4	32.00	3	33.00	36	—	—
38	—	—	—	—	—	—	38
39	4	35.00	3	36.00	—	39	—
40	—	—	—	—	—	—	40
42	4.5	37.50	3	39.00	42	—	—
45	4.5	40.50	3	42.00	—	45	—
48	5	43.00	3	45.00	48	—	—
50	—	—	—	—	—	—	50
52	5	47.00	3	49.00	—	52	—

Metric Screw Threads *(Continued)*

NOMINAL SIZE (mm)	SERIES WITH GRADED PITCHES*				THREAD-DIAMETER PREFERENCE†		
	COARSE	TAP DRILL‡	FINE	TAP DRILL‡	1	2	3¶
55	—	—	—	—	—	—	55
56	5.5	50.50	4	52.00	56	—	—
58	—	—	—	—	—	—	58
60	5.5	54.50	4	56.00	—	60	—
62	—	—	—	—	—	—	62
64	6	58.00	4	60.00	64	—	—
65	—	—	—	—	—	—	65
68	6	62.00	4	64.00	—	68	—

* The pitches shown in bold type are those which have been estimated by ISO as a selected coarse- and fine-thread series for commercial threads and fasteners.
† Select thread diameter from columns 1, 2, or 3 with preference for selection being in that order.
‡ For an approximate 75% thread, use the formula: nominal O.D. minus $0.97 \times$ pitch.

Wire Numbers and Sizes

NO. AWG* MCM	STRANDING†	DIAMETER (in.)	AREA (cmil)	AREA‡ (sq in.)
40	Solid	0.0031	10	0.000008
38	Solid	0.0040	16	0.000013
36	Solid	0.0050	25	0.00002
34	Solid	0.0063	40	0.00003
32	Solid	0.0080	64	0.00005
30	Solid	0.0100	100	0.00008
28	Solid	0.0126	159	0.00012
26	Solid	0.0159	253	0.00020
24	Solid	0.0201	404	0.00032
22	Solid	0.0253	640	0.00050
20	Solid	0.0320	1,023	0.00080
18	Solid	0.0403	1,620	0.0013
16	Solid	0.0508	2,580	0.0020
14	Solid	0.0641	4,110	0.0032
12	Solid	0.0808	6,530	0.0051
10	Solid	0.1019	10,380	0.0081
8	Solid	0.1285	16,510	0.0130
6	7	0.184	26,240	0.027
4	7	0.232	41,740	0.042
3	7	0.260	52,620	0.053
2	7	0.292	66,360	0.067
1	19	0.332	83,690	0.087
0	19	0.372	105,600	0.109
00	19	0.418	133,100	0.137
000	19	0.470	167,800	0.173
0000	19	0.528	211,600	0.219
250	37	0.575	250,000	0.260
300	37	0.630	300,000	0.312
350	37	0.681	350,000	0.364
400	37	0.728	400,000	0.416
500	37	0.813	500,000	0.519
600	61	0.893	600,000	0.626
700	61	0.964	700,000	0.730
750	61	0.998	750,000	0.782
800	61	1.030	800,000	0.833
900	61	1.090	900,000	0.933
1000	61	1.150	1,000,000	1.039
1250	91	1.289	1,250,000	1.305
1500	91	1.410	1,500,000	1.561
1750	127	1.526	1,750,000	1.829
2000	127	1.630	2,000,000	2.087

* Wire numbers from 40 to 0000 are AWG, while those from 250 to 2000 are MCM. MCM ≅ 1000-circular-mil area.
† Solid conductors listed can be produced in stranded configurations.
‡ Area given is that of a circle equal to the overall diameter of a stranded conductor.

Appendix E

Product and Manufacturers' Directories, Catalogs, and Other Sources of Design Information

One drafting technique a designer must master is how to find and understand information about an electrical or electronic component. This information is required for schematics, layouts, bills of materials, and other electrical and electronics drawings. Usually, the technique is gained through experiences of finding out "where to look," "where to go," or "whom to talk to."

Generally, a designer who does not know what specific product to use or what manufacturer makes the product or provides a service will consult one of the product and manufacturers' directories available today. These directories vary from large multivolume sets to single volumes and from a very general to a very specialized scope of products. They are published by technical magazines, specialized publishers, and large publishing companies. Essentially all product and manufacturers' directories have a product index and a manufacturers' index, and some have a trade-name index and/or manufacturers' catalogs. The product index is arranged alphabetically by product classification, and, sometimes, subclassification [e.g., ICs (integrated circuits), transistor array; or interrupters, fault]. Under these product classifications, there is an alphabetical listing of manufacturers that make the product, giving their addresses, phone numbers, and, sometimes, teletype (TWX) numbers and branch offices. Some product indexes give more complete listings than others. The manufacturers' index is an alphabetical list of the manufacturers, and for each manufacturer gives the manufacturer's address, phone number, and sometimes, TWX number, gross sales, management personnel, number of employees, branch offices, and distributors. A trade-name index is an alphabetical listing of trade names of products and their respective manufacturers. Some product and manufacturers' directories are listed below.

Product and Manufacturer's Directories

FOR GENERAL USE:

1. *Thomas Register*
   Thomas Publishing Company
   One Penn Plaza
   New York, NY 10001
   (212) 695-0500

2. *U.S. Industrial Directory*
   Cahners Publishing Company
   8 Stamford Forum
   (P.O. Box 10277)
   Stamford, CT 06904
   (203) 328-2500

3. *Sweets Construction and Reno-*
   vation Catalog File and Plant
   Engineering Extension
   Sweets Division
   McGraw-Hill, Inc.
   1221 Avenue of the Americas
   New York, NY 10020
   (212) 512-2000

4. *Plant Engineering Directory and*
   Specifications Catalog
   Technical Publishing Company
   1301 South Grove Avenue
   Barrington, IL 60010
   (312) 381-1840

FOR ELECTRICAL EQUIPMENT:

1. *Design, Professional Product*
   Bulletin Directory
   Consulting Engineer
   Technical Publishing Co.
   Division of Dun & Bradstreet
   1301 South Grove Ave.
   Barrington, IL 60010
   (312) 381-1840

2. *Electrical Products Yearbook*
   Electrical Construction
   & Maintenance
   McGraw-Hill, Inc.
   1221 Avenue of the Americas
   New York, NY 10020
   (212) 512-2000

3. *Annual Specifiers & Buyers*
   Guide
   Transmission and Distribution
   Cleworth Publishing Company,
   Inc.
   One River Road
   Cos Cob, CT 06807
   (203) 661-5000

4. *Index of Advertisers*
   and Products (CEE)
   Sutton Publishing Company
   707 Westchester Avenue
   White Plains, NY 10604
   (914) 949-8500

FOR ELECTRICAL/ELECTRONIC EQUIPMENT:

1. *Electrical/Electronics Reference*
   Edition
   Design News
   Cahners Publishing Company
   270 St. Paul Street
   Denver, CO 80206
   (303) 388-4511

FOR ELECTRONIC EQUIPMENT.

1. *Electronics Designs Gold Book*
   — Master Catalog & Directory
   of Suppliers to Electronics Man-
   ufacturers
   Hayden Publishing Co., Inc.
   10 Mulholland Drive
   Hasbrouck Heights, NJ 07604
   (201) 393-6000

2. *Electronics Buyers Guide*
   McGraw-Hill, Inc.
   1221 Avenue of the Americas
   New York, NY 10020
   (212) 512-2000

3. *Electronic Engineers Master Catalog*
Hearst Business Communication Inc./UTP Division
645 Stewart Avenue
Garden City, NY 11530
(516) 222-2500

4. *Electronics Industry Telephone Directory*
Harris Publishing Company
2057-2 Aurora Road
Twinsburg, OH 44087
(216) 425-9143

5. *D.A.T.A. Book Electronics Information Series*
D.A.T.A., Inc.
9889 Willow Creek Road
San Diego, CA 92131
(619) 578-7600

6. *IC Master*
Hearst Business Communications Inc./UTP Division
645 Stewart Avenue
Garden City, NY 11530
(516) 222-2500

FOR CONTROL AND INSTRUMENTATION:
1. *INCS Buyers Guide*
Chilton Publishing Company
Chilton Way
Radnor, PA 19089
(215) 964-4000

Other sources of product and manufacturer information are technical and trade magazines and journals, salespeople, distributors, other engineers and designers, and purchasing departments. The authors have even used stockbrokers and Dun & Bradstreet Reports to find the address of a known manufacturer which could not be found elsewhere.

After a designer has selected a product, a manufacturer's or distributor's catalog is obtained within the office from the catalog file or library or from another engineer, or it is obtained from a salesperson. Usually, engineering firms have large and fairly complete catalog files, such as the one shown in Fig. E-1. Although catalogs vary considerably, designers should observe the following concepts when looking at an unfamiliar catalog.

1. Check the issue date or number to see if it is the most up-to-date version.

2. Scan the table of contents, index, and the catalog for an overview or "big picture" of the contents of the catalog.

3. Examine the conditions of sale, discount schedules, and other general information, usually found in the front of the catalog.

4. Then find the product description in the catalog, using the index.

5. Once the product description is located in the catalog, look at the pictures of the product, read all the information about the product, including the notes and footnotes, which are sometimes found at the front or end of the product class

FIG. E-1 Typical engineering catalog file. This one belongs to an electrical contractor. (Boese-Hilburn Electric Service Company.)

section. Observe all the operating parameters and specifications, physical dimensions, available modifications, and costs associated with the product.

Once the product is selected, record the relevant information from the catalog onto the drawings, paying particular attention to getting all necessary ordering information on the bill of materials—such as a general product description, manufacturer, all manufacturer's model and type numbers, operating parameters, general industry class numbers.

Bibliography

Electrical or Electronics Drawing

Raskhodoff, Nicholas M.: *Electronic Drafting and Design,* 3d ed., Prentice-Hall, Inc., Englewood Cliffs, N.J., 1977.

Richter, Herbert W.: *Electrical and Electronic Drafting,* John Wiley & Sons, Inc., New York, 1977.

Snow, Charles W.: *Electrical Drafting and Design,* Prentice-Hall, Inc., Englewood Cliffs, N.J., 1976.

Engineering Drawing or Graphics

French, Thomas E., Carl L. Svensen, Jay D. Helsel, and Byran Urbanick: *Mechanical Drawing,* 9th ed., McGraw-Hill Book Company, New York, 1980.

────── and Charles J. Vierck: *Engineering Drawing and Graphic Technology,* 12th ed., McGraw-Hill Book Company, New York, 1978.

──────, ──────, and Robert J. Foster: *Graphic Science and Design,* 4th ed., McGraw-Hill Book Company, New York, 1984.

Giesecke, F. E., A. Mitchell, and H. C. Spencer: *Technical Drawing,* 6th ed., The Macmillan Company, New York, 1974.

Levins, A. S.: *Graphics with an Introduction to Conceptual Design,* John Wiley & Sons, Inc., New York, 1974.

Luzadder, W. J.: *Fundamentals of Engineering Drawing,* Prentice-Hall, Inc., Englewood Cliffs, N.J., 1978.

Miscellaneous, Electronics or Electrical

Bishop Graphics, Inc.: *The Design and Drafting of Printed Circuits,* McGraw-Hill Book Company, New York, 1979.

Chute, George M., and Robert Chute: *Electronics in Industry,* 5th ed., McGraw-Hill Book Company, New York, 1979.

Clifford, Martin: *The Handbook of Electronic Tables,* Gernsback Library, Inc., New York, 1965.

Deboo, G. J., and C. Burrous: *Integrated Circuits and Semiconductor Devices: Theory and Application,* 2nd ed., McGraw-Hill Book Company, New York, 1977.

Grinich, Victor, and Horace Jackson: *Introduction to Integrated Circuits,* McGraw-Hill Book Company, New York, 1975.

Hall, Douglas V.: *Microprocessors and Digital Systems,* McGraw-Hill Book Company, New York, 1980.

Hamilton, Douglas J., and William Howard: *Basic Integrated Circuit Engineering,* McGraw-Hill Book Company, New York, 1975.

Heumann, G. M.: *Magnetic Control of Industrial Motors,* John Wiley & Sons, Inc., New York, 1954.

Hordeski, Michael: *Illustrated Dictionary of Microcomputer Terminology,* Tab Books, Blue Summit, Pa., 1978.

Keonjian, Edward: *Microelectronics: Theory, Design, and Fabrication,* McGraw-Hill Book Company, New York, 1963.

Kiver, Milton S.: *Transistor and Integrated Electronics,* 4th ed., McGraw-Hill Book Company, New York, 1972.

Krouse, John K.: "CAD/CAM — Bridging the Gap from Design to Production," *Machine Design,* June 12, 1980, pp. 117–125.

Lindsey, Darryl: *The Design and Drafting of Printed Circuits,* Bishop Graphics, Inc., Westlake Village, Calif.

Malvino, A. P.: *Semiconductor Circuit Approximations,* 4th ed., McGraw-Hill Book Company, New York, 1985.

"National Electrical Code®," NFPA, New York, 1984.

New York Institute of Technology: *A Programmed Course in Basic Pulse Circuits,* McGraw-Hill Book Company, New York, 1977.

Markus, John, and Vin Zeluff: *Handbook of Industrial Electronic Control Circuits,* McGraw-Hill Book Company, New York, 1956.

Ramirez, Edward V., and Melvyn Weiss: *Microprocessing Fundamentals: Hardware and Software,* McGraw-Hill Book Company, New York, 1956.

Schuler, Charles A.: *Electronics: Principles and Applications,* McGraw-Hill Book Company, New York, 1979.

Sippl, Charles J.: *Microcomputer Dictionary,* Radio Shack, USA, 1981.

Slurzberg, Morris, and William Osterheld: *Essentials of Communication Electronics,* 3d ed., McGraw-Hill Book Company, New York, 1973.

Technical Manual and Catalog (for PC-board layout), Bishop Graphics, Inc., Westlake Village, Calif.

Voisinet, Donald D.: *Introduction to Computer-Aided Design,* McGraw-Hill Book Company, New York, 1983.

Index

Abbreviations for drawings, 487–490
Active devices, 296
Aids for drafting (*see* Drafting aids)
Aircraft electrical diagrams, 115, 118, 262
Airline diagrams, 124–126
Alphanumeric, 52, 57
American National Standards, 11, 12,
    81, 357, 491, 508
Amplifier circuits, 231–234
    [*See also* Schematic (elementary)
    diagrams]
Appliqués, 89, 166–169
Architectural plans, electrical drawing
    for, 411–439
    drawing symbols, 508–516
    (*See also* specific component)
Architecture, microprocessor, 51, 61,
    207–209
Artwork, 162, 163, 167–169, 179, 286,
    290
Assembler, 53
Assembly drawings, 141–144
Automated machine tools, 351–357
Automotive wiring diagrams, 117

Balloon drawing, 358–361
Bar charts, 461–463
Baseline diagrams, 124–126
BASIC (language), 214, 310, 311
Battery symbol, 81, 82
Bend radii for aluminum, 529
Block diagrams, 52, 203–228, 280, 283,
    354, 366
    (*See also* Flow diagrams)
Branch circuit, 412, 422
Bus ducts, 423–425

Cabling diagrams, 128–134
Capacitor symbols, 82, 83
Catalogs, 543, 544
Cathode-ray tube (CRT), 49, 50, 98, 99
Cellular floor system, 426, 427
Chassis drawings:
    holes and terminals, 138–140
    layout and construction, 133–140
    photo drawings, 144, 145
Chassis manufacture, 136–143
Chassis symbol, 83, 256
Circuit breaker, function of, 322
    air, 384, 385
    oil (OCB), 381
Circuit return symbols, 83, 256

Circuits [*See* Schematic (elementary)
    diagrams]
Color codes, 117, 119, 120, 122–124
Common connection symbol, 83, 496
Compiler, 53
Computer-aided design (CAD), 46–79
Computer-aided manufacturing (CAM),
    77–79, 351–357
Computer control, 351–353
Computer drawings, 3, 16, 17, 70–76
Computer-output, 62–76
Connection diagrams, 113–132
    aircraft, 118, 262
    airline-type, 124–126
    automotive, 117
    cabling, 128–134
    chassis, 133–140
    color coding, 117, 119–125, 517–528
    with elementary diagram, 127
    harness, 128–134
    highway-type, 122–124
    interconnection diagrams, 115–120
    line spacing, 127, 128
    local cabling, 128–131
    pictorial, 116, 132
    point-to-point, 116, 118, 120, 121
    straight-line, 126, 127
    wiring (to-and-from) lists, 127, 133
Connections and crossovers, 83, 84, 239,
    496
Construction and assembly drawings,
    133–144
Contactor function, 322
Contactor symbols, 85, 324–326
Continuous data, 443, 444
Controls, 321–355
    circuit-breaker function, 322
    contactor function, 322
    control switches, 324
    device designations, 329, 330
    ladder diagram, 329–331, 349, 371
    overload protection, 322
    power-generating field, 380–385, 389
    reduced-voltage starting, 322
    relay circuits, 348, 362
    relay function, 362
    speed controller, 366–369
    undervoltage protection, 322
Conversion, 366
Coordination of drawings, 427, 428
Coupling, interstage, 231

Cryogenic switch, 87
Cursor, 57–59
Curves, 451, 455–459
   equations for, 455
   families of, 452
   smooth, 444
   straight-line, 455–459

DC (direct coupling), 232
Decision table, 215, 216
Demand-load table, 428, 429
Dependency, 215–219
Detail drawings, 394, 396, 401
Device designations, 104, 238
Digital watch circuit, 205–207
Digitizer, 54–56
Dimensioning, 135–140
   chassis drawings, 135–138
   hole locations, 138–140
Diode symbols, 89–92
Disconnect switch, 324, 381
Discrete (discontinuous) data, 442–444
Disk, 52, 62, 63
Display console, 58–60
Drafting aids:
   drafting machines, 5, 170
   mechanical lettering devices, 18, 243
   preprinted symbols, 89, 166–169, 245
   standard grids, 128, 165
   templates, 6–9, 14–17, 24, 89, 90
Drafting machines, 5, 170
Drill sizes, 138–140, 532–535
Drilling drawing, 139, 140, 180, 181

Electric power installations, drawings for,
   377–409
Electrical layout:
   building, 417, 419, 421, 426
   industrial controls, 326–328
Electromechanical controls, 329, 334
Electron-tube symbols, 98, 100, 494, 502
Elementary diagrams, 324, 325, 331,
   336, 338, 349
   [See also Schematic (elementary)
   diagrams]
Empirical equations, 455–459
Epitaxial growth, 284, 285
Etched circuits, 282, 285, 286

Feedback, AGC (automatic gain control)
   circuit, 204
Feeder circuits, 414, 422, 423
Feeder lines, 122, 123, 125
Feeder-load tables, 429, 430
Film circuits, 295–302

Fixture schedule, electrical, 420
Flow diagrams, 52, 203–228
   dependency, 215–219
   logic diagrams, 210–215
   microprocessors, 207–209
   (See also Block diagrams)
Foil conductor, copper, widths of, 166,
   526
Font, 1, 17
FORTRAN, 53, 61
Fractions, decimal equivalents, 531
Freehand sketch, 4, 25, 27
French curve, use of, 450, 451
Frequency spectrum, 490
Functions, logic, 210–215

Gate arrays, 303, 304
General arrangement drawing, 396, 397
Glossary of electronics and electrical
   terms, 282, 308, 480–485
Graphical representation of data,
   440–479
   abscissa, 443
   bar charts, 461–463
   continuous versus discrete data,
   442–444
   conversion scale, 466
   curve fitting and drawing, 443, 450, 451
   curve identification, 444
   equations for curves, 455–459
   independent versus dependent variable, 443
   lettering, 450, 451
   logarithmic paper, 455–457
   origin or zero point, 442, 446
   paper, types of, 441
   pictorial graphs, 453–456, 458, 460,
   464, 465
   pie graph, 463, 464
   plotting symbols, 445–448
   polar graphs, 459–461
   principles of, 440
   procedure, 447–450
   scales and captions, 446–451
   selection:
      of scales, 450, 451, 455
      of variables and curve fitting, 443
   semilogarithmic paper, 453, 454,
   457–459
   steps in construction (layout), 447–451
   variables, dependent and independent,
   443
Graphs (see Graphical representation of
   data)

Ground symbol, 83, 256
Grounding, 404

Hard copy, computer, 66, 67
Hardware, 52–68
Harness diagrams, 128–133
Highway diagrams, 122–124
Hockey puck, 55
Hybrid circuits, 295–303

Identification of lines in wiring diagrams,
    115–125
Inductor symbols, 84, 495
Industrial controls, 321–375
Instrument drawing techniques, 1, 5–10
Instrumentation, 357–366
Integrated semiconductor circuits,
    278–295
    capacitor, resistor characteristics of, 297
    diffusion steps, 282–285
    drawing sequence, 285–293
    epitaxial growth, 284, 285
    glossary, 279, 282, 308
    intraconnection pattern, 285, 294
    masking steps, 284
    packaging, 293–296, 302, 303
    schematic diagrams, 291–293
    symbols for schematics, 293
Interactive graphics, 59, 60
Interconnection diagram, 115–120, 334
Interstage coupling, 231
Inversion, 366
Inverter, 211
Irregular curve and splines, 450, 451

JIC (Joint Industrial Council) standard
    identification, 500–503
Joystick, 55

Keyboard, 52, 55, 57, 59

Ladder diagrams, 329–331, 349, 371
Layout(s):
    component, 74, 142, 178
    sheet metal, 136–144
Learning curve, 311–313
Legends:
    for architectural plans, 420
    for graphs, 444, 445, 449
    for printed circuits, 181
Lettering, 11–18, 450, 451
    alphabets, 11–13
    on flow diagrams, 204, 205
    font, 16, 17
    on graphs, 449–451, 464
    guidelines for, 13–15

Lettering *(Cont.)*
    mechanical lettering devices, 16, 18
    on parts list, 15, 420
Light pen, 49, 52
Lighting-load table, 429
Lightning arrester, 381
Line spacing in wiring diagrams, 127, 128
Line weights, typical lines and their uses, 5
Line work, examples of, 2–5, 11, 15, 19
Load centers for buildings, 422–424
Load computation, 428–430
Local cabling in wiring diagrams,
    128–133
Logarithmic paper and scales, 453–457
Logic diagrams, 210–219, 350, 385–391
    AND, 210–212
    control circuits, 347, 350, 389
    in electric power field, 385–391
    EXCLUSIVE OR, 211
    flip-flop, 211, 212
    inverter (NOT), 211
    logic functions, 210–215
    NAND, 211, 213, 236, 304
    negative logic, 213
    NOR, 211, 236
    OR, 211, 213
    symbols, 210–214, 218, 219, 387,
        504–507
Loop, crossover, 251, 272

Machine-tool control, 351–355, 399
Main service entrance, 422–424
Manufacturing layout and time rate,
    136–138
Marking drawing, 180, 181
Mechanical linkage, 253, 254, 262
Menu, 55–60, 215, 311
Metal-oxide semiconductor (MOS)
    technology:
    definition of, 279, 280
    process steps, 315
    schematic diagram, 290, 292
    symbols, 95, 96, 293
Microelectronics, 279–320
    film circuits, 296–302
    (*See also* Integrated semiconductor
      circuits; Microprocessors)
Microprocessors, 207–209, 305–311
    applications of, 305–311
    diagrams, 207–209
    functional diagram, 207, 307
    glossary of terms, 308
    programming, 310, 311
Monolithic (integrated) circuits, 278–293

Motor control, 321–335
Motor control center, 326–328
  layout, 327, 328
  schedule, 327, 328
Motors, types of, 335–343
Mouse, 54, 55
Multiview drawing, 18–21, 137, 139, 341

National Electrical Code® (NEC),
  411–416, 428–430
  areas of coverage, 411
  definitions used in, 412–415
Negative logic, 213
NEMA (National Electrical Manufacturers Association) Standard, 329,
  504–507
NOR circuit, 211
Notes for schematic diagrams, 238, 252,
  258
Numerical control, 75, 139, 351–353

Office building wiring, 419–426
One-line diagram, 68–70, 381, 383
Orthographic views, 18–21, 139, 341, 396
Outlet symbols, 508–515
Overcurrent (overload) protection, 322,
  325

Panel layout, 326–329, 333, 334, 394
  schedule for, 327, 328
Panelboards for buildings, 422, 425
Parts-assembly drawings, 142, 143, 145
PASCAL, 53, 61
Passive devices, 296
Photodrawing for chassis assembly, 144,
  145
Photolithography, 282, 285
Pictorial drawing, 21–30
  dimetric projection, 28
  graphs, 464, 465
  isometric drawing, 22–25
  oblique projection, 26, 27
  perspective drawing, 21, 28, 29, 377
  sections, 30
  types of, 22
  wiring diagrams, 116, 132
Picture tube, 98–100
Pie graph, 463, 464
Pigtail leads, 121–122
Pixel, 50
Plotters, 62–68
  beltbed, 65, 66, 68
  drum, 65, 73
  electrostatic, 66, 67
  flatbed, 64, 65

Plotters (Cont.)
  photoplotters, 65
Plotting symbols, 445, 447, 448
Point-to-point wiring diagrams, 116,
  118, 120, 121
Polar graphs, 459–461
Power semiconductors, 302, 303, 366, 368
Prefixes and units, 238, 241, 252
Printed circuit boards, 159–202
  assembly drawing, 162, 178
  board shapes, 160, 161, 162, 179, 192
  components for, 164, 172, 174, 185
  component spacing, 162–165
  conductor spacing, 166, 185
  double-sided layout, 176–182
  drawing sequence, 160–166, 176
  fastening (mounting), 161, 190–193
  master layout, 75, 80, 163, 179
  standard grids, 165, 185
  through connections, 165, 185
Product directories and catalogs, 541–
  544
Production drawings, 113–145, 179, 187
  construction and assembly drawings,
  133–145
  (See also Connection diagrams;
  Printed circuit boards)
Programmable controllers, 341–350
  block diagram, 344
  description of, 341–344
  example of, 345–350
  program for, 350
  relays in system, 346

Raceways, 424–427
Radii for conduit bends, 431
Raster, 50
RC (resistance-capacitance) coupling, 232
Receptacles, 115, 132, 415, 417, 420
Rectification, 366, 367
Rectifier symbols, 92, 367
Reduced voltage starting, 322
Reference designations, 104, 105, 238,
  252
Relay circuits, 324, 330
  contact designations, 329–331
  function of, 329, 345, 346, 348
  symbols for, 84, 85, 382
Resistor symbols, 85, 330, 331
Riser diagrams, 423–425
Robots, 193, 354, 357
Routing, harness, 127, 131, 133

Schedule:
  lighting fixture, 420
  for panelboard, 327, 328

Schematic (elementary) diagrams, 229–277
  amplifier, 231, 233, 234, 365
  audio amplifier, 251
  basic transistor circuits, 229–237
  calculators, 246, 248–250
  CB receiver, 142, 265
  computer circuits, 234, 235
  coupling, interstage, 231
  digital ICs, 236, 237
  dc motor drive, 367
  electric power generation, 390, 391
  flasher circuit, 239–241
  flip-flop circuits, 235, 241, 273
  flu gas system, 71
  fundamentals for preparation, 244, 245
  hybrid diagram, 261, 262
  interruption and separation, 255–257
  latch circuit, 291, 293
  layout procedure, 239–245
  linkage (mechanical), 253, 254, 262
  NAND function, 211, 213, 236
  NOR function, 211, 236, 237
  notes for drawings, 238, 258, 259
  preamplifier circuit, 258
  prefixes and units, 238, 241, 252
  radar display circuit, 176, 177
  radio-receiver circuit, 272
  reference designations, 104, 105, 238, 252
  robot, 193, 356
  shunt-regulator circuit, 233
  speed controller, 366–369
  standards, 491–503
  transistor-radio circuit, 204, 247, 270
  typical patterns, 230–237
  video distribution amplifier, 246, 248
  voltage sequence, 245, 246
  waveforms, 253
Semilogarithmic paper, 453, 454, 457–459
Sheet metal data, 529, 530
Sheet metal layouts, 133–140
Silicon controlled rectifier (SCR), 73, 366–369
Single-line diagrams, 68–70, 381, 383
Size of symbols, 88, 212, 242, 244
Software, 51–53
Solar energy, 205, 376–379
Specification writing, 400
Speed controller, 366–369
Stage definition of, 203
Standards:
  abbreviations for drawings, 411
  alphabet, 11, 12

Standards *(Cont.)*
  for architectural drawings, 411–416, 508–516
  device (electrical and electronic) symbols, 491–503, 517–526
  instrument symbols and identification, 357–361
  for logic diagrams, 210, 217
  National Electrical Code, 411–416
  for wiring, chassis, etc., 113, 526
Straight-line wiring diagram, 126, 127
Straps, 125
Stylus, 57, 59
Surface mounted devices, 97, 185–190
Switch symbols, 86, 87, 500, 501
Symbols:
  in block diagrams, 201–203
  for control diagrams, 325, 326, 347, 359, 367
  for electrical devices, 81–102, 491–503 (*See also specific symbols, e.g.,* Resistor symbols)
  for logic diagrams, 210–215, 217–219
  sizes for drawing, 88, 212, 242, 244
  standard:
    for architectural wiring, 508–516
    for electronic devices, 491–503

T square, use of, 6, 8
Tablet, graphics, 52, 55, 59, 60
Television tubes, 98–100
Templates:
  for blocks and device symbols, 7, 89, 90, 243
  for lettering, 17
  use of, 8–10
Thick-film circuits, 296–299
Thin-film circuits, 297–301
  device tolerances, 297
  materials for, 296, 297
Threads (screw), 139, 140, 536–539
Three-line diagram, 385, 397
Thyristor, 281
To-and-from diagrams, 127, 133
Transformer coupling, 232
Transformer symbols, 87
Transistor symbols, 93–97
Truth tables, 210–214
Tube symbols, 100, 210–214

Under-floor systems, 426, 427
Units, prefixes and, 252, 253

Vacuum-tube symbols, 98, 100
Variables, dependent and independent, 443, 444

Video-tube symbols, 99, 100

Waveforms, 253
Weights of lines, 5
Wire and sheet metal gages, 530, 540
Wiring diagrams, 113–132
    aircraft, 118
    airline-type, 124–126
    for architectural plans, 417, 419, 421,
       423–426
    automotive, 117
    cabling diagrams, 128–134
    color coding, 117, 119–125, 526–528
    connection and elementary, 127
    harness, 128–134
    highway-type, 122–124

Wiring diagrams *(Cont.)*
    interconnection, 115–120
    line spacing, 127, 128
    local cabling, 128–131
    panel, 326–329, 333, 334
    pictorial, 21, 22–30, 116, 132
    point-to-point, 116, 118, 120, 121
    straight-line, 126, 127
    wire sizes, 540
    wiring (to-and-from) lists, 127, 133
Work station, 55, 59

X-ray photographs, 184, 185

Zooming, 60